Model-Driven Risk Analysis

Mass Soldal Lund · Bjørnar Solhaug · Ketil Stølen

Model-Driven Risk Analysis

The CORAS Approach

Springer

Mass Soldal Lund
Bjørnar Solhaug
Ketil Stølen
SINTEF ICT
P.O. box 124 Blindern
0314 Oslo
Norway
Mass.S.Lund@sintef.no
Bjornar.Solhaug@sintef.no
Ketil.Stolen@sintef.no

ISBN 978-3-642-12322-1   e-ISBN 978-3-642-12323-8
DOI 10.1007/978-3-642-12323-8
Springer Heidelberg Dordrecht London New York

Library of Congress Control Number: 2010936190

ACM Computing Classification (1998): K.6, D.2.9

© Springer-Verlag Berlin Heidelberg 2011
This work is subject to copyright. All rights are reserved, whether the whole or part of the material is concerned, specifically the rights of translation, reprinting, reuse of illustrations, recitation, broadcasting, reproduction on microfilm or in any other way, and storage in data banks. Duplication of this publication or parts thereof is permitted only under the provisions of the German Copyright Law of September 9, 1965, in its current version, and permission for use must always be obtained from Springer. Violations are liable to prosecution under the German Copyright Law.
The use of general descriptive names, registered names, trademarks, etc. in this publication does not imply, even in the absence of a specific statement, that such names are exempt from the relevant protective laws and regulations and therefore free for general use.

Cover design: KünkelLopka GmbH, Heidelberg

Printed on acid-free paper

Springer is part of Springer Science+Business Media (www.springer.com)

Preface

Exposure to risk is inescapable in most domains. People and families, enterprises, governments, private and public organisations, infrastructure providers, service providers, and so forth all encounter risks on an ongoing and frequent basis. The kinds of risks however vary from domain to domain, be it safety, economy, information and ICT security, politics, civil protection, emergency planning, defence, law, health, and so on. The need for understanding and managing risk is self-evident. Risk management is moreover in many cases imposed as a prerequisite, be it by law and legal regulations or from the public opinion, in particular within critical areas that may affect privacy and welfare, or even health and human life. In other cases, the lack of good routines, cultures and processes for managing risk may be a decisive factor for risks to emerge that should or could have been avoided.

In this book, we present CORAS, which is a model-driven approach to risk analysis. Risk analysis is a core part of the overall process of risk management. In order to conduct risk analysis in practice, there is clearly a need for well-defined methods, techniques and guidelines for how to do this, and this is precisely what CORAS offers. Risk analysts, or for that matter anyone with a need for identifying and understanding risks, will in this book find guidance on how to conduct a stepwise, structured and systematic analysis and documentation of risks.

The book also serves as an introduction to risk analysis in general, and as an introduction to the central and well-established underlying concepts and terminology. Practitioners, as well as graduate or undergraduate students, particularly within the IT domain, are therefore main target groups of this book. CORAS is strongly related to international standards on risk management, and this book therefore serves as an introduction to many of the issues that are addressed in these standards.

An important objective of this book is to accompany standardised risk management guidelines and terminology with comprehensive pragmatic support. International standards generally focus on the *what*, but say little or nothing about the *how*. This book is a self-contained contribution not only to understand what risk management, risk analysis and risk related concepts are, but also to learn how to do risk analysis in practice. Extensive use of practical and illustrative examples furthermore facilitates a deep understanding of both the pragmatics and the conceptual aspects.

The comprehensiveness of CORAS is manifested by the three complementary parts of the approach. CORAS consists of a customised language for risk modelling, a tool supporting the language, and a risk analysis method into which the tool-supported risk modelling language is tightly interwoven. It is particularly the specialised support for risk modelling that distinguishes CORAS from other approaches to risk analysis. The CORAS language provides explicit support for the risk analysis steps and tasks, and is furthermore closely related to the underlying risk analysis concepts.

The CORAS approach as presented in this book is the result of work that was initiated in 2001, and that draws upon academic research, empirical studies, thorough experience, as well as close interaction and cooperation with actors from several industrial domains. Along the way, we have benefited greatly from fruitful cooperation with many colleagues, and much work on different aspects of CORAS has already been published in articles, papers, reports and doctoral theses. Several colleagues have also contributed to this book by coauthoring some of the chapters, or by giving valuable criticism, suggestions and feedback, and for this we owe them great thanks.

We are deeply grateful to Ida Hogganvik Grøndahl for her influential doctoral work. Many aspects of the CORAS approach as presented in this book are strongly inspired by her work, in particular the basic CORAS language.

We owe our great thanks to Gyrd Brændeland, Atle Refsdal and Fredrik Seehusen for each coauthoring a chapter in this book, and for their valuable suggestions and comments. Fredrik Seehusen has moreover contributed by being the main developer of the current version of the CORAS tool. Many thanks also to Folker den Braber, Heidi Dahl and Fredrik Vraalsen for their contributions over the past years, and to Olav Ligaarden for helping us with the index and for making valuable suggestions.

Many thanks to Tobias Mahler for his many comments and fruitful criticism, in particular on the chapter on legal aspects. His doctoral work on legal risk management served as a valuable source of inspiration, and we acknowledge the synergies between his work and the work that has led to this book.

We are thankful to Jan Øyvind Aagedal, Iselin Engan, Bjørn Axel Gran, Jan Heim, Siv Hilde Houmb, Tormod Håvaldsrud, Tom Lysemose, Aida Omerovic, Eva Skipnes and Jan Håvard Skjetne, each of which has contributed by valuable suggestions or via fruitful cooperation in CORAS related work.

We are thankful to our colleagues at SINTEF ICT, including our Head of Department Bjørn Skjellaug. Many thanks also to the colleagues that we have worked with in several national and international projects that have been related to CORAS. These people include Demissie Aredo, Gustav Dahll, Theo Dimitrakos, Ivan Djordjevic, Rune Fredriksen, Chingwoei Gan, Eva Henriksen, Erik Mork Knutsen, Monica Kristiansen, Simon Lambert, Katerina Papadaki, Xavier Parent, Athanasios Poulakidas, Dimitris Raptis, Brian Ritchie, Yannis Stamatiou, Nikos Stathiakis, Atoosa Thunem, Erik Wisløff and Bjarte Østvold.

We also recognise the valuable feedback and knowledge acquired from many industrial field trials and commercial risk analyses based on CORAS. In relation to this, we would like to thank Tor Aalborg, Semming Austin, Nils Inge Brubrerg, Petter Christensen, Sten Vidar Eikrem, Håvard Fridheim, Are Torstein Gimnes, David

Goldby, Janne Hagen, Rune Hagen, Tor-Gaute Indstøy, Hege Jacobsen, Ole Jarl Kvammen, Arne Bjørn Mildal, Per Myrseth, Mikkel Skou, Petter Taugbøl, Anne Karin Wahlfjord, Hermann Steen Wiencke and Jon Ølnes.

We are also in debt to the many students who have followed our course INF5150 at the University of Oslo since it was started up in 2001, as well as the to the MSc students who have addressed various aspects of CORAS in their thesis work. In particular, we would like to thank Emese L. Bogya, Jenny Beate Haugen, Vikash Katta, Igor Kodrik, Mihail Korabelnikov, Stig Torsbakken, and Shahbaz Chaudhary Yaqub.

Our work on developing the CORAS approach has benefited from research in joint projects with a number of good partners. The initial CORAS approach was developed within the CORAS project funded by the European Commission that ran from 2001 until 2003. We are thankful to the project coordinator Yves Paindaveine, as well as the project leaders Tom Arthur Opperud and Tony Price, for providing a good environment for fruitful research. We are also grateful to Habtamu Abie who together with Eva Skipnes in 1999 invited us to join the consortium that later started the CORAS project.

Some of the research results that is reported in this book has partly been funded by the Research Council of Norway through the projects COBRA, COMA, DIGIT, EMERGENCY, ENFORCE and SECURIS. The research has also partly been funded by the European Commission through the projects iTrust, MASTER, MODELWARE, SecureChange, S3MS and TrustCoM.

Oslo, Norway
Mass Soldal Lund
Bjørnar Solhaug
Ketil Stølen

Contents

Part I Introductory Overview

1 Introduction 3
   1.1 The Importance of Risk Analysis 3
   1.2 Asset Identification 4
   1.3 Risk Modelling 5
   1.4 The CORAS Approach 5
       1.4.1 The CORAS Language 6
       1.4.2 The CORAS Tool 6
       1.4.3 The CORAS Method 6
   1.5 The Generality of CORAS 7
   1.6 Overall Aim and Emphasis 8
   1.7 Organisation 8
       1.7.1 Part I: Introductory Overview 9
       1.7.2 Part II: Core Approach 9
       1.7.3 Part III: Selected Issues 11
       1.7.4 Appendices 12
   1.8 Colours in CORAS and in this Book 13

2 Background and Related Approaches 15
   2.1 Basic Terminology 15
   2.2 Related Approaches 17
       2.2.1 Risk Analysis Methods 17
       2.2.2 Table-based Risk Analysis Techniques 18
       2.2.3 Tree-based Risk Analysis Techniques 18
       2.2.4 Graph-based Risk Analysis Techniques 19
       2.2.5 Situating CORAS Within this Picture 20

3 A Guided Tour of the CORAS Method 23
   3.1 Preparations for the Analysis 23
   3.2 Customer Presentation of the Target 25
   3.3 Refining the Target Description Using Asset Diagrams 26

	3.4	Approval of the Target Description	31
	3.5	Risk Identification Using Threat Diagrams	33
	3.6	Risk Estimation Using Threat Diagrams	37
	3.7	Risk Evaluation Using Risk Diagrams	39
	3.8	Risk Treatment Using Treatment Diagrams	41

Part II Core Approach

4 The CORAS Risk Modelling Language 47
  4.1 Central Concepts 48
    4.1.1 What is a Threat? 48
    4.1.2 What is a Threat Scenario? 49
    4.1.3 What is a Vulnerability? 51
    4.1.4 What is an Unwanted Incident? 53
    4.1.5 What is an Asset? 55
  4.2 The Diagrams of the CORAS language 56
    4.2.1 Asset Diagrams 56
    4.2.2 Threat Diagrams 58
    4.2.3 Risk Diagrams 60
    4.2.4 Treatment Diagrams 62
    4.2.5 Treatment Overview Diagrams 64
  4.3 How to Schematically Translate CORAS Diagrams into English Prose 65
    4.3.1 How to Translate Asset Diagrams 65
    4.3.2 How to Translate Threat Diagrams 67
    4.3.3 How to Translate Risk Diagrams 69
    4.3.4 How to Translate Treatment Diagrams 69
    4.3.5 How to Translate Treatment Overview Diagrams 70
  4.4 Summary 71

5 Preparations for the Analysis 73
  5.1 Overview of Step 1 73
  5.2 Conducting the Tasks of Step 1 76
  5.3 Summary of Step 1 78

6 Customer Presentation of the Target 81
  6.1 Overview of Step 2 81
  6.2 Conducting the Tasks of Step 2 83
    6.2.1 Presentation of the CORAS Terminology and Method . . . 83
    6.2.2 Presentation of the Goals and Target of the Analysis 86
    6.2.3 Setting the Focus and Scope of the Analysis 89
    6.2.4 Determining the Meeting Plan 91
  6.3 Summary of Step 2 94

7 Refining the Target Description Using Asset Diagrams 95
  7.1 Overview of Step 3 95

	7.2	Conducting the Tasks of Step 3 97
		7.2.1 Presentation of the Target by the Analysis Team 97
		7.2.2 Asset Identification 101
		7.2.3 High-level Analysis 106
	7.3	Summary of Step 3 109
8	**Approval of the Target Description** 111	
	8.1	Overview of Step 4 111
	8.2	Conducting the Tasks of Step 4 113
		8.2.1 Approval of the Target Description 114
		8.2.2 Ranking of Assets 115
		8.2.3 Setting the Consequence Scales 116
		8.2.4 Setting the Likelihood Scale 118
		8.2.5 Defining the Risk Function 120
		8.2.6 Deciding the Risk Evaluation Criteria 122
	8.3	Summary of Step 4 124
9	**Risk Identification Using Threat Diagrams** 125	
	9.1	Overview of Step 5 125
	9.2	Conducting the Tasks of Step 5 128
		9.2.1 Categorising Threat Diagrams 128
		9.2.2 Identification of Threats and Unwanted Incidents 129
		9.2.3 Identification of Threat Scenarios 133
		9.2.4 Identification of Vulnerabilities 137
	9.3	Summary of Step 5 144
10	**Risk Estimation Using Threat Diagrams** 147	
	10.1	Overview of Step 6 147
	10.2	Conducting the Tasks of Step 6 149
		10.2.1 Likelihood Estimation 150
		10.2.2 Consequence Estimation 154
		10.2.3 Risk Estimation 157
	10.3	Summary of Step 6 163
11	**Risk Evaluation Using Risk Diagrams** 165	
	11.1	Overview of Step 7 165
	11.2	Conducting the Tasks of Step 7 167
		11.2.1 Confirming the Risk Estimates 167
		11.2.2 Confirming the Risk Evaluation Criteria 168
		11.2.3 Providing a Risk Overview 169
		11.2.4 Accumulating Risks 170
		11.2.5 Estimating Risks with Respect to Indirect Assets 173
		11.2.6 Evaluating the Risks 182
	11.3	Summary of Step 7 185

12 Risk Treatment Using Treatment Diagrams 187
  12.1 Overview of Step 8 187
  12.2 Conducting the Risk Treatment 188
    12.2.1 Grouping of Risks 189
    12.2.2 Treatment Identification 191
    12.2.3 Treatment Evaluation 196
  12.3 Summary of Step 8 203

Part III Selected Issues

13 Analysing Likelihood Using CORAS Diagrams 207
  13.1 Using CORAS Diagrams to Calculate Likelihood 208
    13.1.1 Specifying Likelihood Using CORAS Diagrams 208
    13.1.2 Rules for Calculating Probability in CORAS Diagrams . . 210
    13.1.3 Rules for Calculating Frequency in CORAS Diagrams . . . 222
    13.1.4 Likelihood as Probability or Frequency 226
    13.1.5 Generalisation to Intervals and Distributions 227
  13.2 Using CORAS Diagrams to Check Consistency 229
  13.3 Using CORAS to Analyse Scenarios with Logical Connectives . . 233
    13.3.1 Using CORAS to Analyse Scenarios with Logical
           Conjunction 233
    13.3.2 Using CORAS to Analyse Scenarios with Logical
           Disjunction 236
  13.4 How to Structure a Threat Diagram to Exploit the Potential for
       Likelihood Analysis 237
    13.4.1 Enabling Application of Rules by Composition 237
    13.4.2 Enabling Application of Rules by Decomposition 239
  13.5 Summary 243

14 The High-level CORAS Language 245
  14.1 Referring Elements and Referenced Diagrams 246
    14.1.1 Threat Scenarios 247
    14.1.2 Unwanted Incidents 250
    14.1.3 Risks 251
    14.1.4 Treatment Scenarios 253
  14.2 Likelihoods in High-level CORAS 257
    14.2.1 Reasoning About the Likelihoods in a High-level Diagram . 260
    14.2.2 Reasoning About the Likelihoods in a Referenced Diagram 261
    14.2.3 Analysing the Relation Between the Likelihoods of a
           Referring Element and the Likelihoods in the Referenced
           Diagrams 263
  14.3 Consequences in High-level CORAS 264
  14.4 Risk Levels in High-level CORAS 266
  14.5 How to Schematically Translate High-level CORAS Diagrams
       into English Prose 267

Contents

	14.5.1 Referring Elements	267
	14.5.2 Referenced Diagrams	270
14.6	Example Case in High-level CORAS	271
	14.6.1 Threat Diagram	272
	14.6.2 Risk Diagram	275
	14.6.3 Treatment Diagram	277
14.7	Summary	279

15 Using CORAS to Support Change Management 283
  15.1 Classification of Changes 283
    15.1.1 Target of Analysis 284
    15.1.2 Scope and Focus 285
    15.1.3 Environment 285
    15.1.4 Assumptions 285
    15.1.5 Parties and Assets 286
    15.1.6 Context 286
    15.1.7 Changes in our Knowledge 287
  15.2 Managing Change 287
    15.2.1 Maintenance Perspective 288
    15.2.2 Before-after Perspective 290
    15.2.3 Continuous Evolution Perspective 294
  15.3 Summary 296

16 The Dependent CORAS Language 297
  16.1 Modelling Dependencies Using the CORAS Language ... 298
    16.1.1 Dependent CORAS Diagrams 299
    16.1.2 Representing Assumptions Using Dependent CORAS Diagrams 300
    16.1.3 How to Schematically Translate Dependent CORAS Diagrams into English Prose 303
  16.2 Reasoning and Analysis Using Dependent CORAS Diagrams ... 305
    16.2.1 Assumption Independence 307
    16.2.2 Assumption Simplification 308
    16.2.3 Target Simplification 309
    16.2.4 Assumption Consequence 310
  16.3 Example Case in Dependent CORAS 311
    16.3.1 Creating Dependent Threat Diagrams 311
    16.3.2 Combining Dependent Threat Diagrams 313
  16.4 Summary 316

17 Using CORAS to Analyse Legal Aspects 319
  17.1 Legal Risk 319
  17.2 Uncertainty of Legal Aspects 321
    17.2.1 Legal Uncertainty 322
    17.2.2 Factual Uncertainty 323
    17.2.3 Combining Legal and Factual Uncertainty 324

	17.3 Modelling Legal Aspects Using the CORAS Language 326
	17.3.1 Legal CORAS Diagrams 326
	17.3.2 How to Schematically Translate Legal CORAS Diagrams into English Prose 328
	17.4 Analysing Legal Aspects through the Eight Steps of CORAS . . . 330
	17.5 Summary 337

18 The CORAS Tool 339
    18.1 Main Functionality of the CORAS Tool 339
    18.2 How to Use the CORAS Tool During Risk Analysis 341
        18.2.1 Initial Modelling Before a Meeting 341
        18.2.2 On-the-fly Modelling During a Meeting 342
        18.2.3 Revising and Analysing Diagrams After a Meeting 344
    18.3 Integration with Other Tools 344
    18.4 Summary 345

19 Relating CORAS to the State of the Art 347
    19.1 Risk Modelling 347
    19.2 Risk Analysis Methods 350
    19.3 Likelihood Analysis 351
    19.4 High-level Risk Modelling 353
    19.5 Change Management 354
    19.6 Dependency Analysis 355
    19.7 Legal Risk Management 356

Appendix A The CORAS Language Grammar 359
    A.1 Basic CORAS 359
        A.1.1 Meta-model 359
        A.1.2 EBNF Grammar 364
        A.1.3 Examples 366
    A.2 High-level CORAS 370
        A.2.1 Meta-model 370
        A.2.2 EBNF Grammar 374
        A.2.3 Examples 377
    A.3 Dependent CORAS 383
        A.3.1 Meta-model 383
        A.3.2 EBNF Grammar 385
        A.3.3 Example 386
    A.4 Legal CORAS 387
        A.4.1 Meta-model 387
        A.4.2 EBNF Grammar 388
        A.4.3 Example 389

Appendix B The CORAS Language Semantics 391
    B.1 Basic CORAS 391
        B.1.1 Elements 392

		B.1.2	Relations	392
		B.1.3	Diagrams	393
		B.1.4	Examples	394
	B.2	High-level CORAS		398
		B.2.1	Referring Elements	398
		B.2.2	Relations	399
		B.2.3	Referenced Diagrams	402
		B.2.4	Examples	403
	B.3	Dependent CORAS		410
		B.3.1	Border	410
		B.3.2	Dependent Diagrams	411
		B.3.3	Example	411
	B.4	Legal CORAS		412
		B.4.1	Elements	412
		B.4.2	Relations	412
		B.4.3	Example	415

Appendix C The CORAS Guidelines 417

- C.1 Step 1: Preparations for the Analysis 418
- C.2 Step 2: Customer Presentation of the Target 418
  - C.2.1 Step 2a: Presentation of the CORAS Terminology and Method 419
  - C.2.2 Step 2b: Presentation of the Goals and Target of the Analysis 419
  - C.2.3 Step 2c: Setting the Focus and Scope of the Analysis 420
  - C.2.4 Step 2d: Determining the Meeting Plan 420
- C.3 Step 3: Refining the Target Description Using Asset Diagrams . . . 421
  - C.3.1 Step 3a: Presentation of the Target by the Analysis Team . . 422
  - C.3.2 Step 3b: Asset Identification 422
  - C.3.3 Step 3c: High-level Analysis 422
- C.4 Step 4: Approval of the Target Description 423
  - C.4.1 Step 4a: Approval of the Target Description 423
  - C.4.2 Step 4b: Ranking of Assets 424
  - C.4.3 Step 4c: Setting the Consequence Scales 424
  - C.4.4 Step 4d: Setting the Likelihood Scale 424
  - C.4.5 Step 4e: Defining the Risk Function 424
  - C.4.6 Step 4f: Deciding the Risk Evaluation Criteria 425
- C.5 Step 5: Risk Identification Using Threat Diagrams 425
  - C.5.1 Step 5a: Categorising Threat Diagrams 426
  - C.5.2 Step 5b: Identification of Threats and Unwanted Incidents . 427
  - C.5.3 Step 5c: Identification of Threat Scenarios 427
  - C.5.4 Step 5d: Identification of Vulnerabilities 428
- C.6 Step 6: Risk Estimation Using Threat Diagrams 428
  - C.6.1 Step 6a: Likelihood Estimation 429
  - C.6.2 Step 6b: Consequence Estimation 429

        C.6.3 Step 6c: Risk Estimation 430
   C.7 Step 7: Risk Evaluation Using Risk Diagrams 430
        C.7.1 Step 7a: Confirming the Risk Estimates 431
        C.7.2 Step 7b: Confirming the Risk Evaluation Criteria 431
        C.7.3 Step 7c: Providing a Risk Overview 432
        C.7.4 Step 7d: Accumulating Risks 432
        C.7.5 Step 7e: Estimating Risks with Respect to Indirect Assets . 432
        C.7.6 Step 7f: Evaluating the Risks 433
   C.8 Step 8: Risk Treatment Using Treatment Diagrams 433
        C.8.1 Step 8a: Grouping of Risks 434
        C.8.2 Step 8b: Treatment Identification 434
        C.8.3 Step 8c: Treatment Evaluation 435

Appendix D The CORAS Terminology 437

Appendix E Glossary of Terms 445
   E.1 Logic 445
   E.2 Sets 445
   E.3 Likelihoods 445
   E.4 Likelihood Intervals 446
   E.5 Deductions 447
   E.6 Extended Backus-Naur Form 447
   E.7 Semantics 448
   E.8 Miscellaneous 448

Acronyms 449

References 451

Index 455

Part I
Introductory Overview

Chapter 1
Introduction

This book is about risk analysis. The term "risk" is known from many fields. On an almost daily basis, we face references to "contractual risk", "economic risk", "operational risk", "environmental risk", "health risk", "political risk", "legal risk", "security risk", and so forth. In order to identify and assess risks, we may conduct risk analysis. The exact nature of an analysis, however, vary considerably depending on the nature of the risks we address. We classify risk analysis approaches into two main categories:

- Offensive approaches: Risk analysis concerned with balancing potential gain against risk of investment loss. Typically, the greater the potential gain the more we are prepared to risk. This kind of risk analysis is for example relevant within finance and political strategy making.
- Defensive approaches: Risk analysis concerned with protecting what is already there. For example, when we build a nuclear power plant we want to protect the surrounding environment, or if you are a business owner you would like to protect the information about your customers to avoid loosing them to a competitor.

This book is mainly about defensive risk analysis, and more explicitly about a particular approach to defensive risk analysis called CORAS.[1]

1.1 The Importance of Risk Analysis

The need for understanding, managing and mitigating risks is apparent in a wide range of domains and areas, including economy, defence, civil protection, information security, health and politics. The variety of stakeholders that are exposed to risks and the organisations or actors that need to manage risk are also wide-ranging, and include businesses, customers, individuals, corporations, governmental and non-governmental organisations, and so forth.

---

[1] CORAS is the name of our approach, and is written in capital letters; it is not, and has never been, an acronym.

The continuous exposure to greater or lesser risks is hardly more evident than in the information society of today, as documented by several annual and biennial surveys [12, 15, 65, 73, 80]. The heavy dependence on computerised information systems and networks means that the importance of systems and information security has never been higher [15]. End-users face the risks of Internet fraud, identity theft, spam, phishing attacks, and so forth. Security incidents occur on a daily basis within most enterprises, where virus attacks, insider abuse of networks and theft of mobile devices are among the most frequent. Whereas many of the incidents are easily recovered from, the average yearly losses caused by computer security incidents are significant; the respondents of a CSI survey reported an average yearly loss of USD 289,000 in 2008 [11]. The potential impact of a single incident may, however, be massive. In January 2007, TJX Companies Inc. lost at least 45 million sets of credit card details, and in January 2009, Heartland Payment Systems exposed data on 130 million credit card users, both incidents due to hacker attacks. In October 2009, hard drives sent for repair exposed data on 76 million US Army veterans [73]. Incidents of that kind can of course not be measured in terms of monetary loss alone as loss of credibility and customer trust can put enterprises out of business.

Security incidents can have disastrous impact. In 2003, the Davis-Besse nuclear plant in Ohio was penetrated by the Slammer worm which disabled a safety monitoring system for nearly five hours [63]. The so-called Northeast Blackout of 2003 was a failure in the electrical power grid that was partly caused by a software bug. The incident left 50 million people in North America without power [64], and affected among other things transportation, communication, industry and water supply.

A comprehensive forensic investigation of data breach cases [4] reported that attacks tend to be not highly difficult, and that they are largely opportunistic rather than targeted. Due partly to this "nearly all breaches would likely have been prevented if basic security controls had been in place at the time of the attack". The report moreover claims that "efforts to locate, catalogue, track, and assess the risk of data stored in and flowing through information assets are highly beneficial in reducing the likelihood of data compromise". A further security breaches survey [15] reveals that "companies that carry out risk assessment are four times as likely to detect identity theft as those that do not", and "twice as likely to detect unauthorised access by staff or attacks on network traffic". The survey also found "a clear correspondence between carrying out a formal risk assessment and the clarity of senior management's understanding", and that risk assessments increase both the understanding of the commercial implications of security decisions and the security awareness. Nevertheless, as much as 52% of companies do not carry out any formal security risk assessment [15]. The report on the survey therefore recommends the use of risk assessment to target the security investments at the most beneficial areas, and to understand the security threats by drawing on the right knowledge sources.

1.2 Asset Identification

In order to defend something, it is important to know exactly what we are defending. Moreover, in order to build up a defence for those things we are defending, it seems sensible to start by identifying the potential ways in which these things may be harmed. This has motivated the invention of what we refer to as asset-driven risk analysis methods. An asset is anything of value to the party on whose behalf the analysis is conducted. In other words, the assets are the things a defensive risk analysis aims to identify means to protect. Although an asset in principle is anything of value, a risk analysis will typically focus only the most valuable ones. Assets may be physical like the "Gold reserve of the Bank of England", the "Health of the inhabitants of Lichtenstein", the "Opera House of Sydney" or the "Online Banking Service of Deutche Bank". Assets may also be conceptual like the "Reputation of Airbus" or the "Trust in French Government services".

In asset-driven risk analysis, the assets to be protected are identified early on as part of the characterisation of the system, service, business or process to be analysed, in the following often referred to as the target of the analysis. Moreover, everything that happens thereafter in the analysis is driven by these assets. In many risk analyses, the target of analysis is broad and the amount of documentation large. To have a clear picture of the assets to protect helps in distinguishing the more relevant documentation from the less so. Not all methods for defensive risk analysis are asset-driven. Within some fields, like environmental risk or safety the assets are more or less given and the identification of assets is no real issue in the analysis process. In other fields, like security, asset identification is an integrated task in any approach to risk analysis. The CORAS method on which we focus in this book is highly asset-driven.

1.3 Risk Modelling

In order to analyse something, we need a clear picture of what this something is. Typically, the customer of a risk analysis will provide piles of documentation, such as technical specifications, program listings, procedures and policies. Its level of quality, the extent to which it is up to date, and its level of abstraction will normally vary a lot. In most cases, when conducting a risk analysis, we are in a situation where some of the customer's documentation may be used as basis for the analysis while other aspects have to be produced or adapted during the analysis.

Modelling the behaviour of the target as it exists or is planned to be built is, however, not the only aspect of modelling in relation to a risk analysis; modelling what can go wrong is even more important. In fact, this is what risk analysis is all about. We then speak of risk modelling. It is the specialised support for risk modelling that keeps CORAS apart from other approaches to risk analysis, and this aspect is a major topic in this book.

1.4 The CORAS Approach

CORAS basically consists of three artefacts, namely a language, a tool and a method. We often refer to the three artefacts together as the CORAS approach.

1.4.1 The CORAS Language

The CORAS language is a customised language for risk modelling. The language is diagrammatic. It uses simple graphical symbols and relations between these to facilitate diagrams that are easy to read and that are suitable as a medium for communication between stakeholders of diverse backgrounds. In particular, CORAS diagrams are meant to be used during brainstorming sessions where the discussion is documented along the way.

The core part of the language is referred to as the basic CORAS language. It offers five different kinds of diagrams, each supporting a particular stage in a risk analysis process. Further modelling support is provided by three extensions:

- The high-level CORAS language supports abstraction and easily comprehensible overviews of large risk models.
- The dependent CORAS language supports documentation of assumptions and dependencies as well as modular reasoning.
- The legal CORAS language supports documentation of legal aspects and their impact.

1.4.2 The CORAS Tool

The CORAS tool supports the CORAS language and is basically a graphical editor for making any kind of CORAS diagram. It is well-suited for creating risk models on-the-fly during brainstorming sessions. The tool furthermore facilitates the documentation and presentation of risk analysis results.

1.4.3 The CORAS Method

The CORAS method is a method for asset-driven defensive risk analysis into which the tool-supported risk modelling language is tightly interwoven. The method comes with detailed guidelines explaining how to conduct the various stages of a CORAS risk analysis in practice.

A risk analysis using CORAS consists of eight steps that comprise every stage from the initial preparations through deriving the eventual results and conclusions. Each step is divided into concrete, practical subtasks with a clearly defined objective. The practical guidelines explain how to fulfil the objectives, in particular by using the features of the CORAS language, but also by applying specific risk analysis techniques.

1.5 The Generality of CORAS

CORAS is a general approach to risk analysis and has been applied to a large variety of risk analysis targets and concerns within numerous domains. Most of the examples in this book are from the domain of IT security, and this choice of domain has been made for two reasons. The first reason is that much of the existing literature on risk analysis and risk management has, because of its legacy, a strong focus on safety and mechanical systems. Risk analysis is an emerging field within the context of IT security, and we see the need for literature that presents risk analysis in this setting, and that provides thorough explanations of the usage of risk analysis in this domain. The second reason is that we have focused on providing large, continuous examples throughout the book. We believe that this gives the reader a deeper and more coherent understanding of the risk analysis process than smaller, fragmentary examples would give. Consequently, there is less room for providing examples from a variety of domains.

While the focus of the examples in this book is on IT security, they do to some degree also illustrate the generality of CORAS by taking into account issues that go beyond information security. Pure IT security analyses generally focus on three main assets, namely confidentiality, integrity and availability of information. In order to demonstrate risk analysis in a wider and more general setting, we exemplify in this book how assets such as trust and reputation may be taken into account. We furthermore address legal issues, both by demonstrating how compliance with laws and legal regulations can be included as assets in their own respect, and by extending the method and the modelling language to facilitate the explicit modelling and reasoning about legal norms that can be an important source of risk.

The many examples throughout the book also give hints about and demonstrations of the use of CORAS within domains such as safety, physical protection and power supply. Although these further domains of applicability are only quite briefly addressed, there is neither in principle nor in practice anything that prevents the use of CORAS in a full fledged risk analysis within these areas.

The generality of CORAS is ensured by the generality of its underlying risk related concepts, the generality of the asset-driven risk analysis method, and the generality of the risk modelling language. Whether the domain is security, safety, law, civil protection, emergency planning, defence, health, and so forth, the basic principles and the practical approach remain the same. What is required, and what CORAS offers within the category of defensive risk analysis in general, is a coherent approach to systematically identify the core assets and the related risks, as well as adequately documenting the results.

The adequacy and suitability of a given risk analysis method for conducting a specific analysis depend on the objectives and purposes of the analysis. In particular, risk analyses may be conducted to various levels of depth and broadness, irrespective of the domain in question. We classify risk analysis projects into four categories:

1. Analyses involving less than 50 man-hours from the analysis team. CORAS has not been developed for this kind of analyses. Diagrams of the CORAS language

may be useful for documentation or communication purposes, but for this kind of "quick and dirty" analyses a more checklist-oriented approach is required.
2. Analyses involving between 50 and 150 man-hours from the analysis team. CORAS can be used in such projects, but some shortcuts must be made.
3. Analyses involving from 150 to 300 man-hours from the analysis team. This is the main segment addressed by CORAS as described in this book.
4. Analyses involving more than 300 man-hours from the analysis team. Also here CORAS may be used, but special measures must be undertaken to take care of issues related to scalability and detail. For example, if the reason for the many hours is that the analysis should be conducted down to a very fine granularity, it might be necessary to combine CORAS with special purpose description and analysis techniques for different aspects of the target.

1.6 Overall Aim and Emphasis

Our overall aim is to deliver a high-quality textbook on how to conduct asset-driven risk analysis using CORAS. In particular, we have aimed at supporting risk analysts in conducting structured and stepwise risk analyses supported by customised diagrams for information gathering and risk modelling as found in the CORAS risk modelling language. The book is also meant to serve as an introduction to risk analysis in general, as well as an introduction to central risk concepts and notions and their relationships.

CORAS has many features in common with other risk analysis methods. The CORAS approach is based on the ISO 31000 [36] standard on risk management published by the International Organization for Standardization. ISO 31000 is a generalisation of the preceding Australian/New Zealand standard for risk management AS/NZS 4360 [75]. Much of what is specified in these standards is also relevant for CORAS. Nevertheless, this will not be an important issue in this book. The main emphasis is on those aspects of the asset-driven CORAS approach that are specific for CORAS; in particular, its modelling language and the way models are used to drive the analysis process. The presentation of the CORAS approach is meant to be self-contained and do not rely on background knowledge in other risk analysis methods and terminologies.

The main target groups of the book are IT practitioners and students of IT at a graduate or undergraduate level.

1.7 Organisation

The book is divided into three main parts. The introductory part, Part I, introduces and demonstrates the central concepts and notations used in CORAS. The presentation is largely example-driven. In Part II, we go through the same material once more, but in much more detail. We start by presenting the basic CORAS risk modelling language. Then we give a thorough presentation of the CORAS method. We devote a separate chapter to each of the eight steps of the method. After having completed Part II, the reader should know enough to start experimenting with the method in practice. In Part III, we focus on a set of problematic issues that is often encountered in real-life risk analyses and to which CORAS offer helpful advice and support. We furthermore present the CORAS tool, which is a special-purpose editor for the CORAS language, before we close Part III by relating CORAS to the state of the art within risk analysis.

In addition to the three main parts, there are Appendices defining the syntax and semantics of the CORAS risk modelling language, providing the full guideline for the CORAS method, as well as providing an overview of the CORAS terminology and other key terms used throughout this book.

1.7.1 Part I: Introductory Overview

In this part of the book, we motivate and explain the basic principles of risk analysis in general and the asset-driven risk analysis as advocated by CORAS in particular. The presentation touches on most of the central issues in CORAS, but without explaining everything in full detail.

Chapter 1: Introduction.
 This chapter motivates the need for CORAS. It also describes our aims and objectives for the book. Finally, it presents the structure of the book.
Chapter 2: Background and Related Approaches.
 This chapter introduces basic concepts required for the guided tour. This includes the basic notions and ideas of risk analysis. It also provides an overview of alternative approaches and describes in what way these differ from CORAS. A more detailed and technical comparison of the various contributions of the CORAS approach to the state of the art is provided in Chap. 19.
Chapter 3: A Guided Tour of the CORAS Method.
 This chapter presents a guided tour of the CORAS method. We follow two analysts in their interaction with an organisation by which they have been hired to carry out a risk analysis. The analysis is divided into eight main steps, and the chapter devotes a separate section to each of them. The chapter focuses in particular on the use of the CORAS risk modelling language as a means for communication and interaction.

1.7.2 Part II: Core Approach

In this part of the book, we describe two of the three main artefacts of which CORAS consists: The basic CORAS risk modelling language and the CORAS method.

Chapter 4: The CORAS Risk Modelling Language.

This chapter provides a more detailed presentation of the basic CORAS risk modelling language, its syntax and semantics. The chapter furthermore introduces and explains the central concepts of a risk analysis and shows how these concepts are reflected in the CORAS language.

Chapter 5: Preparations for the Analysis.

The CORAS method is divided into eight steps, and this chapter is devoted to the first of these, namely the initial preparations for a risk analysis.

Chapter 6: Customer Presentation of the Target.

This chapter focuses on the second step of the CORAS method which is the introductory meeting with the customer on the behalf of which the analysis is conducted. The main item on the agenda for this meeting is to get the representatives of the customer to present their overall goals of the analysis and the target they wish to have analysed, as well as setting the focus and scope of the analysis.

Chapter 7: Refining the Target Description Using Asset Diagrams.

The third step of the CORAS method as described in this chapter also involves interaction with representatives of the customer. However, this time the analysis team will present their understanding of what they learned at the first meeting and from studying documentation that has been made available to them by the customer. Based on interaction with the customer the analysts will also identify the main assets to be protected and conduct a rough, high-level analysis to identify major enterprise level threat scenarios and vulnerabilities that should be investigated further.

Chapter 8: Approval of the Target Description.

The fourth step of the CORAS method involves presenting a more refined description of the target to be analysed, including assumptions and preconditions being made. Typically, the analysts describe the target using a formal or semi-formal notation. It may be some in-house notation of the customer or UML-related notations like class diagrams, use-case diagrams, sequence diagrams or activity diagrams. At the end of the meeting the description of the target should either be approved or at least the analysts should have a detailed description of changes required for the customer to accept the description of the target. The step furthermore includes defining the scales that will be used for estimating likelihoods, consequences and risk levels, as well as deciding the risk evaluation criteria for each asset. This analysis step concludes the context establishment.

Chapter 9: Risk Identification Using Threat Diagrams.

This chapter focuses on the fifth step of the CORAS method, the risk identification. To identify risks, CORAS makes use of structured brainstorming. Structured brainstorming is a step-by-step walk-through of the target of analysis and is carried out as a workshop. The main idea of structured brainstorming is that since the analysis participants represent different competences, backgrounds and interests, they will view the target from different perspectives and consequently identify more, and possibly other, risks than individuals or a more homogeneous group would have managed. The brainstorming is documented on-the-fly

1.7 Organisation   11

as CORAS threat diagrams. Each threat diagram describes the threats, vulnerabilities, scenarios and incidents of relevance for the risks in question.

Chapter 10: Risk Estimation Using Threat Diagrams.

This chapter is concerned with the sixth step of the CORAS method, namely how to estimate likelihoods and consequences for the identified unwanted incidents. This is typically done in a separate workshop. The values are used to compute the risk value.

Chapter 11: Risk Evaluation Using Risk Diagrams.

This chapter explains how risk are evaluated according to step seven of the CORAS method. The objective is to decide which of the identified risks are acceptable, and which of the risks must be further evaluated for possible treatment. Whether or not the risks are acceptable is determined by using the already defined risk evaluation criteria and the results of the risk estimation.

Chapter 12: Risk Treatment Using Treatment Diagrams.

This chapter is concerned with the eighth and last step of the CORAS method, namely the identification and analysis of treatments. The risks that are found to be unacceptable are evaluated to find means to reduce them. A treatment should contribute to reduced likelihood and/or consequence of an incident. Since treatments can be costly, they are assessed with respect to their cost-benefit, before a final treatment plan is made.

1.7.3 Part III: Selected Issues

In the third part of the book, we address a number of special issues for which the core approach as described in the previous part is not sufficient on its own.

Chapter 13: Analysing Likelihood Using CORAS Diagrams.

The estimation and proper treatment of likelihoods is a major problem in many risk analyses. In this chapter, we explain how the CORAS language can be used to estimate likelihoods and check the consistency of different likelihood estimates.

Chapter 14: The High-level CORAS Language.

This chapter introduces a number of useful high-level constructs for the CORAS risk modelling language. These features are also supported by the CORAS tool and they are a prerequisite to structure large models.

Chapter 15: Using CORAS to Support Change Management.

A risk analysis is based on a description that represents the target of analysis at a given point in time, as well as a set of assumptions about the target and its environment. The results of a risk analysis, including the documented risk picture, are therefore valid only under this description and these assumptions. This chapter presents methods for how to take into account changes that the target of analysis and its surroundings may undergo, and how to update and correct the risk picture accordingly.

Chapter 16: The Dependent CORAS Language.
> This chapter extends the basic CORAS language with constructs to facilitate the documentation of and reasoning about dependencies on the assumptions that we make in a risk analysis. The assumptions may, for example, be about the environment, by which we mean the surrounding things of relevance that may affect or interact with the target of analysis. The new features are supported by the CORAS tool.

Chapter 17: Using CORAS to Analyse Legal Aspects.
> This chapter has special focus on legal issues connected to risk analysis. It distinguishes between factual risks (which are the risks treated in the book up to this point) and legal risks, and extends the CORAS approach to also deal with the latter. Support for analysing and documenting the legal aspects are provided through extensions of the basic CORAS language that also are supported by the CORAS tool.

Chapter 18: The CORAS Tool.
> This chapter presents the CORAS tool. We give a description of the functionality of the tool, and describe how to use it during risk analyses with particular focus on risk identification.

Chapter 19: Relating CORAS to the State of the Art.
> This chapter gives a detailed and more technical comparison of the various contributions of CORAS to the state of the art. The presentation is structured according to the structure of the presentation of CORAS in Part II and Part III of this book.

1.7.4 Appendices

In this part of the book, we give the complete definition of the CORAS risk modelling language as well as the CORAS method.

Appendix A: The CORAS Language Grammar.
> This appendix defines the grammar of the five basic kinds of CORAS diagrams, as well as high-level CORAS, dependent CORAS and legal CORAS. The grammar is defined both by means of an Object Management Group (OMG) style meta-model, and by means of a textual representation using the Extended Backus-Naur Form (EBNF) grammar.

Appendix B: The CORAS Language Semantics.
> This appendix defines the semantics of CORAS diagrams in terms of a schematic and unambiguous translation into English prose.

Appendix C: The CORAS Guidelines.
> This appendix presents the methodological guidelines for the CORAS method.

Appendix D: The CORAS Terminology.
> This appendix defines the terminology for the CORAS method.

Appendix E: Glossary of Terms.
> This appendix gives an overview and explanation of the terms, notations, operators, and so forth that are used throughout this book.

1.8 Colours in CORAS and in this Book

The CORAS language makes use of colours to improve readability and comprehension. The colouring has, however, been defined in such a way that the diagrams are easily understandable also in black and white. To keep down the retail price of the printed book, only the figures defining the syntax of the language are in colour; the rest is printed in black and white.

Chapter 2
Background and Related Approaches

The term "risk" is used in a variety of contexts and domains. We see references to notions like "risk management", "risk analysis", "risk evaluation", "risk treatment", and so fourth, and there are all kinds of methods, processes, frameworks, techniques and tools that are supposed to aid us in dealing with risks. The extent to which the various approaches differ or complement each other is often unclear. The problem partly lies in the lack of precise definitions. Something presented as a method in one context may be regarded as a framework, methodology or technique in another, and risk may be the same as hazard in one domain, and used in the meaning of threat in another. In the first section of this chapter, we make an attempt to sort out such issues, or at least precisely define the meaning of the most important of these terms in this book. Additional terminology is defined in later chapters, and in Appendix D we provide the full CORAS terminology.

The chapter is divided into two main sections. In the first of these, we define our basic terminology; in the second, we present other approaches to risk analysis and relate CORAS to these.

2.1 Basic Terminology

According to [36, 37], *risk management* is coordinated activities to direct and control an organisation with regard to risk. The diagram in Fig. 2.1 has been adapted from [36] and illustrates the seven subprocesses of which the overall risk management process consists. The risk management process is defined as the systematic application of management policies, procedures and practices to the activities of communicating, consulting, establishing the context, and identifying, analysing, evaluating, treating, monitoring and reviewing risk.

The five subprocesses in the middle of Fig. 2.1 constitute what we refer to as the process of *risk analysis*. The purpose of the subprocesses is as follows:

- *Establish the context* is to identify stakeholders and vulnerabilities, and to decide what parts of the system, process or organisation will receive attention.

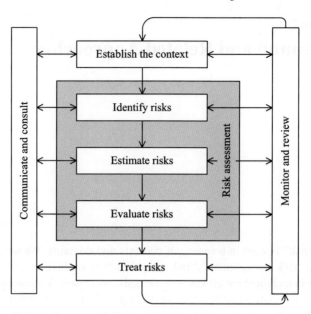

Fig. 2.1 The overall risk management process

- *Identify risks* is to identify potential threats and threat scenarios/incidents that may constitute risks.
- *Estimate risks* is to estimate likelihoods and consequences for the risks that have been identified.
- *Evaluate risks* is to prioritise risks according to their severity/risk level in order to identify the ones that will be subject to treatment.
- *Treat risks* is to find proper treatments.

The two remaining subprocesses *communicate and consult* and *monitor and review* connect the risk analysis process to the rest of the business, system or organisation. The first focuses on interaction with internal and external stakeholders, while the latter addresses the continuous reviewing and monitoring with respect to all aspects of risk.

In this book, we distinguish between risk analysis methods and risk analysis techniques. A risk analysis method provides detailed methodological advice on how to carry out each of these five subprocesses, while a risk analysis technique is more narrow in the sense that it addresses only some aspects of the risk analysis process. A risk analysis method typically makes use of one or more risk analysis techniques.

As pointed out in Chap. 1, we focus on defensive risk analysis; risk analysis concerned with protecting what is already there in the form of assets. Moreover, the CORAS approach is asset-driven in the sense that assets are identified early on and everything that happens thereafter in the risk analysis process is driven by these assets. The Unified Modeling Language (UML) [59] class diagram in Fig. 2.2 models the relationship between core concepts of asset-driven risk analysis.

2.2 Related Approaches

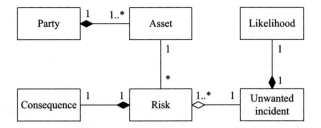

Fig. 2.2 The basic risk terminology

> *Remark* The associations between the elements in a UML class diagram have cardinalities specifying the number of instances of one element that can be related to one instance of the other. The hollow diamond symbolises aggregation and the filled composition. Elements connected with an aggregation can also be part of other aggregations, while composite elements only exist within the specified composition.

An *asset* is something to which a party assigns value and hence for which the party requires protection. A *party* is an organisation, company, person, group or other body on whose behalf the risk analysis is conducted. An *unwanted incident* is an event that harms or reduces the value of an asset, and a *risk* is the likelihood of an unwanted incident and its consequence for a specific asset.

2.2 Related Approaches

As mentioned above, we distinguish between risk analysis methods and risk analysis techniques. The former typically make use of one or more of the latter. In the following, we present two risk analysis methods, OCTAVE and CRAMM. They are both well-established and in many respects related to the CORAS approach in spirit. Thereafter, we present a number of risk analysis techniques classified according to whether they use table-based, tree-based or graph-based documentation. Finally, we align CORAS to this overall picture. The reader is referred to Chap. 19 for a more detailed and technical comparison of the various contributions of the CORAS approach to the state of the art.

2.2.1 Risk Analysis Methods

Operationally Critical Threat, Asset, and Vulnerability Evaluation (OCTAVE) [1] is a risk-based strategic assessment and planning method for security. The method has tree phases. The first phase is dedicated to the identification of critical assets and threat profiles, the second to key components and vulnerabilities and the third and final phase to risks and risk mitigation. During an analysis, several workshops with structured brainstorming sessions are organised to obtain the required information. The results are documented in tables and by using a form of threat trees.

CCTA Risk Analysis and Management Method (CRAMM) is based on the UK Government's Risk Analysis and Management Method [5]. In this approach, risk analysis is understood as the identification and assessment of security risks, while risk management is concerned with identifying appropriate treatments for such risks. The method has three phases dedicated to asset identification, threat and vulnerability identification, and countermeasure identification, and the required information is usually collected by means of interviews. The documentation is made using a standardised CRAMM table format. CRAMM may help an organisation to achieve compliance with ISO 17799 [35], and the outcome is furthermore compliant with the mandatory documentation needed to achieve ISO 27001 [34] certification.

2.2.2 Table-based Risk Analysis Techniques

Hazard and Operability (HazOp) analysis [33] is a well-known risk identification technique that in one form or another is integrated into most risk analysis methods. A HazOp analysis gathers a number of participants that identifies hazards towards a target system through a kind of structured brainstorming. The brainstorming is structured around system documentation and a set of guidewords that the HazOp leader uses to formulate questions about the components of the system. The results are presented in HazOp tables that document the hazards, the relations to system components, the keywords used for identifying the hazards, and so fourth. HazOp can be tailored to fit any situation; for example, in [66], the method is used to assess risks related to software.

Failure Mode Effect Analysis/Failure Mode Effect and Criticality Analysis (FMEA/FMECA) [9] is a bottom-up approach that is particularly suitable for detecting a system's possible failure modes and for determining their consequences. The failure identification is normally organised as a brainstorming that is structured according to the system's functional descriptions. Failure modes for each component are identified and their propagation to other components are explored. Finally, the criticality and probability of the failure modes are assessed. The findings are documented in FMEA/FMECA tables that list the potential failure modes for each component, as well as their effects, criticality and other relevant information.

2.2.3 Tree-based Risk Analysis Techniques

Fault Tree Analysis (FTA) [30] is a risk analysis technique based on so-called fault trees. Variations of FTA and fault trees are well known and widely used within

2.2 Related Approaches

risk analysis. The fault tree notation enables structuring of the order of and decomposition of events, and is particularly useful when quantitative statistical data are available. Fault trees may for example be used to model the findings of HazOp analyses [74]. The top node of a fault tree represents an unwanted incident, or failure, and the different events that can lead to the top event are modelled as intermediate nodes or leaf nodes. The probability of the top node is calculated based on the probability of the leaf nodes and the logical gates representing logical conjunction and logical disjunction. Fault trees can be used both qualitatively to specify the different paths that lead to the unwanted incident, as well as quantitatively to estimate the likelihood of the top node incident.

Event Tree Analysis (ETA) [31] uses a tree notation to represent the outcome (or consequences) of an event and the probability of the various consequences. In the same manner as for fault trees, the event trees are both qualitative (shows the outcomes from an event) and quantitative (estimates the likelihood of each outcome). When constructing an event tree it is normal to use a binary split from the initial event towards the final consequences (success/failure). The event tree lets the modeller specify every detail about the expected outcome of an unwanted incident. This also includes the barriers, or the mechanisms that prevent the consequences of an unwanted incident from escalating, and what the outcome will be if the barriers fail to work.

Attack trees [69] are a modelling notation that aims to provide a formal and methodical way of describing the security of a system based on the attacks it may be exposed to. The notation uses a tree structure similar to FTA, with an attack goal as the top node and different ways of achieving that goal as leaf nodes.

2.2.4 Graph-based Risk Analysis Techniques

Cause-Consequence Analysis (CCA) [57] employs diagrams that combine the features of both the fault trees and the event trees. When constructing a cause-consequence diagram, the starting point is an unwanted incident. From this incident, the diagram is developed backwards to find its causes (fault tree) and forwards to find its consequences (event tree). A cause-consequence diagram illustrates the chain of events from the very beginning, including the initiators of unwanted incidents, to their final consequences towards assets.

A Bayesian network [7, 18] is a directed, acyclic graph that may be used, for example, to predict the number of faults in a software component. The nodes of the graph represent contributing factors and the edges represent dependencies between the nodes. Each node is characterised by a probability distribution over its possible values depending on the values of its parent nodes. For any manipulation of the probabilities of the nodes, the effects both forwards towards child nodes and backwards towards parent nodes can be computed. A Bayesian network can be used both quantitatively and qualitatively. If the Bayesian network is analysed qualitatively, it provides relations between causes and effects. When analysed quantitatively, it serves as a mathematical model for computing probabilities, which depends not only on the probabilities of the leaf nodes like in FTA, but on all the nodes in the graph.

Markov analysis [27, 32, 43] is a stochastic mathematical analysis method that looks at events and sequences events. A Markov analysis considers the system as a number of states, and assigns probabilities to the transitions between these states. The states are modelled graphically, and statistical calculations or simulations are performed to determine the probability of the system reaching particular states or similar properties of the state model. Markov analysis may be used to analyse the reliability of systems that have a high degree of component dependencies. It is also well suited to analyse systems that may fail partially or experience degraded states. Markov models can be suitable for showing the operation modes of a system and changes in these, but less so for representing chains of events related to, for example, a security attack.

2.2.5 Situating CORAS Within this Picture

As mentioned in the preface of this book, the initial version of the CORAS approach was invented and delivered as the main result of a European research project with the same name that ran from 2001 until 2003. What we present as the CORAS approach in this book has been further developed and refined in many respects since then. A major influence has been our experiences from using CORAS in numerous industrial full scale risk analyses; these risk analyses have been within a wide range of domains including banking, telecommunication, power-supply, defence, industrial process control and web-based services. Another major influence has been technical developments in more recent research projects like SECURIS, including the influential PhD-research of Ida Hogganvik [23], COBRA, COMA, DIGIT, EMERGENCY, and ENFORCE funded by the Research Council of Norway, as well as the projects iTrust, MASTER, MODELWARE, SecureChange, S3MS and TrustCoM funded by the European Commission.

As already explained, the CORAS approach consists of three main artefacts: (1) A customised diagrammatic language for risk modelling; (2) a tool supporting the language; (3) a method for asset-driven defensive risk analysis into which the tool-supported risk modelling language is tightly interwoven.

There are many relationships between the risk analysis methods and techniques presented above and the CORAS approach. For example, it may be argued that fault trees resembles CORAS diagrams. However, fault trees focus more on the logical decomposition of an incident into its constituents, and less on the causal relationship between events which is the emphasis in CORAS. Event trees are also related to the CORAS notation. Event trees focus on illustrating the (forward) consequences of an event and the probabilities of these. CORAS diagrams on the other hand may be developed in any direction; they can be developed from the assets towards the threats, from the threats towards the assets, or by starting in middle working in all directions, more in the style of CCA.

2.2 Related Approaches

The structure of CORAS diagrams have similarities to Bayesian networks since both are directed acyclic graphs with the edges representing causal dependencies. On the other hand, the models used for analysing likelihoods are different. Bayesian networks focus on powerful probability analysis and require completeness in the specification of contributing factors, while likelihood analysis in CORAS is simpler but more focused towards analysis in situations where we only have incomplete data. Markov analysis is not directly comparable to CORAS. The focus of Markov analysis is on operation and failure modes in state-based systems, and is more suitable for reliability analysis than risk analysis as such. CORAS diagrams, on the other hand, are used to specify scenarios consisting of causal chains of events.

CORAS diagrams can easily be related to table-based documentation. For example, as we will see in later chapters, when we follow the path from a threat to an asset in a CORAS diagram we basically get the information we need to fill in one row in a HazOp table.

Although the CORAS language has similarities with other approaches to risk documentation, it is nevertheless unique in its support for the whole risk analysis process, from asset identification to risk treatment. It also differs from other approaches in that it has been developed to facilitate communication and interaction during structured brainstorming sessions involving people of heterogeneous backgrounds [24, 25]. To this end, the CORAS language makes use of graphical symbols, or icons, that are closely related to the underlying risk analysis concepts, and that are intended to be easily comprehensible. It also offers a schematic procedure allowing the translation of any fragment of a CORAS diagram into a paragraph in English.

Another unique feature of CORAS is the tight interweaving of the CORAS language into everything that takes place during the risk analysis process. The CORAS method offers detailed guidelines for how to fulfil the goals of the various steps and tasks in general, and how to fulfil these goals making efficient use of the CORAS language in particular. Furthermore, to handle the issue of scalability in relation to risk documentation, the CORAS language offers a set of constructs that can be used to structure individual CORAS diagrams. The CORAS approach also offers calculi for various kinds of likelihood reasoning, it provides advice on how to deal with change with respect to analysis documentation and comes with specific support for handling legal issues in relation to risk analysis.

Chapter 3
A Guided Tour of the CORAS Method

This chapter presents a guided tour of the CORAS method. As illustrated by Fig. 3.1, the CORAS method is divided into eight steps. The first four of these steps are introductory in the sense that we use them to establish a common understanding of the target of the analysis, and to make the target description that will serve as a basis for the subsequent risk identification. The introductory steps include documenting all assumptions about the environment or setting in which the target is supposed to work, as well as making a complete list of constraints regarding which aspects of the target should receive special attention, which aspects can be ignored, and so forth. The remaining four steps are devoted to the actual detailed analysis. This includes identifying concrete risks and their risk level as well as identifying and assessing potential treatments for unacceptable risks.

In the following sections, we go through each of the eight steps of the CORAS method by means of a running example from the telemedicine domain. We follow two analysts in their interaction with an organisation by which they have been hired to carry out a risk analysis. They conduct the analysis according to the eight steps of the CORAS method.

3.1 Preparations for the Analysis

The purpose of Step 1 is to do the necessary initial preparations prior to the actual startup of the risk analysis. This includes to roughly set the scope and focus of the analysis so that the analysis team can make the necessary preparations. It also includes informing the customer of its responsibilities regarding the analysis. We now introduce our example.

Example 3.1 In one region of the country, an experimental telemedicine system has been set up. A dedicated network between the regional hospital and several primary

---

This chapter is an adaptation of the guided tour to the CORAS method presented in [10].

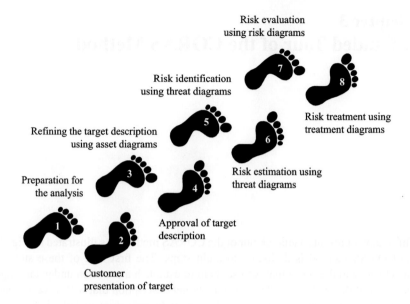

Fig. 3.1 The eight steps of the CORAS method

health care centres (PHCC) allows a general practitioner (GP) to conduct a cardiological examination of a patient (at the PHCC) in cooperation with a cardiologist located at the hospital. During an examination, both of the medical doctors have access to the patient's health record, and all data from the examination is streamed to the cardiologist's computer.

The National Ministry of Health is concerned whether the patient privacy is sufficiently protected, and hires a risk analysis consultancy company to conduct a risk analysis of the cardiology system with particular focus on privacy. The consultancy company appoints a team of two consultants to do the job. They are in the following referred to as "the analysts" and assigned the roles of risk analysis leader and risk analysis secretary, respectively.

As a first step, the analysis leader organises a preparatory meeting with a representative from the ministry. At this meeting, the analysis leader is briefed and provided with documentation and background information. In particular, the representative from the ministry hands over the existing privacy regulations that the system should comply with. The analysis leader highlights the importance of having a fixed interaction point for the analysts at the ministry throughout the analysis. In particular, the ministry is asked to appoint one person that will serve as contact point and also be present at all important meetings throughout all of the subsequent steps of the analysis. The analysis leader also presents a plan for the analysis. It is decided that the effort on behalf of the analysts should be 200 man-hours and the plans are adjusted to fit with that. They also agree that the analysis should be completed within three months and a tentative meeting schedule is worked out. The analysis leader presents the objectives of the different meetings and what the objectives imply when it comes to selecting participants to the meetings. For example, during risk identification it is important to involve technical expertise, while the presence of decision makers is essential at the initial meetings to help define the focus and scope of the analysis.

3.2 Customer Presentation of the Target

Step 2 involves an introductory meeting. The main item on the agenda for this meeting is to get the representatives of the customer to present their overall goals of the analysis and the target they wish to have analysed. Hence, during the second step, the analysts will gather information based on the customer's presentations and discussions.

Before starting to identify and analyse potential risks to something, it is necessary to know exactly what this something is. What is the scope of the analysis, and what are the assumptions that we may make? In other words, we need to know what we are supposed to protect before we can start finding the threats against it and how it may be harmed, as well as how it should be protected. It is furthermore essential that the parties of the risk analysis and the analysts agree on a common terminology and how it should be used. They also need to arrive at a joint understanding of what should be the target of analysis, the assets to protect, the scope and focus, as well as all assumptions being made.

Example 3.2 A meeting is organised where, in addition to the analysts and a representative from the ministry, the IT manager of the regional hospital and a general practitioner from one of the PHCCs participate.

This meeting is where the overall setting of the analysis is decided, and the first step is taken towards establishing the target description that will be used later in the analysis. The meeting starts with the risk analysis leader giving a brief presentation of the method to be used, what the customer (the National Ministry of Health) can expect from the analysis, and a proposed meeting schedule. The analysis leader reminds the representative of the ministry of the responsibilities with respect to providing necessary information and documentation about the target in question, as well as allocating people with suitable background to participate at the scheduled meetings and workshops.

The IT manager then presents the telemedicine system intended as target. As part of the presentation, she draws the picture shown in Fig. 3.2. From the picture, we see that speech and other data from the examination of a patient is streamed over a dedicated network, while access to the patients' health record (stored in a database at the regional hospital) is given through an encrypted channel over the Internet. Next in line after the IT manager is the medical doctor from the PHCC. She talks about her personal experiences from using the system.

After the presentations, a discussion on the scope and focus of the analysis follows. The representative of the ministry emphasises that they are particularly worried about the confidentiality and integrity of the health records and other medical data, first and foremost for the sake of the patients' health, but also because of the

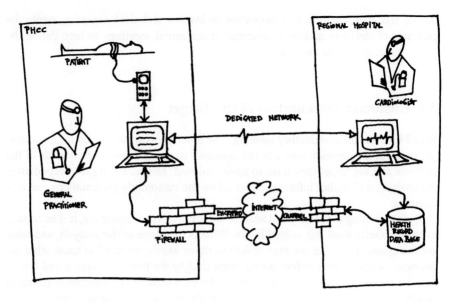

Fig. 3.2 Picture of target

public's trust in the national health care system. For the medical doctor, the most important thing is the patients' health and well-being, and hence also the availability and integrity of the telemedicine system. The IT manager explains that they have already made a security analysis of the health record database and the encrypted access, so she is confident that this part of the system is secure and reliable. After some discussion, the representative of the ministry decides that the focus will be on confidentiality and integrity of medical data, and the availability of the service, but that the access to the health record database is outside the scope of analysis.

As the last point on the agenda, the participants set up a plan for the rest of the analysis with dates for meetings and delivering of reports, as well as indications of who should attend the various meetings.

3.3 Refining the Target Description Using Asset Diagrams

The objective of Step 3 is to arrive at a more correct and refined understanding of the target and the objectives of the customer. Also this step typically involves a meeting between the analysts and the representatives of the customer. The meeting is divided into three parts: (1) presentation of the target as understood by the analysts; (2) asset identification; (3) high-level risk analysis.

The purpose of the presentation of the of the target by the analysts is to correct misunderstandings on behalf of the analysts and to settle issues in need of clarification. The asset identification involves pinpointing the most important valuables of the parties of the analysis. The parties typically include the customer, but may also

3.3 Refining the Target Description Using Asset Diagrams

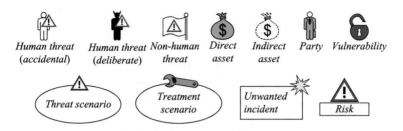

Fig. 3.3 Symbols of the CORAS risk modelling language

be other relevant stakeholders with respect to the target in question. The assets are the things or entities that these parties want to protect, and are the real motivation for conducting the risk analysis in the first place. The identified assets are documented using so-called asset diagrams. Asset diagrams are one of five kinds of diagrams offered by the CORAS risk modelling language. The other four play important roles in later steps of the CORAS method as we will see. Common for all five kinds of diagrams is that they make use of partly overlapping subsets of the graphical symbols presented in Fig. 3.3. In the case of asset diagrams, the subset consists of the two symbols for asset, and the one for party.

The main purpose of the high-level analysis is to get an overview of the main threats and risks with respect to the identified assets, in particular, at an enterprise level and from the perspective of the decision makers. The high-level analysis helps the analysts in identifying the aspects of the target that have the most urgent need for in-depth analysis, and hence makes it easier to define the exact scope and focus of the full analysis.

Example 3.3 The meeting starts with the analysis leader presenting the analysts' understanding of the target to be analysed. The analysts have formalised the information presented by the customer at the previous meeting, as well as the documentation received in the mean time. It was decided to use UML for this formalisation. The UML class diagram of Fig. 3.4 shows the relevant concepts and how they relate to each other, while the UML collaboration diagram of Fig. 3.5 illustrates the physical organisation of the target. Furthermore, the medical doctor's description of use has been captured as a UML activity diagram as shown in Fig. 3.6. During this presentation, the participants representing the customer make corrections and eliminate errors, so that the result is a target description that all parties can agree upon. In the class diagram and the collaboration diagram, the analysis leader has also indicated what he understands is the scope of the analysis.

After agreeing on a target description, the analysis moves on to asset identification. An asset is something in or related to the target to which the customer or other party of the analysis assigns great value. Based on the discussion at the introductory meeting, the analysis leader has prepared the initial *CORAS asset diagram* of Fig. 3.7 to help specifying the scope of the analysis. The asset diagram shows the National Ministry of Health as the party on whose behalf the assets are identified, and its four assets *Health records, Provision of telecardiology service, Patients'*

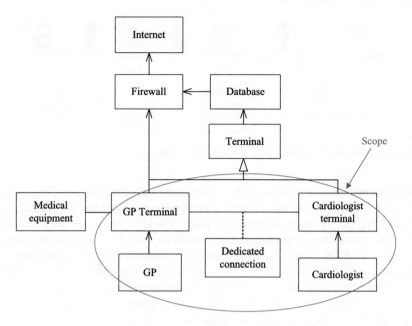

Fig. 3.4 Class diagram showing the target concepts

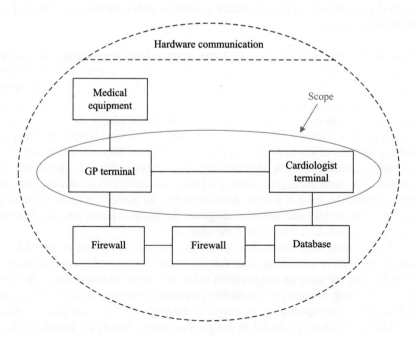

Fig. 3.5 Collaboration diagram illustrating the physical communication lines

3.3 Refining the Target Description Using Asset Diagrams

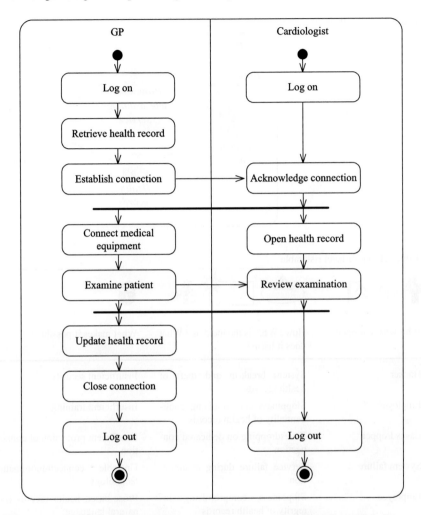

Fig. 3.6 Activity diagram describing the parallel processes of the GP and the cardiologist

health and *Public's trust in system*. The arrows show dependencies between the assets, such that for example, harm to *Health records* may cause harm to *Public's trust in health care system*.

The analysis leader explains that *Public's trust in system* is represented with a slightly different symbol than the other three because this is a so-called indirect asset. He goes on to explain that an asset is indirect if, with respect to the target of analysis, it is harmed only through harm to other assets. The remaining assets are direct. In Steps 4, 5 and 6, the indirect asset may to a large extent be ignored since risks with respect to this asset can be identified by identifying risks with respect to the direct assets. The analysts still need to provide a risk picture for the indirect asset during the risk evaluation of Step 7. It follows from the diagram that risks to

Fig. 3.7 Asset diagram

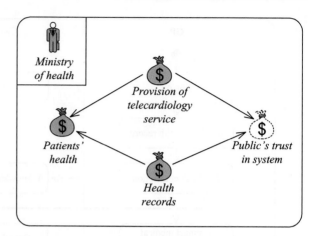

Table 3.1 High-level risk table

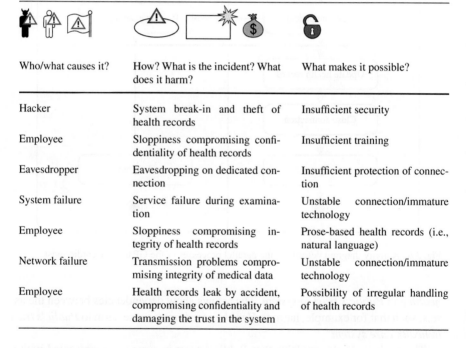

Who/what causes it?	How? What is the incident? What does it harm?	What makes it possible?
Hacker	System break-in and theft of health records	Insufficient security
Employee	Sloppiness compromising confidentiality of health records	Insufficient training
Eavesdropper	Eavesdropping on dedicated connection	Insufficient protection of connection
System failure	Service failure during examination	Unstable connection/immature technology
Employee	Sloppiness compromising integrity of health records	Prose-based health records (i.e., natural language)
Network failure	Transmission problems compromising integrity of medical data	Unstable connection/immature technology
Employee	Health records leak by accident, compromising confidentiality and damaging the trust in the system	Possibility of irregular handling of health records

Public's trust in system are identified via risks to *Health records* and *Provision of telecardiology service*. The asset *Patients' health* may also be harmed via harm to the latter two, but may also be harmed directly in other ways within the scope of the analysis and is therefore specified as a direct asset.

After agreeing on the assets, the analysts conduct a high-level analysis. The short brainstorming should identify the most important threats and vulnerabilities with respect to the identified assets, without going into great detail. In this case, the customer is concerned about hackers, eavesdroppers, system failure and whether the security mechanisms are sufficient. These threats and vulnerabilities do not necessarily involve major risks, but give the analysis leader valuable input on where to start the analysis. The analysis secretary documents the results by filling in the high-level risk table shown in Table 3.1. The symbols above the three columns indicate what kind of information should be documented.

3.4 Approval of the Target Description

Step 4 also typically involves a separate meeting, but may alternatively be conducted by email or other means of communication. The main objective of Step 4 is to agree on the description of the target to be analysed, including scope, focus and all assumptions, and for the customer to approve the description. Important aspects of the target documentation are definitions of scales for likelihoods and consequences as well as risk evaluation criteria. The formulation of these aspects are subtasks of Step 4.

We often need multiple consequence scales, which are used when it is difficult or inappropriate to measure or describe damage to all assets according to the same scale. It is easier, for example, to measure income in monetary values than to do the same for company brand. There should only be one likelihood scale for the analysis based, for example, on time-intervals such as years, weeks and hours, or on probabilities. The last activity of the approval step is to decide upon the risk evaluation criteria. These criteria characterise the minimal level of risk required for risks to deserve a detailed evaluation for possible treatment. Step 4 should not terminate before the full documentation as prepared by the analysts has been approved by the customer.

Example 3.4 The analysis leader has updated his presentation from the last meeting based on input and comments that he received from the representatives of the customer during Step 3, and the target and asset descriptions are now to be finally approved. Based on the discussions at the previous meetings and the issues identified in the high-level analysis, the customer and the analysis team decided to narrow the scope of the analysis, and agree upon the following target definition:

> The target of analysis is the availability of the telecardiology service, and confidentiality and integrity of health records and medical data in relation to use of the service and related equipment.

An unwanted incident is an event that when it occurs harms or reduces the value of at least one of the identified assets. A risk is a characterisation of the severity of an unwanted incident with respect to a single asset. If an unwanted incident harms more than one asset, we get a separate risk for each of the harmed assets. Typically the customer is forced to accept some risks, either because of shortage of resources, conflicting concerns or because the treatment costs will be greater than the benefits. As a first step towards distinguishing risks that can be accepted from those that

Table 3.2 Asset table

Asset	Importance	Type
Health records	2	Direct asset
Provision of telecardiology service	3	Direct asset
Public's trust in system	2	Indirect asset
Patients' health	1	Direct asset

Table 3.3 Likelihood scale

Likelihood value	Description	Definition
Certain	Five times or more per year	$[50, \infty) : 10y = [5, \infty) : 1y$
Likely	Two to five times per year	$[20, 50) : 10y = [2, 5) : 1y$
Possible	Less than twice per year	$[5, 20) : 10y = [0.5, 2) : 1y$
Unlikely	Less than once per two years	$[1, 5) : 10y = [0.1, 0.5) : 1y$
Rare	Less than once per ten years	$[0, 1) : 10y = [0, 0.1) : 1y$

cannot, the representatives of the customer are asked to rank the assets according to their importance (1 = very important, 5 = minor importance) and fill in the asset table as shown in Table 3.2.

Having finished the asset table, they go on to define the likelihood scale and the consequence scales. A likelihood is a general description of the frequency or probability for incidents to occur, and the likelihood scale defines the values that will be used when assigning likelihood estimates to unwanted incidents. A consequence is a description of the impact of unwanted incidents on the assets in terms of degree of damage, and the consequence scale defines the values that will be used when estimating the impact of unwanted incidents.

The analysts initiate the discussion by suggesting a scale of likelihood based on the following rule of thumb: The lowest likelihood *rare* is set to be maximum one occurrence during the target's lifetime; the remaining intervals have an increasing number of expected events until the maximum possible number of incidents per year is reached. Table 3.3 gives the likelihood scale defined for the target of analysis. The likelihood *possible*, for example, denotes less than twice a year, which is defined by the precise interval from 5 to 20 occurrences per 10 years, as shown in the table. By using the same scale for all scenarios and incidents, it is possible to extract combined likelihood values as shown later in the risk estimation step.

Because incidents may have different impact depending on which asset is harmed, they decide to make a separate consequence scale for each of the direct assets. Table 3.4 shows the consequence scale defined for the asset *Health records* in terms of number of health records that are affected. If desired, the consequence description for an asset may include more than one measure. For example, *major* could be the number of disclosed health records, the number of deleted records, and so forth.

Table 3.4 Consequence scale for *Health records*

Consequence value	Description
Catastrophic	1000+ health records are affected
Major	101–1000 health records are affected
Moderate	11–100 health records are affected
Minor	1–10 health records are affected
Insignificant	No health records are affected

Table 3.5 Risk evaluation matrix

		Consequence				
		Insignificant	Minor	Moderate	Major	Catastrophic
Frequency	Rare					
	Unlikely					
	Possible					
	Likely					
	Certain					

Finally, the representatives of the customer define the risk evaluation criteria. The risk evaluation criteria assert whether a risk to an asset should be evaluated further or not. A risk that is not accepted according to the risk evaluation criteria may nevertheless have to be accepted as a result of the cost-benefit analysis conducted when deciding how to respond to the conclusions from the risk analysis. They define these criteria by means of a risk evaluation matrix for each asset. The risk analysis leader draws the matrix for the asset *Health records* on a blackboard. It has likelihood and consequence values as its axes so that a risk with a specific likelihood and consequence will belong to the intersecting cell. Based on a discussion in the group, the risk analysis leader marks the cells in the matrix as either *acceptable* or *unacceptable* (i.e., *must be evaluated*) by filling the cells with the colour green or red, respectively. The resulting risk evaluation matrix is shown in Table 3.5. The participants decide to use these criteria for the other assets as well.

After all this has been approved by the customer, including the target description with the target models, the analysts have the framework and vocabulary they need to start identifying threats (a potential cause of an unwanted incident), vulnerabilities (weaknesses which can be exploited by one or more threats), unwanted incidents and risks.

3.5 Risk Identification Using Threat Diagrams

Step 5 is organised as a workshop gathering people with expertise on the target of analysis. The goal is to identify as many potential unwanted incidents as possible, as well as threats, vulnerabilities and threat scenarios.

To do this identification, we make use of a technique called structured brainstorming. Structured brainstorming may be understood as a structured walk-through of the target of analysis and is carried out as a workshop. The main idea of structured brainstorming is that since the participants of the analysis represent different competences, backgrounds and interests, they will view the target from different perspectives and consequently identify more, and possibly other, risks than individuals or a more homogeneous group would have managed.

The findings of the brainstorming are documented using CORAS threat diagrams, which are the second kind of diagrams offered by the CORAS risk modelling language.

Example 3.5 The analysis leader challenges the participants to work with questions like: What are your biggest concerns with respect to your assets? (Threat scenarios and unwanted incidents.) Who/what may initiate threat scenarios and unwanted incidents? (Threats.) What makes this possible? (Vulnerabilities.) The answers are documented by the secretary on-the-fly using CORAS threat diagrams that are displayed to the participants.

The analysis leader has used this technique on numerous occasions before. He does not employ exactly the same procedure in every case, but adapts it to fit the target domain. He often finds it useful to include checklists and "best practices" for a specific technology or domain. In this case he needs IT experts and medical personnel (general practitioners) to participate in the brainstorming, but some will only participate when their competences are needed for specific scenarios. Since people may be involved at different stages of the analysis, it is essential that information gathered during this session is documented in a simple and comprehensive way.

The analysis leader uses the target models approved during Step 4 as input to the brainstorming session. The models, as exemplified by Figs. 3.4, 3.5 and 3.6, are assessed in a stepwise and structured manner and the identified unwanted incidents are documented on-the-fly.

A set of initial, preliminary threat diagrams has been prepared by the analysis team on the basis of the high-level analysis table shown in Table 3.1. Three of these threat diagrams are shown in Figs. 3.8, 3.9 and 3.10. These may represent a starting point for discussion. The analysis team has structured the three diagrams according to the different kinds of threats: human accidental, human deliberate and

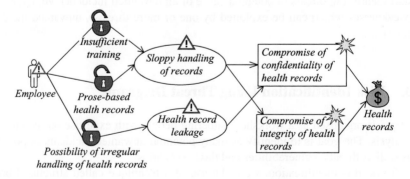

Fig. 3.8 Initial threat diagram for accidental actions

3.5 Risk Identification Using Threat Diagrams

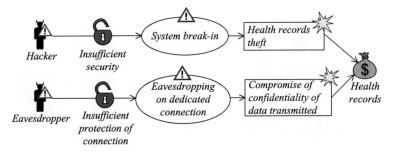

Fig. 3.9 Initial threat diagram for deliberate actions

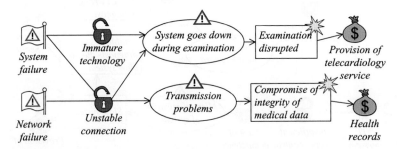

Fig. 3.10 Initial threat diagram for non-human threats

non-human threats. To what extent these diagrams are presented to the participants of the brainstorming session is dependent on the situation. In any case, these initial threat diagrams provide helpful guidance to the analysis leader with respect to what aspects on which to focus.

The threat diagram in Fig. 3.8 shows how a combination of insufficient training or prose-based health records together with sloppiness may compromise the integrity and confidentiality of the patients' health records. The system also allows for irregular handling of health records with the result that an employee accidentally may cause a leakage of records. A confidentiality or integrity breach may harm the health record in the sense that it no longer is secret or correct, respectively. In the extreme consequence, a faulty health record may affect the patients' health.

In the threat diagram of Fig. 3.9 that describes deliberate harmful actions caused by humans, the participants have identified two main threats, namely hacker and eavesdropper. A hacker may exploit insufficient security mechanisms to break into the system and steal health records. An eavesdropper is someone who, due to insufficient protection of communication lines, may gather data that is transmitted and thereby compromise its confidentiality.

The participants also worry about threats like system failure and network failure, as documented in Fig. 3.10. They fear that unstable connections or immature technology are vulnerabilities that may lead to system crashes during examination or transmission problems. A transmission problem may interfere with the data that is stored in the system and leave the health records only partly correct.

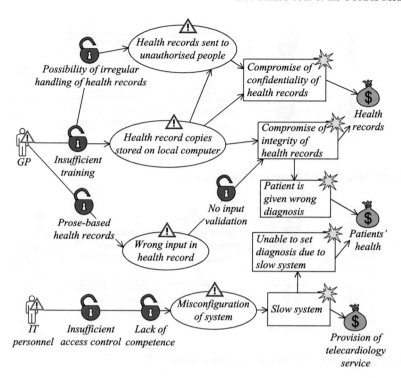

Fig. 3.11 Final threat diagram for accidental actions

During the brainstorming session, the initial threat diagrams are expanded with new information on-the-fly. The diagrams may require updating and polishing after the session has ended. The threat diagram of Fig. 3.8 illustrating incidents caused by accidental actions of employees receives much attention among the participants and develops into the diagram shown in Fig. 3.11. In the following, we concentrate on just this one, and do not explore the other two threat diagrams further.

The participants decide that the threat *Employee* must be split into *general practitioner (GP)* and *IT personnel* since they may cause different incidents. If the GP has too little security training, she may store copies of health records on a local computer. This may compromise the integrity of the records and in the worst case lead to an erroneous diagnosis of a patient. The same incidents may also occur if the GP enters wrong information into the patients' health record. The system allows for irregular handling of health records which opens for the possibility of accidentally sending records to unauthorised people. This would compromise the confidentiality of the health record. The policy of the IT personnel with respect to access control has been very "loose". They explain this with their responsibility for making critical updates in emergencies and that they do not have the time to wait for personnel with correct access rights to show up. An unfortunate consequence of this is that personnel without the required competence sometimes become responsible for critical changes. This may lead to misconfiguration of the system, which again may slow the system down. If the system is too slow it may be impossible to set a patient's diagnosis, and also the ability of providing a telecardiology service may be compromised.

3.6 Risk Estimation Using Threat Diagrams

When the threat scenarios, unwanted incidents, threats and vulnerabilities are properly described in threat diagrams it is time to estimate likelihoods and consequences. This is the main task of Step 6 which is also typically conducted as a structured brainstorming. The likelihoods and consequences are needed in order to compute the risk values which are used to decide whether risks are acceptable or should be further evaluated for possible treatment.

The participants of the brainstorming session provide likelihood estimates based on their judgements or give advice with respect to how they may be determined from historical data that they are aware of. Since risk values are calculated from the likelihoods of unwanted incidents, and not threat scenarios, the unwanted incidents are the main focus of the likelihood estimation. However, if the likelihood of an unwanted incident is hard to determine or very uncertain, we may try to deduce the value from the likelihoods of the threat scenarios and unwanted incidents to which they are directly related. The documentation of information about the likelihoods of threat scenarios is useful also because it shows the most important sources of risks. This gives a more detailed risk picture and furthermore serves as a basis for determining where to direct treatments.

Consequences are estimated for each relation from an unwanted incident to an asset. The consequence values and the likelihood values are taken from the consequence scale of the asset and the likelihood scale, respectively, as defined during Step 4.

Example 3.6 The analysis leader organises the risk estimation as a separate workshop with a structured brainstorming where the starting point is the threat diagrams from the previous workshop. He knows that in this workshop it is especially important to include people with diverging backgrounds and competences, such as users, technical experts and decision makers. The participants of the analysis decide that "most likely" estimates will provide more realistic risk values than "worst case" estimates. First, they provide as many estimates as possible for the threat scenarios, which in turn facilitates the estimation of the likelihood of the unwanted incidents. Second, the consequences of the unwanted incidents for each harmed asset are estimated. The estimates are documented by annotating the diagrams as shown in Fig. 3.12.

There are different ways of computing the likelihood of an incident that may be caused by more than one threat scenario. If the estimates are suitable for mathematical calculations a computerised tool may be used. Since the likelihood scale in our case is in the form of intervals, the analysis leader decides to use an informal method that is quite straightforward and transparent. The threat scenario *Health records sent*

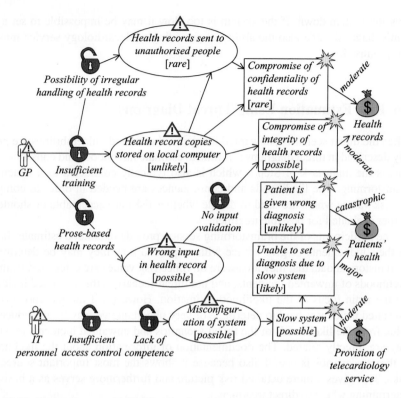

Fig. 3.12 Threat diagram with likelihood and consequence estimates

Table 3.6 Combined likelihood estimates

Threat scenario	Likelihood	Unwanted incident	Combined likelihood
Health records sent to unauthorised people	Rare $([0, 1) : 10y)$	Compromise of confidentiality of health records	$[0, 1) : 10y + [1, 5) : 10y = [1, 6) : 10y$
Health record copies stored on local computer	Unlikely $([1, 5) : 10y)$		It is decided that *unlikely* is the best fit

to unauthorised people and Health record copies stored on local computer can both lead to Compromise of confidentiality of health records. Table 3.6 shows the rough estimation of the combined likelihood. The technique is informal, but suitable for the creative setting of a structured brainstorming. It is of course important that the combined estimates reflect reality, meaning that the combined estimates should be presented to the participants for validation or adjustment.

As shown by Table 3.6, the aggregated likelihood of the two threat scenarios is $[1, 6) : 10y$ which is overlapping both *unlikely* and *possible*. Since the aggregated interval hardly intersects with *possible*, and since it clearly gravitates towards *unlikely*, the latter is chosen to represent the aggregation.

However, although the aggregation of the likelihoods of the two threat scenarios yields *unlikely*, it may still not be that this value correctly represents the likelihood of the unwanted incident *Compromise of confidentiality of health records*. This is because the storage of health records on a local computer not necessarily leads to the compromise of the records. In fact, the participants in the brainstorming group reject the suggested estimate for *Compromise of confidentiality of health records*, arguing that the likelihood is less than *unlikely*. The value is therefore adjusted to *rare*, as documented in the threat diagram of Fig. 3.12.

3.7 Risk Evaluation Using Risk Diagrams

Step 7 involves giving the customer the first overall risk picture. This will typically trigger some adjustments and corrections of the information documented so far. The objective of the risk evaluation is to determine which of the identified risks that must be considered for possible treatment based on the risk estimation of the previous step, as well as the risk evaluation criteria.

The risk evaluation furthermore includes the estimation and evaluation of the risks with respect to the indirect assets. Because the indirect assets are harmed only through harm to the direct assets, the relevant unwanted incidents with likelihoods are already identified. What remains is to determine the consequence of the harm to the direct assets on the related indirect assets. For the purpose of this, we need to define a consequence scale for each of the indirect assets, and we need to define their risk evaluation criteria.

Example 3.7 The analysis leader shows the asset diagram of Fig. 3.7 and explains that every risk with respect to the direct assets *Provision of telecardiology service* and *Health records* may represent a risk with respect to the indirect asset *Public's trust in system*. A consequence scale similar to the one in Table 3.4 with values ranging from *insignificant* to *catastrophic* is defined. The customer furthermore decides to use the risk evaluation criteria defined in Table 3.5 also for the indirect asset.

Each unwanted incident that harms one or both of the two relevant direct assets represents a risk with respect to the indirect asset. The analysis leader presents each of the relevant unwanted incidents in turn in order to have the participants to decide the consequence for the indirect asset. Figure 3.13 shows the consequence estimations for two of the unwanted incidents. The analysis leader explains that the full documentation of the risks with respect to the indirect asset is already given in the threat diagrams for the direct assets.

Once all the relevant unwanted incidents have been identified, and their likelihoods as well as consequences for both direct and indirect assets have been estimated, we are ready to evaluate the risks.

Fig. 3.13 Harm to indirect assets

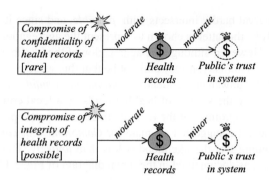

Table 3.7 Risk evaluation matrix with risks

		Consequence				
		Insignificant	Minor	Moderate	Major	Catastrophic
Frequency	Rare			CC1, CC1(I)		
	Unlikely					PR1
	Possible		CI1(I), SS1(I)	CI1, SS1		
	Likely				SS2	
	Certain					

Example 3.8 In this case, the risks are evaluated by plotting them into the risk evaluation matrix. From the five unwanted incidents in the threat diagram, the analysis secretary extracts five risks with respect to the direct assets. *CC1: Compromise of confidentiality of health records*, which may affect health records. *CI1: Compromise of integrity of health records*, which may also harm health records. The latter may also lead to the risk *PR1: Patient is given wrong diagnosis*, which may harm the patient's health. *SS1: Slow system* may affect the provisioning of the telecardiology system, and also lead to the risk *SS2: Unable to set diagnosis due to slow system*, which may affect the patients' health. Only *CC1* is within acceptable risk level; the remaining risks need further evaluation.

Three of the unwanted incidents in Fig. 3.12 moreover represent risks with respect to the indirect asset. The risk *CC1(I): Compromise of confidentiality of health records* may affect *Public's trust in system*, and the analysis secretary uses the suffix *(I)* to convey that the risk is with respect to the indirect asset. The other risks with respect to the indirect asset are named by the same convention.

Table 3.7 positions the risks within the risk evaluation matrix. The customer is at this point also invited to reconsider the risks to the indirect asset based on the risks to the direct assets, but our customer is now of the opinion that it is sufficient to focus on the direct ones in this particular analysis.

The analysis leader invites the customer to adjust likelihood and consequence estimates, as well as risk evaluation levels, to make sure that the results reflect reality as much as possible.

The participants request an overview of the risks. They want to know who or what is initiating them and which assets they harm. In response the analysis leader presents the risks, including the risks with respect to the indirect assets, with their associated risk values in terms of CORAS risk diagrams. The final diagram regarding

Fig. 3.14 Risk diagram

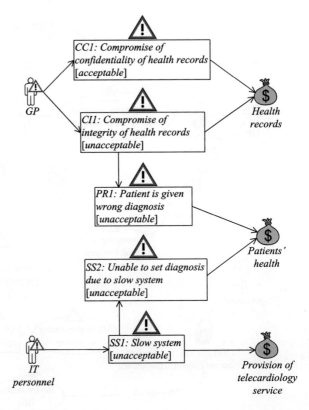

the direct assets for risks accidentally caused by employees is shown in Fig. 3.14. Since the *CC1* is within the acceptable risk level it will not be considered in the treatment phase of Step 8 of the CORAS method. The same is the case for the risks with respect to the indirect assets.

3.8 Risk Treatment Using Treatment Diagrams

Step 8 is devoted to treatment identification, as well as addressing cost-benefit issues of the treatments. A main task of Step 8 is the treatment identification using CORAS treatment diagrams, which is also often organised as a workshop. The risks that are not acceptable are all addressed in order to find means to reduce their likelihood and/or consequence. Since treatments can be costly, they are assessed with respect to cost-benefit, before a final treatment plan is made. The initial treatment diagrams are similar to the final threat diagrams except that unwanted incidents are replaced by the risks from the risk diagram.

Example 3.9 The analysis leader presents preliminary treatment diagrams showing all the unacceptable risks, ready to be filled in with treatments. He knows that

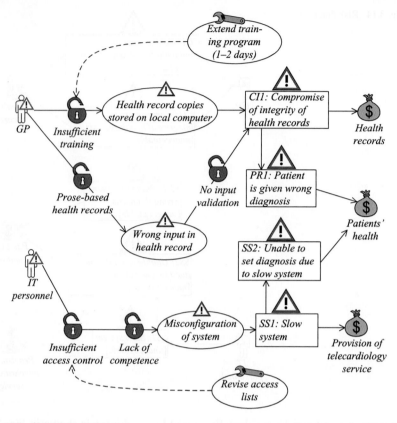

Fig. 3.15 Treatment diagram

participants of analyses often find it most intuitive to address vulnerabilities when looking for treatments. Hence, he highlights the possibility of treating other parts of the target as well, such as threats or threat scenarios. The participants involve in a discussion of potential treatments, and decide which ones will reduce the risks to acceptable levels. On some occasions, when the discussion gets slightly out of scope, the analysis leader suggests treatments taken from best-practice descriptions for network solutions and cryptography to help the discussion back on track. The diagrams are annotated with the identified treatment options indicating where they will be implemented. Finally, the following treatments are suggested and annotated in the treatment diagram of Fig. 3.15:

- Extend the training program for practitioners with 1–2 days, with a special focus on security aspects.
- Revise the list of people having access to conduct maintenance.

When the final results from the analysis are presented to the customer, an overview of the risks and the proposed treatments is useful. In our case, the treatment overview diagram of Fig. 3.16 is used for this purpose.

3.8 Risk Treatment Using Treatment Diagrams

Fig. 3.16 Treatment overview diagram

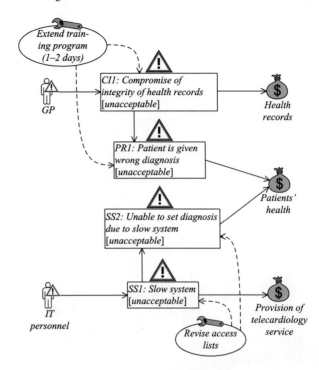

3.3 Rack Treatment Using Treatment Diagrams

Fig. 3.16. Treatment Overview diagram

Part II
Core Approach

Chapter 4
The CORAS Risk Modelling Language

As we saw in the guided tour, there is a lot to keep track of in a risk analysis—assets, threats, risk, and so forth—and not least the relationships between them: Which threats threatens which assets, and what vulnerabilities do they exploit? What risks originate from which threats, and what are their treatments? With the huge amount of information that is gathered, discussed, analysed, and written down during a risk analysis, it is easy to get lost. But not only that; people with different backgrounds, roles, responsibilities, and interests are involved, and these people do not speak the same language. The management people speak "economian", the users speak "GUI", and the developers speak one of the many dialects of "technical". And even if the challenges related to these different kinds of "speech" are overcome, one still remains: In the end, a report from the risk analysis should be produced, and this report should be readable also for people that did not take part in the analysis.

Experience from risk analysis in practice shows that we have to take special measures to avoid Babylonian confusion in the analysis sessions. There are also strong indications that the same is true when it comes to communicating the risk analysis results outside the group that conducted the analysis.

Our response to these challenges is to introduce a common language for the participants of risk analyses (and readers of risk analysis reports). The language is called the *CORAS risk modelling language* (or the *CORAS language* for short). The CORAS language is graphical, or diagram-based. In the CORAS method, the purpose of the CORAS language is threefold:

1. Provide structure to the information gathered during risk analysis and help the participants keep the appropriate level of abstraction and detail.
2. Aid communication between the participants of a risk analysis by giving them a common vocabulary (set of terms and concepts) and a notation with a precise meaning for expressing risk related information.
3. Provide a standard format for documenting the risk analysis which increases readability and helps emphasising the most important findings.

In the guided tour of Chap. 3, we have already seen several examples of diagrams made with the CORAS risk modelling language. In this chapter, we give the comprehensive introduction to the language. After reading this chapter, the reader should

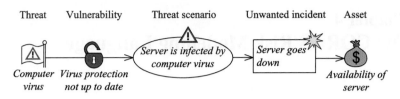

Fig. 4.1 CORAS diagram

be sufficiently familiar and comfortable with the language to understand CORAS diagrams and be able to start experimenting with threat and risk modelling. We start in Sect. 4.1 by introducing the most central concepts of the language: *threat*, *threat scenario*, *vulnerability*, *unwanted incident*, and *asset*. After that, in Sect. 4.2, we present the five basic kinds CORAS of diagrams, namely *asset diagrams*, *threat diagrams*, *risk diagrams*, *treatment diagrams*, and *treatment overview diagrams*. In Sect. 4.3, we show how the correct interpretation of any CORAS diagram, that is, the precise meaning of the diagrams, can be derived by a schematic and unambiguous translation into a paragraph in English.

4.1 Central Concepts

Figure 4.1 shows a typical description of a risk scenario in CORAS. This example expresses that the virus protection of the target is out of date, and that a virus might exploit this so that the server is infected. This may in turn cause the server to go down, which again will harm the availability of the server. The diagram is a threat diagram, which is the kind of diagram that is most used in CORAS, and contains a specimen of each of the most central concepts of the language: *threat*, *threat scenario*, *vulnerability*, *unwanted incident*, and *asset*. In the following, we will go through each of these concepts and explain in detail how they should be understood and how they are used in the CORAS language. The CORAS language contains other elements and diagrams as well, as we will come back to in Sect. 4.2.

4.1.1 What is a Threat?

In every day life, the word "threat" is ambiguous and may have several different meanings. In CORAS terminology, we say that a *threat* is something that can cause an effect. A threat is the cause of an unwanted incident, not the unwanted incident itself. It might sometimes be useful to think of a threat as an initiator of events that may harm the system we are analysing.

Definition 4.1 A *threat* is a potential cause of an unwanted incident.

4.1 Central Concepts

Fig. 4.2 Threats

Human threat (deliberate) — Hacker
Human threat (accidental) — System administrator
Non-human threats — Power failure, Computer virus

Hackers represent one class of threats that it is easy to think of, and that is likely to show up in nearly every risk analysis of information systems. Hackers are in many ways good examples of threats, because we will never confuse a hacker as the cause of an unwanted incident with the unwanted incident itself. But this must not lead us to think that all threats are human beings with malicious intents.

Another class of threats that is likely to show up in risk analyses of information systems are computer viruses. This shows that a threat does not have to be human. It does not even have to be an artefact (as computer viruses are), but can equally well be events outside of our control, like power failures or natural disasters. It is also important to remember that a threat does not have to be malicious; people with good intentions making mistakes may also be represented as threats.

In the CORAS language, we have three symbols for threats, designed to help remind us of the three main classes of threat: malicious human threats, accidental human threats, and non-human threats. The three symbols are *Human threat (deliberate)*, *Human threat (accidental)*, and *Non-human threat*, as illustrated in Fig. 4.2. Threats are usually given short descriptive names, as in the figure.

It might sometimes be difficult to determine exactly who or what the threat in a threat scenario is. Is it the Trojan horse infecting the computer or the hacker who made the Trojan horse? Is it the power failure or is it the tree falling over the power line? Is it the employee who forgets the laptop in a restaurant or the crook misusing the information found on the forgotten laptop? In such cases, we should consider what the scope of our analysis is, what is within and what is outside of our control, whether there is intent involved, and whether the system we are analysing is a direct target of the threat.

This means that if the hacker uses a Trojan horse in a directed attack against the target of the analysis, the hacker should probably be the threat. If the Trojan horse is just "in the wild" and might affect our target, then probably the Trojan horse should be the threat. If our target is a power line, then the falling tree should be the threat, but in most other cases it should be the power failure. If the laptop is stolen by a crook going specifically after our target, then the crook should probably be the threat. If the crook just misuses any interesting information he gets his hands on, then probably the employee should be the threat, and so forth.

4.1.2 What is a Threat Scenario?

Above, we characterised a threat as something that may cause harm or initiate unwanted incidents. We can therefore say that threats are *what* threatens assets. In order to answer the question of *how* the threats may cause harm or initiate unwanted incidents, we describe *threat scenarios*.

> **Definition 4.2** A *threat scenario* is a chain or series of events that is initiated by a threat and that may lead to an unwanted incident.

By this definition, a threat scenario is a characterisation of a sequence (or chain) of events that is initiated by a threat and that may result in an incident that harms one or more of the assets of the target under analysis.

The actual events may be of a very technical nature, and detailed knowledge of the target may be needed in order to comprehend them. Even if it sometimes may be useful to specify the threat scenarios in full detail, this is usually not required. Usually, we are content with an overview of the full situation with respect to the target in question and the assets. Often a threat scenario is more general than a sequence of events. We may therefore think of threat scenarios also as patterns describing collections of event sequences.

We try to characterise the patterns of these sequences with short explanatory texts. In this way, we seek to establish the full picture without drowning in details. After establishing the big picture, we may choose to describe some of the threat scenarios in more detail in order to understand and analyse them more deeply. This can, for example, be done by decomposing them into a structure of more detailed threat scenarios or by using other description techniques like sequence diagrams to specify their behaviour more carefully.

The threat scenario of the diagram in Fig. 4.1 describes that the server is infected by computer virus. All the technical details of exactly how this happens is suppressed, because this is not the information we want to convey. This will in any case vary from virus to virus and only a computer security expert may fully understand it.

The level of detail in the threat scenarios is usually a matter of scope, target and context of the analysis. In a technical analysis such as an analysis of the security of a specific application, the threat scenario can be quite detailed and technical. In analyses at an enterprise level, for example, an analysis of the information management in an organisation, the threat scenarios will be general and contain little technical details.

It is, when this is said, not a goal to put all information into a single threat scenario, and there is nothing wrong in splitting them up. In the CORAS language, we are allowed to have one threat scenario initiating another threat scenario so that we get series of threat scenarios. Further, we can have one threat scenario initiating several other threat scenarios and several threat scenarios initiating the same threat scenario. An example is shown in Fig. 4.3.

How much information to place in one threat scenario is to some degree a matter of preference, but it is still possible to devise some rules of thumb. One rule of thumb is that if a network of threat scenarios is too big to fit on a standard piece of paper (or on a PowerPoint slide), the description of the scenarios is probably too detailed. Another rule of thumb is that threat scenarios with several lines of text should be split into more scenarios, especially if the text contains the connectives

4.1 Central Concepts

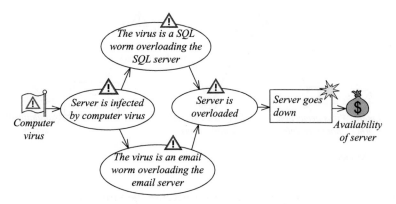

Fig. 4.3 Network of threat scenarios

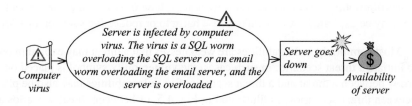

Fig. 4.4 Large threat scenario

"and" or "or". As an example, we prefer the diagram shown in Fig. 4.3 to the diagram of Fig. 4.4. Even though it should not be considered a strict rule, it is useful to think of threat scenarios in series as connected by "and" and scenarios in parallel as connected by "or". By following this rule, the network of threat scenarios in Fig. 4.3 should read as follows, which essentially is the description of the threat scenario in Fig. 4.4:

> Server is infected by computer virus and the virus is a SQL worm overloading the SQL server, or the virus is an email worm overloading the email server and the server is overloaded.

How to structure risk related information in threat scenarios is furthermore a matter of which pieces of risk related information that we need to understand and reason about separately. By extracting specific scenarios and representing them as separate threat scenarios, we can analyse them to understand their individual contribution to the general risk picture.

4.1.3 What is a Vulnerability?

A *vulnerability* is a weakness or shortcoming of the target of analysis that allows a threat to initiate a threat scenario, or that allows a threat scenario to lead to an unwanted incident or another threat scenario. In the diagram of Fig. 4.1, for example,

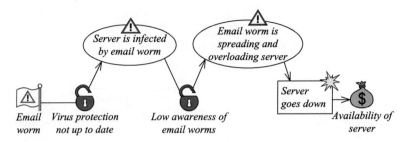

Fig. 4.5 Vulnerabilities

the computer virus may infect the server because the virus protection is not up to date.

> **Definition 4.3** A *vulnerability* is a weakness, flaw or deficiency that opens for, or may be exploited by, a threat to cause harm to or reduce the value of an asset.

Most commonly, vulnerabilities are weaknesses that a threat may exploit in order to initiate a threat scenario. In such cases, the vulnerability symbol is placed on the arrow between a threat and a threat scenario. But vulnerabilities may also be placed between threat scenarios, as illustrated in Fig. 4.5, or between threat scenarios and unwanted incidents,

A vulnerability may be something the target has or something it lacks; in any case it should be made explicit when modelling threat scenarios, as vulnerabilities are the prerequisites for threat scenarios to happen. It might sometimes be difficult to distinguish vulnerabilities from threat scenarios. There are two rules that can help us in deciding what we express as vulnerabilities and what we express as threat scenarios. The first is that vulnerabilities often are features or properties of the target, while threat scenarios are occurrences of events. Insufficient virus protection, for example, is a property of the target, but that the system is infected by computer virus is an event that is happening. The second rule is that a vulnerability is something that is exploited in order for something to happen; the computer virus exploits the insufficient virus protection in order to make the virus infection happen.

A question we can ask at this point is where a vulnerability should be placed in a chain from a threat, through one or more threat scenarios, and to an unwanted incident. The answer to this is simply where it fits best into the reading of the diagram. The diagram in Fig. 4.5 we read:

> An email worm exploits the not up to date virus protection to infect the server. Due to low awareness of email worms, the worm is spreading and overloading the server, which causes the server to go down.

If we had put both vulnerabilities between the threat and the first threat scenario we would get:

> An email worm exploits the not up to date virus protection and the low awareness of email worms to infect the server. The worm is spreading and overloading the server, which causes the server to go down.

4.1 Central Concepts

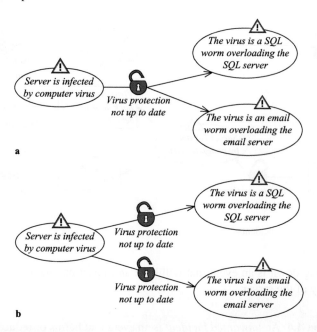

Fig. 4.6 A fragment of a diagram; **a** with shorthand notation for arrows with vulnerabilities, **b** without using the shorthand

On the other hand if both vulnerabilities were placed between the last threat scenario and the unwanted incident the diagram would have read:

> An email worm infects the server. The worm is spreading and overloading the server. Because the virus protection is not up to date and the awareness of email worms are low, this causes the server to go down.

The first of these three paragraphs seems to be somewhat more meaningful that the last two. The reason for this is that we have found good places to put the vulnerabilities in the path from the threat to the unwanted incident.

Formally, vulnerabilities are annotations to the arrows and not independent elements like threats and threat scenarios. This is the reason why there is no arrowhead on the lines going to vulnerabilities. Despite this we sometimes apply a shorthand notation where arrows branch or meet in vulnerabilities. This is illustrated by the fragments of a diagram in Fig. 4.6, where the upper fragment (a) is a shorthand for the lower fragment (b).

4.1.4 What is an Unwanted Incident?

An *unwanted incident* is an event that results in harm to one or more assets of the target under analysis. In the diagram of Fig. 4.1, the unwanted incident *Server goes down* has an arrow pointing at the asset *Availability of server*. This means that the

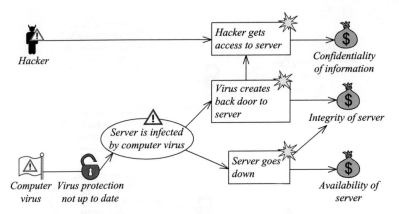

Fig. 4.7 Unwanted incidents

server going down is an event that will have an impact on, and cause harm to, the availability of the server.

Definition 4.4 An *unwanted incident* is an event that harms or reduces the value of an asset.

In Fig. 4.7, we see a more complicated picture. We see that a threat scenario might lead to more than one unwanted incident, and that an unwanted incident may lead to new unwanted incidents. Further, we see that an unwanted incident may be initiated directly by a threat and that the same unwanted incident can impact more than one asset. If we saw the need, we could also have placed vulnerabilities directly before the unwanted incident, for example on the arrow going from *Hacker* to *Hacker gets access to server*, or on the arrow going from *Server is infected by computer virus* to *Virus creates back door to server*. We could also have placed vulnerabilities on the arrow going from the unwanted incident *Virus creates back door to server* to the unwanted incident *Hacker gets access to server*. When an unwanted incident takes place, the assets to which it is connected are harmed by definition. We therefore never assign vulnerabilities to the arrows connecting unwanted incidents to assets.

There are two more rules concerning unwanted incidents. The first is that an unwanted incident always has an arrow going to an asset; if an incident does not harm any asset, it is by definition not an unwanted incident. The second rule is that an unwanted incident must be initiated by a threat scenario, another unwanted incident or a threat. In sum, we can say that an unwanted incident must always have at least one incoming arrow and at least one outgoing arrow.

When unwanted incidents are events and threat scenarios are chains of events leading to unwanted incidents, how do we know which events should be unwanted incidents and which should be part of threat scenarios? In other words, when do we decide that an event is to be taken out of a threat scenario and given the status of unwanted incident? In order to answer that question, we simply ask whether or not the event causes harm to at least one of the assets we have defined for the analysis. If an event does not cause harm to any of the assets it should not be represented as an unwanted incident. Likewise, if an event in a threat scenario damage at least one of the assets, it should be extracted from the threat scenario and represented as an unwanted incident.

4.1.5 What is an Asset?

The last of the central concepts we treat in this section is the *asset*. In every example in this chapter, we have seen unwanted incidents causing harm to assets, so obviously an asset is something to which harm may be caused.

> **Definition 4.5** An *asset* is something to which a party assigns value and hence for which the party requires protection.

This definition implies that an asset is something the party in question wants to protect. A party is a person, organisation, or other stakeholder who has interests in the target of analysis. For the time being, we assume that the party of a risk analysis is identical to the customer of the analysis, that is, the person or organisation commissioning the analysis. (We will come back to parties in Sect. 4.2.1.) When a party orders a risk analysis, the party obviously has something it wants to protect, and this something has value for the party. The things or entities with value for the party, and that the party wants to protect, are what we call assets.

Again, we are dealing with a very general concept. The reason for using such a general definition is that we do not want to put constraints on what the parties involved in a risk analysis may want to protect. Assets can be physical objects, such as computers, or objects we do not necessarily think of as physical, such as databases. Furthermore, assets may be desirable properties of the target or parts of the target, such as availability, confidentiality and integrity, as we saw in the example of Fig. 4.7, or they may be relations to the outside world, such as reputation, goodwill, and compliance with laws and regulations. In addition, assets may of course be pure values, such as bank account balances or stock prices.

The process of identifying and assigning value to assets are treated in detail in Chap. 7. The important thing to remember here is that asset is a general concept of value, and that different assets may have different value. The latter is also the reason why it is important to represent the assets in the CORAS diagrams; we cannot know what the consequences of an unwanted incident are unless we know which assets are impacted by the unwanted incident. How much harm, expressed as loss of asset value, an unwanted incident causes is obviously dependent on the value of the assets it harms.

4.2 The Diagrams of the CORAS language

So far in this chapter, we have only seen examples of one kind of CORAS diagrams, namely the diagrams we refer to as *threat diagrams*. Usually, we distinguish between five kinds of basic CORAS diagrams: *asset diagrams*, *threat diagrams*, *risk diagrams*, *treatment diagrams*, and *treatment overview diagrams*. These different diagrams are actually overlapping. The distinction between them is more of a pragmatic nature, and we have given them different names because they are used in different parts of the analysis for different purposes; asset diagram are used in asset identification, threat diagrams are used in risk identification and risk estimation, risk diagrams are used in risk evaluation, and treatment diagrams and treatment overview diagrams are used in treatment identification. This means that asset diagrams become the starting point for threat diagrams, threat diagrams become the starting point for risk diagrams, and so forth.

In this section, we go through each of the five kinds of diagrams of the CORAS risk modelling language in full detail. In the previous section, the main focus was on the basic concepts and how they are represented by separate constructs in the CORAS language. We refer to these constructs as the *elements* of a CORAS diagram. In the following, also the arrows of the diagrams, often referred to as the diagram *relations*, will get a proper treatment. We continue with presenting the language by means of examples and intuitive explanations. In Sect. 4.3, we explain how any diagram or diagram fragment can be schematically translated into a paragraph in English. A more formal treatment of the language is presented in Appendices A and B. In Appendix A, we give a formal syntax, and in Appendix B we formally define the structured semantics capturing the meaning of the diagrams. For more information on the process of making CORAS diagrams, we refer to Chaps. 7 through 12.

4.2.1 Asset Diagrams

An *asset diagram* is, as the name indicates, a diagram mainly used for the purpose of defining and documenting the assets of relevance for the analysis. The elements of asset diagrams are *assets* and *indirect assets*. Asset diagrams also have a special symbol called *party*.

> **Definition 4.6** A *party* is an organisation, company, person, group or other body on whose behalf the risk analysis is conducted.

An example asset diagram is shown in Fig. 4.8. In the figure, we see the party in a compartment in the upper left corner. All the assets in the diagram belong to that party. Usually, there is only one party of a risk analysis, namely the client or customer of the analysis. But it might also be the case that the customer of the analysis wants the risk analysis to also take into account the interests and concerns of other parties. Examples of this may be a bank that orders a risk analysis of its

4.2 The Diagrams of the CORAS language

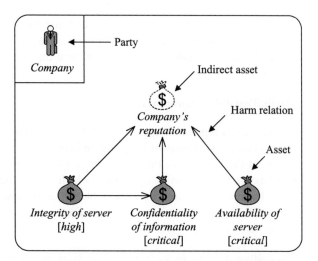

Fig. 4.8 Example asset diagram

online banking solution and includes its customers as a party, or a governmental agency that orders a risk analysis of its databases and includes the public as a party. In such cases, we make one asset diagram for each of the parties of the analysis, thereby specifying and documenting the assets of each party separately.

As shown in the asset diagram of Fig. 4.8, we can explicitly specify the value that the party in question assigns to each asset. The asset *Integrity of server*, for example, is assigned the asset value *high*. In the practical setting of a risk analysis, we use these values to guide and prioritise the analysis, and it often suffices to determine the relative order of importance between the assets. Ranking assets according to importance is a topic of Chap. 8.

Different parties will usually have different concerns and therefore also different assets. Even when assets of different parties are seemingly the same, they may be assigned different values, they may have different scales of consequences of incidents, and the impact of a given incident on the assets may differ. It might even be that the unwanted incidents causing harm are not the same, even though two assets of different parties on the surface look the same. Throughout the analysis, and especially in the estimation of risk level we need to consider the party of an asset. In order to make the ownership of assets explicit, we keep the assets of different parties in separate diagrams.

The party present in an asset diagram assigns value to, and requires protection of, the assets specified in the diagram. One of the assets in the diagram of Fig. 4.8 looks different from the other assets, namely *Company's reputation*. The white asset symbol with the dashed line identifies *Company's reputation* as an indirect asset. An indirect asset is an asset that, with respect to the target and scope of the analysis, is harmed only via harm to other assets.

We refer to the arrows connecting the assets as *harm relations*. The harm relation expresses that an asset can be harmed through harm to another asset. *Confidentiality of information*, for example, can be harmed through harm to *Integrity of server*, and *Company's reputation* can be harmed through harm to *Confidentiality of informa-*

tion. This means that an unwanted incident that causes harm to *Integrity of server* may also cause harm to *Confidentiality of information* and an unwanted incident that causes harm to *Confidentiality of information* may also cause harm to *Company's reputation*.

Definition 4.7 e_1 *harm* e_2: Harm to e_1 may result in harm to e_2.

Company's reputation is an indirect asset. This means that, with respect to the target in question, *Company's reputation* is only harmed through harm to the other assets. In the diagram of Fig. 4.8, there is a harm relation going from each of the other assets to *Company's reputation*, which means that harm to these may also cause harm to *Company's reputation*. Notice, importantly, that this alone does not imply that *Company's reputation* is an indirect asset, since it could be that this asset is also harmed directly without harm to any of the other assets being caused first.

Definition 4.8 An *indirect asset* is an asset that, with respect to the target and scope of the analysis, is harmed only via harm to other assets.

Definition 4.9 A *direct asset* is an asset that is not indirect.

4.2.2 Threat Diagrams

Figure 4.9 shows a threat diagram. The diagram is the same diagram as in Fig. 4.7, but with more details. In explaining this diagram, we first give names to the arrows. Arrows going from a threat to a threat scenario or an unwanted incident are called *initiates relations*.

Definition 4.10 e_1 *initiates* e_2: e_1 exploits some set of vulnerabilities to initiate e_2 with some likelihood.

Arrows from a threat scenario to an unwanted incident or between two threat scenarios or two unwanted incidents are called *leads-to relations*.

Definition 4.11 e_1 *leads-to* e_2: e_1 leads to e_2 with some likelihood, due to some set of vulnerabilities.

Arrows going from an unwanted incident or a threat scenario to an asset are called *impacts relations*.

Definition 4.12 e_1 *impacts* e_2: e_1 impacts e_2 with some consequence.

Both initiates relations and leads-to relations can be annotated with vulnerabilities and likelihoods.

4.2 The Diagrams of the CORAS language

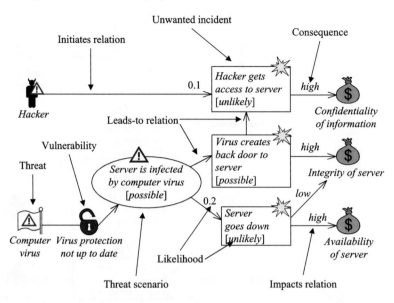

Fig. 4.9 Example threat diagram

Definition 4.13 *Likelihood* is the frequency or probability of something to occur.

Impacts relations can be annotated with consequence values.

Definition 4.14 *Consequence* is the impact of an unwanted incident on an asset in terms of harm or reduced asset value.

From the diagram of Fig. 4.9, we can for example read that the computer virus exploits a virus protection that is not up to date to initiate infection of the server. Further, the diagram says that in the case of an infection of the server there is a likelihood of 0.2 for this leading to the server going down, and that the server going down impacts the availability of the server with the consequence *high*.

In the diagram, also the threat scenario and the unwanted incidents have been assigned likelihoods. The threat scenario is given the likelihood *possible* and the unwanted incident *Server goes down* the likelihood *unlikely*. We can therefore read from the diagram that threat scenario *Server is infected by computer virus* occurs with likelihood *possible* and that the unwanted incident *Server goes down* occurs with likelihood *unlikely*.

Note that the likelihoods *possible* and *unlikely*, and the consequences *high* and *low* may be qualitative values, while the 0.1 and 0.2 assigned to one of the initiates relations and one of the leads-to relations in the diagram are quantitative values. The CORAS language supports the use of both quantitative and qualitative values. The values will not, however, be the same in every analysis; they should be defined as

Fig. 4.10 Example risk diagram

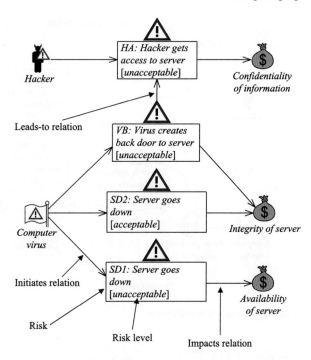

part of the risk analysis prior to the modelling activities. Defining these values is a topic of Chap. 8.

4.2.3 Risk Diagrams

When an incident is unwanted it inflicts damage, and this potential constitutes a risk. Another way to perceive a risk is to think of it as an incident causing harm to an asset. Each risk encompass a likelihood and a consequence. There is therefore one risk for each pair of unwanted incident and asset. The resulting risks are shown in *risk diagrams*. An example, based on the threat diagram of Fig. 4.9, is shown in Fig. 4.10. As can be seen in the diagram, it is customary to give a unique identity to each risk.

> **Definition 4.15** A *risk* is the likelihood of an unwanted incident and its consequence for a specific asset.

In Fig. 4.9, the unwanted incident *Server goes down* is assigned the likelihood *unlikely*, and the impacts relation relating this unwanted incident to the asset *Availability of server* is assigned the consequence *high*. Likewise, the other unwanted incidents and impacts relations are assigned likelihoods and consequences, respectively. Each pair of an unwanted incident and an impacts relation represents a risk.

4.2 The Diagrams of the CORAS language

This is the reason why we have the two risks *SD1: Server goes down* and *SD2: Server goes down* in the diagram, and why we distinguish each risk with a unique identity. The risks originate from the same unwanted incident, but also from two different impacts relations relating the unwanted incident to two different assets. Each risk is given a *risk level* that is calculated from the likelihood and consequence of the unwanted incident, where the consequence as annotated on the impacts relation.

Definition 4.16 *Risk level* is the level or value of a risk as derived from its likelihood and consequence.

The different risk levels, as well as the function for calculating the risk levels from likelihoods and consequences, should, as the likelihood and consequence values, be defined as part of the risk analysis prior to the modelling. How the risk levels and the risk function are defined is explained in Chap. 8. In the example, we have only used two risk levels, *acceptable* and *unacceptable*, and decided that likelihood *unlikely* or *possible* together with consequence *high* yields risk level *unacceptable*, while likelihood *unlikely* together with consequence *low* yields the risk level *acceptable*.

In addition to risks, a risk diagram shows the threats and the assets that are involved. In threat diagrams, threats initiate unwanted incidents or threat scenarios that lead to unwanted incidents. In risk diagrams, we have the initiates relations going directly to risks. Such a relation shows what threat was the original initiator of the chain of events leading to the unwanted incident that constitutes the risk. In the risk diagram in Fig. 4.10, for example, *Computer virus* initiates the risk *VB: Virus creates back door to server*. The reason for this is that there is a path from the threat *Computer virus* to the unwanted incident *Virus creates back door to server* in the threat diagram of Fig. 4.9 on which the risk diagram is based.

Similar to the way a risk diagram shows the initiators of risks, it also shows the assets that are affected by the risks. This is achieved by the *impacts* relations. In the diagram of Fig. 4.10, we have an impacts relation going from the risk *VB: Virus creates back door to server* to the asset *Integrity of server*, meaning that *VB: Virus creates back door to server* impacts *Integrity of server*. This impacts relation is analogous to the impacts relation going from the unwanted incident *Virus creates back door to server* to *Integrity of server* in Fig. 4.9, with the difference that is does not have a consequence value as this is embedded in the risk level of the risk.

In the same way that an unwanted incident may lead to another unwanted incident in a threat diagram, a risk can lead to another risk in a risk diagram. This is specified by the *leads-to* relation. In Fig. 4.10 the risk *VB: Virus creates back door to server* leads-to the risk *HA: Hacker gets access to server*. This is because in the threat diagram of Fig. 4.9, the unwanted incident *Virus creates back door to server* leads-to the unwanted incident *Hacker gets access to server*.

From the above explanations of risk diagrams and their relationship to threat diagrams, it should be clear that the risk diagram corresponding to a threat diagram to a large extent may be schematically constructed from the threat diagram.

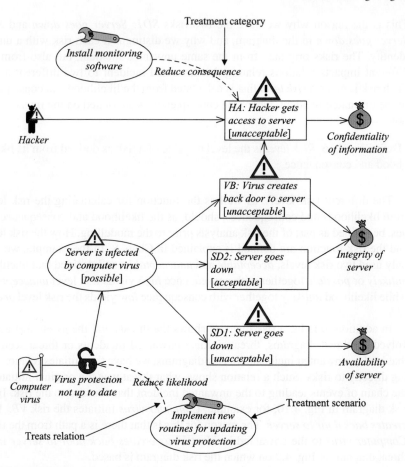

Fig. 4.11 Example treatment diagram

4.2.4 Treatment Diagrams

Treatment diagrams are a kind of extended threat diagrams that are used to document treatments to risks. A treatment diagram based on the threat diagram of Fig. 4.9 is shown in Fig. 4.11.

The first thing to observe in this diagram is that the unwanted incidents are replaced by the risks from the risk diagram in Fig. 4.10. The treatment diagram of Fig. 4.11 can therefore be seen as a merge of the threat diagram of Fig. 4.9 and the risk diagram of Fig. 4.10. The unwanted incident *Server goes down*, for example, has been replaced by the two risks *SD1* and *SD2* that both originate from this unwanted incident. Furthermore, the consequences on the impacts relations are redundant and therefore removed. Optionally, the risks with acceptable risk values and the paths leading up to them can be removed from the diagram, as these risks are not in need for treatments. The second thing to observe is the *treatment scenarios*.

4.2 The Diagrams of the CORAS language

Definition 4.17 A *treatment scenario* is the implementation, operationalisation or execution of appropriate measures to reduce risk level.

A treatment scenario describes a way of dealing with a risk. A risk is the result of series of potential events and weaknesses, as well as the fact that we have assets we want to protect. A risk can therefore be dealt with in many different ways by addressing any of the factors involved in a risk. In the example, we are dealing with the risks by dealing with the vulnerability *Virus protection not up to date* and by directly dealing with the risk *HA: Hacker gets access to server*. This is shown by the treatment scenarios *Implement new routines for updating virus protection* and *Install monitoring software* that are related to the vulnerability and to the risk, respectively. This kind of relation is referred to as the *treats* relation.

Definition 4.18 e_1 *treats* e_2: e_1 reduces risk by providing treatment to e_2.

A treats relation can optionally be annotated with one of five *treatment categories*. The treatment categories that are provided are a general, although not exhaustive, categorisation of treatment scenarios, and describe in what way a treatment scenario treats a risk.

Definition 4.19 A *treatment category* is a general approach to treating risks. The categories are:

- *Avoid*: Avoid risk by not continuing the activity that gives rise to it.
- *Reduce consequence*: Reduce risk level by reducing the harm of unwanted incidents to assets.
- *Reduce likelihood*: Reduce risk level by reducing the likelihood of unwanted incidents to occur.
- *Transfer*: Share the risk with another party or other parties.
- *Retain*: Keep risk at current level by informed decision.

In the example diagram of Fig. 4.11, the treats relations are annotated with the treatment categories *Reduce consequence* and *Reduce likelihood*. This means that treatment scenarios reduce the risk levels of the risks by reducing the likelihood and the consequence of the unwanted incident, respectively.

The treatment scenario *Install monitoring software* is related directly to a risk by a treats relation annotated with *Reduce consequence*. Thus, the treatment scenario reduces the harm inflicted on the asset. The treatment scenario *Implement new routines for updating virus protection* reduces the likelihood of a vulnerability being exploited, and indirectly through this reduces the likelihood of the risks.

It is not uncommon that most treats relations point at vulnerabilities. For this reason, most treatment scenarios will usually fall into the *Reduce likelihood* category. We can, however, direct treatment scenarios to any of the (non-relation) elements in a CORAS diagram. If untrained employees are a threat, for example, training the employees can be a relevant treatment scenario. And if we monitor certain critical activities, this may be represented as a treatment scenario for a threat scenario. If

Fig. 4.12 Example treatment overview diagram

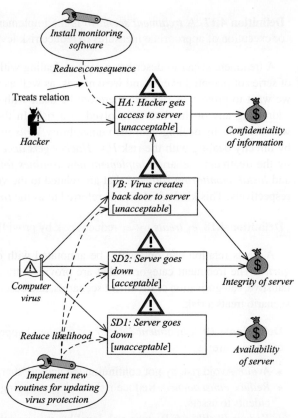

a barrier prevents an unwanted incident from happening, this may be a treatment scenario for the risk originating from the unwanted incident. Buying an insurance transfers risk, and may be seen as a treatment scenario directed towards an asset.

4.2.5 Treatment Overview Diagrams

As we saw above, a risk diagram provides a condensed view of the identified risks; in a way it is a collapsed version of a threat diagram. A *treatment overview diagram* is similar to a risk diagram, and can likewise be seen as a collapsed version of a treatment diagram. An example treatment overview diagram is shown in Fig. 4.12. We can easily see that this is the risk diagram from Fig. 4.10 with the treatment scenarios from the treatment diagram of Fig. 4.11. As for treatment diagrams, the risk with acceptable risk value can optionally be removed along with the diagram paths that go through only acceptable risks.

The diagram does not introduce any new elements, but presents an overview of the treatment scenarios documented in the treatment diagram. A treatment scenario treats a risk if it reduces its risk value.

4.3 How to Schematically Translate CORAS Diagrams into English Prose

The CORAS language has been developed to facilitate the specification of easily comprehensible risk models, to facilitate communication between people of heterogeneous backgrounds, and to ensure that the communication and the results are commonly understood. A crucial property of the CORAS language in this respect is the easily recognisable graphical icons, as well as the immediate relationship between these language constructs and the core concepts of a risk analysis. However, visualisation and diagrammatic representation of information is not enough to ensure a correct and common interpretation; for any language to be well understood it should have a well-defined semantics. In other words, the language should be equipped with a precise and unambiguous explanation of the meaning of any syntactically correct expression. In our case, a syntactically correct expression is a CORAS diagram or a fragment of a CORAS diagram.

In this section, we describe from a practical point of view how we may schematically translate any CORAS diagram, or fragment thereof, into English prose. The reader is referred to Appendix B for a more formal treatment and for additional examples.

The translation is defined for all kinds of CORAS diagrams, and is conducted as a straightforward, systematic mapping of diagram elements into sentences in English. More specifically, an arbitrary CORAS diagram consisting of n elements and m relations between elements is translated into a paragraph in English consisting of $n + m$ sentences. That is, the translation yields one sentence for each element and one sentence for each relation. In the following, we explain and exemplify how to translate each kind of CORAS diagram.

4.3.1 How to Translate Asset Diagrams

The elements of asset diagrams are party, direct asset and indirect asset. The translation of these elements straightforwardly yields sentences of the following form, respectively:

> ... is a party.
> ... is a direct asset.
> ... is an indirect asset.

The annotation of the element is inserted into the open field to complete the sentence. The party of the asset diagram of Fig. 4.8, for example, is named *Company*, and the translation of this particular element is therefore as follows:

> *Company* is a party.

In the same manner, the translation of the direct asset *Availability of server* yields the following sentence:

> *Availability of server* is a direct asset.

The only relation in asset diagrams is the harm relation, which is translated into a sentence of the following form:

> Harm to ... may result in harm to ....

The description of the asset at the source of the harm relation is inserted into the open field to the left, and the description of the asset at the target of the harm relation is inserted into the open field to the right. The source of a relation is the diagram element in which the relation begins, and the target is the diagram element to which the relation points. The translation of the harm relation from the direct asset *Availability of Server* to the indirect asset *Company's reputation* in the asset diagram of Fig. 4.8 therefore yields the following sentence:

> Harm to *Availability of server* may result in harm to *Company's reputation*.

Notice, importantly, that the kind of the source element and the kind of the target element are not given in the translation of the relations. This information is redundant since it is already represented in the translation of the elements.

In addition to the harm relation, there is in each asset diagram an implicit relation between the party and each of the assets. These relations are made explicit in the semantics, and therefore captured in the translation of asset diagrams. The schematic translation depends on whether the asset value is explicitly specified. For each asset that is not annotated with an asset value, the implicit relation between a party and an asset in an asset diagram is translated into a sentence of the following form:

> ... assigns value to ....

The description of the party is inserted into the open field to the left, and the description of the asset is inserted into the open field to the right. For each asset that is annotated with an asset value, on the other hand, the relation is translated into a sentence of the following form:

> ... assigns the value ... to ....

The assigned asset value is then inserted into the open field in the middle.

Consider again the asset diagram of Fig. 4.8. Ignoring the two direct assets *Integrity of server* and *Confidentiality of server*, the translation of this diagram yields the following sentences:

> *Company* is a party.
> *Availability of server* is a direct asset.
> *Company's reputation* is an indirect asset.
> *Company* assigns the value *critical* to *Availability of server*.
> *Company* assigns value to *Company's reputation*.
> Harm to *Availability of server* may result in harm to *Company's reputation*.

Notice that everything that is written in italics represents the descriptions that are inserted into the diagram. Everything not in italics represents the CORAS language constructs.

It may from the example seem redundant to include the implicit relation between the party and each of the assets in the translation. However, as there may be more than one party in an analysis, we need to take into account not only that the different

4.3 How to Schematically Translate CORAS Diagrams into English Prose 67

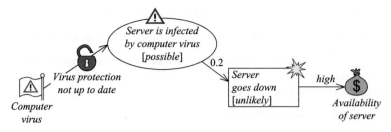

Fig. 4.13 Threat diagram

parties may have different assets, but also that they can assign different values to the same parts or aspects of the target in question.

4.3.2 How to Translate Threat Diagrams

In order to explain the semantics of threat diagrams, we demonstrate the translation of the example diagram given in Fig. 4.13. This diagram is a fragment of the threat diagram shown in Fig. 4.9.

Every kind of language element and relation is translated in the same way independent of the kind of CORAS diagram in which it occurs. The assets in threat diagrams are therefore translated in the same way as for asset diagrams.

There are three kinds of threat elements, namely deliberate human threat, accidental human threat and non-human threat. These are translated into sentences of the following form, respectively:

> ... is a deliberate human threat.
> ... is an accidental human threat.
> ... is a non-human threat.

For example, the non-human threat in the diagram of Fig. 4.13 is named *Computer virus*, and the translation of this particular element is therefore as follows:

> *Computer virus* is a non-human threat.

Vulnerabilities are translated into sentences of the following form:

> ... is a vulnerability.

Accordingly, the translation of the vulnerability in the threat diagram of Fig. 4.13 is:

> *Virus protection not up to date* is a vulnerability.

Threat scenarios and unwanted incidents may also be annotated with likelihoods. A threat scenario with a likelihood annotation is therefore translated into a sentence of the following form:

> Threat scenario ... occurs with likelihood ... .

The translation of the threat scenario of Fig. 4.13 is as follows:

> Threat scenario *Server is infected by computer virus* occurs with likelihood *possible*.

The rules for translating diagram elements take into account that annotations such as likelihoods or risk levels are optional. A threat scenario for which the likelihood is not specified is therefore translated into:

> Threat scenario ... occurs with undefined likelihood.

The translation of unwanted incidents, with or without likelihood annotations, is similar to the translation of threat scenarios. Translating the unwanted incident of Fig. 4.13 therefore yields:

> Unwanted incident *Server goes down* occurs with likelihood *unlikely*.

The translation of the various relations vary somewhat depending on the annotations. An initiates relation, for example, may or may not be annotated with a likelihood and may or may not be annotated with one or more vulnerabilities. The basic form is nevertheless the same; it describes the type of relation, the source of the relation, and the target of the relation. If an initiates relation is not annotated with vulnerabilities, the translation yields sentences of the following form, depending on whether a likelihood is specified:

> ... initiates ... with undefined likelihood.
> ... initiates ... with likelihood ....

The translation of an initiates relation that is annotated with one vulnerability and for which the likelihood is not specified yields:

> ... exploits vulnerability ... to initiate ... with undefined likelihood.

The initiates relation of Fig. 4.13 is therefore translated into the following sentence:

> *Computer virus* exploits vulnerability *Virus protection not up to date* to initiate *Server is infected by computer virus* with undefined likelihood.

The translation of the leads-to relation is similar. The leads-to relation of Fig. 4.13, for example, is translated into the following sentence:

> *Server is infected by computer virus* leads to *Server goes down* with conditional likelihood 0.2.

Notice that the rules for the schematic translation of CORAS diagrams take into account not only that initiates relations and leads-to relations may or may not be annotated with vulnerabilities; they also take into account the number of vulnerabilities. The noun "vulnerability" is therefore translated to its singular form in case there is only one vulnerability, and it is translated into its plural form in case there are arbitrary many of them, but more than one.

The impacts relation has an asset as target, and may be annotated with a consequence. The impacts relation without annotation and the impacts relation with annotation are translated into sentences of the following forms, respectively:

> ... impacts ....
> ... impacts ... with consequence ....

4.3 How to Schematically Translate CORAS Diagrams into English Prose

Fig. 4.14 Risk diagram

The impacts relation of Fig. 4.13 is therefore translated into the following sentence:

Sever goes down impacts *Availability of server* with consequence *high*.

4.3.3 How to Translate Risk Diagrams

We ignore in the presentation of the semantics of risk diagrams the elements and relations already covered in relation to asset diagrams and threat diagrams.

The language element that represents risks may or may not be annotated with a risk value. A risk is therefore translated into a sentence of one of the two following forms:

Risk ... occurs with undefined risk level.
Risk ... occurs with risk level ....

In the second case, the risk level is inserted into the open field to the right, whereas the description of the risk is inserted into the open field to the left. The translation of the risk in Fig. 4.14 is therefore:

Risk *Server goes down* occurs with risk level *unacceptable*.

The semantics of the risk diagram of Fig. 4.14 as whole is then:

Computer virus is a non-human threat.
Risk *Server goes down* occurs with risk level *unacceptable*.
Availability of server is a direct asset.
Computer virus initiates *Server goes down*.
Server goes down impacts *Availability of server*.

4.3.4 How to Translate Treatment Diagrams

As to the semantics of treatment diagrams, we only need to explain the translation of treatment scenarios and treats relations; the rest has already been covered above.

A treatment scenario yields a sentence of the following form:

... is a treatment scenario.

The treatment of the vulnerability *Virus protection not up to date* shown in Fig. 4.11 is therefore translated into the following sentence:

Implement new routines for updating virus protection is a treatment scenario.

Fig. 4.15 Treatment overview diagram

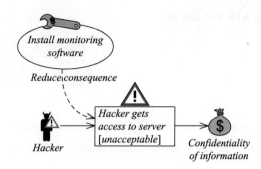

The treats relation has a treatment scenario as source and the element it treats as target. The treats relation can furthermore optionally be annotated with the treatment category to which the treatment belong. Any treatment relation that is not annotated with a treatment category is translated into a sentence of the following form:

> ... treats ....

The description of the treatment scenario is inserted into the open field to the left, and the description of the diagram element it treats is inserted into the field to the right.

There are five treatment categories as defined in Definition 4.19, and a separate translation is defined for each of them. The treatment *Implement new routines for updating virus protection* in Fig. 4.11, for example, belongs to the category *Reduce likelihood* as shown by the annotation on the treats relation. This treats relation is translated into the following sentence:

> *Implement new routines for updating virus protection* decreases the likelihood of *Virus protection not up to date*.

4.3.5 How to Translate Treatment Overview Diagrams

A treatment overview diagram is a risk diagram annotated with treatment scenarios. We have therefore in the above already explained how to translate each element and relation of a treatment overview diagram into English prose. To exemplify the translation of a treatment overview diagram into English, we therefore simply give the full translation of the diagram shown in Fig. 4.15. The translation is conducted by using the rules we have introduced above and yields the following paragraph:

> *Hacker* is a deliberate human threat.
> Risk *Hacker gets access to server* occurs with risk level *unacceptable*.
> *Confidentiality of information* is a direct asset.
> *Install monitoring software* is a treatment scenario.
> *Hacker* initiates *Hacker gets access to server*.
> *Hacker gets access to server* impacts *Confidentiality of information*.
> *Install monitoring software* decreases the consequence of *Hacker gets access to server*.

Table 4.1 Summary of elements of the CORAS language

Language element	Definition
Party	An organisation, company, person, group or other body on whose behalf the risk analysis is conducted
Asset	Something to which a party assigns value and hence for which the party requires protection
Indirect asset	An asset that, with respect to the target and scope of the analysis, is harmed only via harm to other assets
Direct asset	An asset that is not indirect
Threat	A potential cause of an unwanted incident
Threat scenario	A chain or series of events that is initiated by a threat and that may lead to an unwanted incident
Vulnerability	A weakness, flaw or deficiency that opens for, or may be exploited by, a threat to cause harm to or reduce the value of an asset
Unwanted incident	An event that harms or reduces the value of an asset
Likelihood	The frequency or probability of something to occur
Consequence	The impact of an unwanted incident on an asset in terms of harm or reduced asset value
Risk	The likelihood of an unwanted incident and its consequence for a specific asset
Risk level	The level or value of a risk as derived from its likelihood and consequence
Treatment scenario	The implementation, operationalisation or execution of appropriate measures to reduce risk level
Treatment category	A general approach to treating risks; the categories are avoid, reduce consequence, reduce likelihood, transfer and retain

Table 4.2 Summary of relations in the CORAS language

Relation	Definition
e_1 *harm* e_2	Harm to e_1 may result in harm to e_2
e_1 *initiates* e_2	e_1 exploits some set of vulnerabilities to initiate e_2 with some likelihood
e_1 *leads-to* e_2	e_1 leads to e_2 with some likelihood, due to some set of vulnerabilities
e_1 *impacts* e_2	e_1 impacts e_2 with some consequence
e_1 *treats* e_2	e_1 reduces risk by providing treatment to e_2

4.4 Summary

One of the challenges of risk analysis is communicating the results of the analysis, both final and interim results, to other people. In order to help this communication, the CORAS method comes with a special purpose graphical modelling language for the modelling of threats, unwanted incidents, assets, risks, treatments, and other information uncovered and documented in a risk analysis.

The language consists of five kinds of diagrams, each tailored to facilitate specific tasks in a risk analysis. The diagrams consist of elements and relations that capture essential aspect of the risk picture under analysis. A summary of the language elements is provided in Table 4.1 and a summary of the relations in Table 4.2.

Chapter 5
Preparations for the Analysis

Before we start conducting the actual risk analysis, it is important that both we, as the analysis team, and the customer are properly prepared. The purpose of the first step is to do the necessary preparations. We should gather basic information about the target of analysis, the assets for which protection is required, the main worries with respect to these assets, as well as the desired scope and level of detail of the analysis. The customer should after this preparatory step be aware of its responsibilities in relation to the analysis and what we expect from the customer. We should agree with the customer on a tentative time schedule for the analysis, and a contact person representing the customer should be appointed, as well as a contact person representing the analysis team.

5.1 Overview of Step 1

A main objective with Step 1 is to roughly sketch the scope and focus of the analysis such that we can make the necessary preparations for the actual analysis tasks. An equally important objective is to inform the customer of its responsibilities for the risk analysis to be carried out.

An overview of the tasks of Step 1 is given in Table 5.1. We conduct the step by interacting with a representative of the customer, preferably as a face-to-face meeting. Alternatively, we may interact remotely, for example by phone or other means of correspondence.

The completion of the tasks of the preparatory step should result in a tentative time schedule and an agreement regarding the duration of the analysis. The customer is throughout the analysis responsible for gathering the meeting attendants representing the customer or representing other relevant stakeholders. The personnel that should participate at the meetings and workshops that are held during the analysis vary depending on the tasks that are to be conducted. When we go through each of the analysis steps and their tasks in the following seven chapters, we specify the recommended participants. The UML class diagram depicted in Fig. 5.1 shows the roles that are commonly filled by the participants.

Table 5.1 Overview of Step 1

Preparations for the analysis

- **Objective**: Gather basic information about the customer, the purpose and domain of the analysis, and the size of the analysis; ensure that the customer and other involved parties are prepared for their roles and responsibilities; appoint a contact person from the customer; agree on a tentative time schedule
- **How conducted**: Held as an interaction between the analysis team and customer representatives; the interaction is preferably held as a face-to-face meeting, but may be conducted by, for example, phone or other forms of interaction; the analysis team may subsequently make preparations by gathering data about the target of analysis, such as information and statistics about typical risks
- **Input documentation**: None
- **Output documentation**: Any relevant information about the target provided by the customer; information that is gathered by the analysis team about the domain and the target that is addressed; a tentative time schedule

Remark UML *class diagrams* are used to represent a classification of model elements such as entities and concepts. The class diagram in Fig. 5.1 depicts participants in a risk analysis and their roles. Each element is represented by a rectangle and the relations between elements are represented by edges between them. The large, hollow arrowheads specify specialisations. In Fig. 5.1, for example, *Analysis leader*, *Analysis secretary* and *Decision maker* are all special cases of *Analysis role*. The white diamonds on edges represent so-called aggregations. An aggregation specifies a whole-part relation between an aggregate (a whole) and its constituent parts. *Analysis leader* and *Analysis secretary*, for example, are parts of *Analysis team*. The numbers annotated at the ends of edges show multiplicities. *Analysis team*, for example, is formed by one *Analysis leader*, one *Analysis secretary* and an arbitrary number of *Analysis member*. The asterisk denotes zero, one or more. The multiplicity next to *Analysis role* on the relation from *Analysis participant* shows that each analysis participant fills arbitrary many analysis roles, but at least one.

From the class diagram in Fig. 5.1, we see that there are arbitrary many so-called parties for each risk analysis, but at least one. A party of a risk analysis is a stakeholder with respect to which the analysis is held. In other words, a party is an organisation, a person, a group of persons or the like whose interests are the concern of the analysis. A party of an analysis has something of value that needs protection, and the objective of the analysis is to understand and evaluate the risk picture from the viewpoint of the party. In a risk analysis addressing a computer system of a company, for example, the company itself is typically a party. What are the risks for the company given this computer system and how is it used and maintained? How can the company be damaged in case there is a security breach or a technical problem in the system? How severe are the consequences of an unwanted incident for the company?

5.1 Overview of Step 1

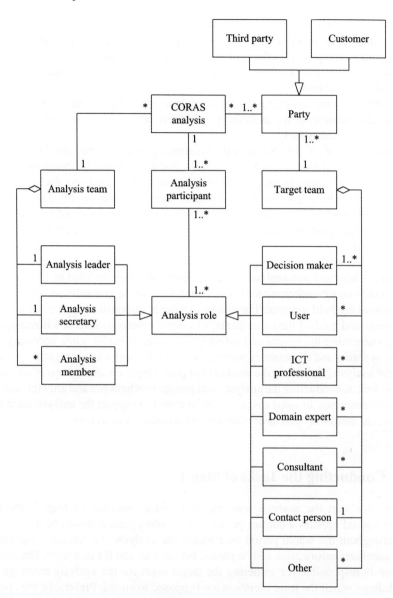

Fig. 5.1 Overview of risk analysis participants and roles

It is sometimes necessary that we take into account more than one party in an analysis and that we look at the risk picture from several perspectives. A company may, for example, want to understand how the company owners or shareholders can be affected by risks towards the system in question, and they may want to take into account the perspective of their customers. The various parties of a risk analysis often have various interests in the target of the analysis. Given a security breach within a computer system of a company, the company itself may be worried about confidentiality of sensitive information, whereas the company shareholders are concerned about the possible effects on the revenue, and the customers may be worried about the dependability of the company with respect to providing services that are critical to the customers.

In practice there is often only one party in a risk analysis, and the party is commonly the customer of the analysis. We therefore often use the notions of party and customer interchangeably.

The parties of a CORAS risk analysis forms a target team, the members of which represent the parties at the various meetings and workshops. As shown by the class diagram, a risk analysis requires the participation of at least one decision maker, and there must be at least one contact person. The other roles are filled by personnel with expert knowledge on various parts or aspects of the target of analysis. Note that one participant may fill several roles and that one role may be filled by several participants.

For a risk analysis, there is one analysis team. The analysis team consists of one analysis leader and one analysis secretary. Both roles may be filled by the same person, which may be adequate for smaller analyses, but we generally recommend that these roles are held separate. The analysis leader is responsible for leading the analysis tasks and guiding the participants, whereas the analysis secretary is responsible for documenting the results and assisting the analysis leader when necessary. The analysis leader and the analysis secretary commonly work closely together throughout the analysis; they share the workload of preparing the various steps of the analysis, as well as conducting the analyses and evaluations between and after the sessions with the customer. In some cases, it may be useful to support the analysis team with additional members such as domain experts, evaluators or trainees.

5.2 Conducting the Tasks of Step 1

Before we start the analysis with the introductory meeting of Step 2, the customer should appoint a contact person. The contact person should be available to us throughout the whole period over which the analysis is conducted, and ensure that necessary information can be passed between us and the customer. The contact person is responsible for gathering the target team for the analysis meetings and workshops which the team members are supposed to attend. Preferably, the contact person should be involved in all of the analysis steps and meetings in order to ensure that at least one person from the target team has full overview of the analysis and its course.

The contact person is commonly the person that represents the customer at the preparatory meeting. An hour or so is usually sufficient time for the customer to give us a briefing about the target of analysis and the customer's objectives for the analysis, and for us to inform about what we expect from the customer throughout the analysis. The contact person must ensure that the meetings and workshops are scheduled and that the required personnel are called in and made available.

5.2 Conducting the Tasks of Step 1

From our side, it is the analysis leader that is the contact person representing the analysis team. The analysis leader should therefore be available to the customer throughout the analysis process.

We cannot expect the customer to think of and remember all relevant details and issues that should be addressed during this preparatory meeting. We should therefore prepare follow-up questions about the target of analysis. After the meeting, we should have basic knowledge about the following.

- What is the target of analysis?
- What is it that the parties seek to protect?
- What kind of users, technologies, networks, work processes, businesses, organisational units, and so forth are relevant within the target?
- What are the main worries?
- What are the most relevant threats?
- Are there areas on which there will be particular focus?
- What is the desired size of the analysis in terms of man-hours?

We should also clarify what is the context of the analysis, and what the customer and the parties of the analysis expect from it. The context may be the initial background of the analysis, for example, that an enterprise needs to increase the security awareness in response to new threats or other changes in the environment. The context also includes the premises of the analysis, for example that the analysis results are intended for a particular addressee. Such an addressee may, for example, be decision makers that need better grounds for decision-making. The context may also be the time horizon of the analysis, for example, that the analysis should be restricted to issues of relevance within the scope of the next three years.

Definition 5.1 The *context* of the analysis is the premises for and the background of the analysis. This includes the purposes of the analysis and to whom the analysis is addressed.

How the analysis results are intended to be used is important knowledge when deciding on the appropriate reporting format for the analysis. In some cases, we work out a comprehensive and detailed report that documents and explains the results of the risk analysis. In other cases, the parties only need the results to be reported as a presentation to the relevant addressees such as decision makers, business partners, employees, or the like.

In the following, we introduce the case that will serve as a running example throughout the presentation of the CORAS method in this and the following chapters.

Example 5.1 AutoParts is originally a mail order company that started out about three decades ago. The company's business is to sell spare parts and accessories for a wide range of car makes and vehicle models. It is distributing catalogues by mail that present its products and is usually shipping the goods to the customers by cash on delivery mail. The company furthermore cooperates with various sub-suppliers and other suppliers within the same business sector that distribute some of their products through AutoParts. The sub-suppliers are mostly authorised car breakers providing spare parts that are not in AutoParts portfolio.

Six or seven years ago AutoParts started publishing the catalogues on the web and provided a primitive mechanism (a web form sending email to the sales department) for the ordering of products via its web site. This did not, however, change the way the orders were processed internally, and the orders coming in through the web site were essentially treated in the same way as ordinary paper form orders. During the last five years or so, the company has experienced a steady increase in the orders received though the web site and a considerable decrease in the paper order forms. At the same time, they have developed better software for keeping track of their inventory. The maintenance of the web site and the continuous development of the inventory software have also resulted in a quite large in-house software development group.

AutoParts has decided to make a transition from the manual system to an automated online store. The most important requirements to the new system are the following:

- Integration of the online store with the inventory software in such a way that manual punching of orders is not longer necessary and so that the warehouse automatically gets a notice when a purchase is made.
- Inclusion of a payment mechanism that allows the purchaser to do in advance payment with a credit card, so that cash on delivery mailing is no longer necessary.
- Offering sub-suppliers of AutoParts and other suppliers within the same business sector the possibility to sell their products directly through the online store of AutoParts with a commission on the sales.

The new system was developed by AutoParts' in-house development team. It has undergone the final pilot testing and is ready to launch, but AutoParts has decided it wants to do a risk analysis of the system before launching. Of particular concern for the management is that the web application is connected to both their customer database, their inventory database and their online store software. In addition the management fears that the new, and more complex, system will be less stable than the old, primitive one. Finally, the management is worried that the processing and storage of personal data submitted by the customers during registration to the system do not comply with data protection laws.

AutoParts has hired the consultancy company CORAS Risk Consulting to do the analysis. CORAS Risk Consulting carries out the security risk analysis in cooperation with AutoParts following the CORAS method. The analysis is estimated to 250 man-hours from the analysis team over a time span of two months.

5.3 Summary of Step 1

Step 1 of the CORAS risk analysis method is summarised by the overview presented in Table 5.2.

5.3 Summary of Step 1

Table 5.2 Summary of Step 1

Preparations for the analysis

Tasks:	People that should participate:
• The customer briefs the analysis team on the target and the objectives, as well as the desired size of the analysis • The analysis leader informs the customer about the customer's responsibilities throughout the analysis • A contact person from the customer is appointed • A tentative time schedule for the analysis is determined	• Analysis leader (required) • Analysis secretary (optional) • Representatives of the customer

Modelling guideline:
- Modelling is not part of this step

Chapter 6
Customer Presentation of the Target

In order to conduct a risk analysis, we need a clear picture of what the customer or the parties of the analysis really want. This includes understanding what are the objectives with the analysis, as well as getting a good grip of the features and architecture of the organisation or system and the processes that the customer wants to analyse. Step 2 of a CORAS analysis is concerned with obtaining this understanding.

Typically, this step involves a face-to-face meeting with the target team. The target team is personnel representing the parties of the analysis, and it may involve third parties such as consultants and other stakeholders. We may alternatively conduct the step as a video conference, or even by phone or through the exchange of written documentation. The latter is not something we recommend, but is a possible alternative. In the following, we assume a meeting, either face-to-face or as a video conference.

6.1 Overview of Step 2

It is during Step 2 of the analysis that we decide on the overall setting for the analysis and make the first move towards establishing the target description that we later use as a basis for the risk analysis. An overview of the tasks is given in Table 6.1. The table also indicates how we should perform each of the tasks.

It is crucial for the efficiency and the correctness of a risk analysis that the participants are able to communicate and exchange knowledge and opinions in a precise manner, and that the target team is well informed about the tasks that are to be conducted and their objectives. As a first task of the current step, we therefore present to the target team the terminology and the method we use.

Operating with a fixed and well defined terminology is important for several reasons. First, since we base the risk analysis on discussions and workshops, it is of vital importance that the participants use a common and precise vocabulary. The target team members typically have various backgrounds and fill different roles

Table 6.1 Overview of Step 2

Customer presentation of the target

- **Objective:** Achieve an initial understanding of what the parties wish to have analysed and what they are most concerned about; decide on the focus, scope and assumptions of the analysis; establish a detailed plan for the analysis
- **How conducted:** The analysis step is conducted as an interaction between the analysis team and representatives of the parties, preferably as a face-to-face meeting; the analysis team introduces the CORAS method before representatives of the parties present the goal and target of the analysis; through plenary discussions, the focus, scope and assumptions are determined, and the plan for the analysis is decided
- **Input documentation:** Any relevant information about the target provided by the customer or the parties in advance; information that is gathered by the analysis team about the type of system that is addressed
- **Output documentation:** Informal drawings and sketches describing the target, the focus and the scope of the analysis; a plan for the analysis with meeting dates, list of participants and dates for delivery of documentation and analysis report
- **Subtasks:**
  a. Presentation of the CORAS terminology and method
  b. Presentation of the goals and target of the analysis
  c. Setting the focus and scope of the analysis
  d. Determining the meeting plan

in the systems or organisations they represent, and a common terminology facilitates communication and reduces the chance of misunderstandings. Second, since the CORAS terminology is customised for risk identification and analysis, the terminology serves to assist the various tasks of the analysis process and it supports the participants in precisely communicating their opinions and views among each other. Third, since the CORAS terminology is accommodated to the CORAS risk modelling language, we can more easily and precisely transfer the findings to the models that we create.

When we present the CORAS method, we focus particularly on making the target team aware of the agenda and goals for today. The presentation of the method also prepares the target team for the tasks that lie ahead; it motivates the various activities and it places each task in the overall setting of the CORAS analysis.

Before starting identifying and analysing potential risks to something, we need to know exactly what this something is. The second task of this step is therefore a presentation held by one or more representatives of the parties where the target of the analysis is introduced. To ensure the success of this task, it is important that we as the analysis team carefully describe our expectations and needs. The parties must be prepared to present relevant aspects, such as the objectives and goals of the analysis, system architecture, applications, work processes and users. The parties should also be prepared to explain what are the main assets of the target, and what are their main worries with respect to these.

The third task is to establish the focus and scope of the analysis, as well as deciding on the assumptions on which the analysis is held. This normally includes a discussion among the participants of the target team, moderated and guided by us. We must make sure that what the target team arrives at is feasible given the resources made available for the analysis. Since conducting risk analyses is time and resource consuming, it is decisive that we from the very beginning determine what should be the main issues of the analysis, at what level of detail the analysis should be conducted, what should be scope, and what assumptions we are allowed to make. The scope must be balanced against the desired level of detail.

During a risk analysis, we typically make several assumptions and decisions about the system, organisation or processes under analysis, and also about the surroundings and the environment. We may, for example, assume that the power supply is reliable or that there are no deliberate human threats within the target. In the end, the conclusions from the risk analysis are valid only under these assumptions, and it is therefore crucial that we are specific about the assumptions we make and that they are properly documented. Moreover, as the target and its surroundings may change over time, the assumptions, and hence also the analysis results, may in the future no longer be valid. Knowing the assumption of the analysis is necessary for determining if and when a new analysis is required for the identification and analysis of new risks or other changes in the current risk picture.

As the final task of this step, we decide on a plan for the remaining meetings and workshops, including dates to the extent this is possible. We also indicate who should be present at each meeting and workshop. A schedule for the delivery of relevant documentation, preliminary reports and particularly the finalised risk analysis report is decided.

6.2 Conducting the Tasks of Step 2

In this section, we go through the tasks of Step 2 in more detail and describe how we conduct each of them. A successful completion of the tasks requires involvement of decision makers of the parties. Involvement of personnel with first-hand knowledge about the target of analysis, such as users or technical expertise may also be helpful.

6.2.1 Presentation of the CORAS Terminology and Method

Our presentation of the CORAS terminology and method, an overview of which is given in Table 6.2, should be prepared for being displayed to the target team. Before we give the presentation, we should emphasise the importance of and motivation for operating with a fixed and well-defined terminology throughout the risk analysis.

When presenting the CORAS terminology, we include the notions of asset, threat, vulnerability, threat scenario, unwanted incident, risk and treatment. We display the precise definition of each of the notions and carefully explain the meaning of each of them. We furthermore characterise the relationships between the notions, since the notions can only be precisely understood by understanding how they are related. The latter is important also because many of the core tasks during the risk

Table 6.2 Overview of Step 2a

Presentation of the CORAS terminology and method

- **Objective:** The target team understands the CORAS terms and their relationships, and is familiarised with the steps of the CORAS method
- **How conducted:** Presentation given by the analysis team; the precise definition of each of the CORAS terms is presented and the relationships between the terms are explained; the CORAS symbols illustrating the terms are displayed and a basic CORAS diagram is presented and explained; an overview of the CORAS method is displayed and each step of the method is briefly presented; the analysis team explains to the target team what can be expected from the analysis, and emphasises the responsibilities of the target team with respect to providing the necessary information and documentation, as well as gathering personnel with suitable background to participate in the various steps of the analysis
- **Input documentation:** None
- **Output documentation:** None

Fig. 6.1 CORAS risk analysis terms

analysis process is to identify and understand how the various threats, vulnerabilities, assets, risk, and so forth of the target of analysis are related. The presentation of the CORAS terminology can with advantage be accompanied with the presentation of a basic, illustrative example.

Example 6.1 The customer AutoParts and the consultancy company CORAS Risk Consulting decided to conduct Step 2 of the analysis as a separate introductory meeting scheduled for three hours. The consultancy company is represented by the analysis leader and the analysis secretary. The participants representing AutoParts are the chief executive officer, the leader of the team developing the automated online store, and the security officer of AutoParts. The security officer is also the appointed contact person for this analysis.

The analysis team has prepared a presentation introducing the CORAS terminology and method, and the first twenty minutes of the meeting is devoted to this task.

The analysis leader holds the presentation, and the terminology is introduced by showing the overview in Fig. 6.1 in one of the slides. For each term, the associated symbol for the CORAS risk modelling language is depicted, and the analysis leader presents the definition of the term. She explains the meaning of each term and carefully clarifies the relationship between them.

6.2 Conducting the Tasks of Step 2

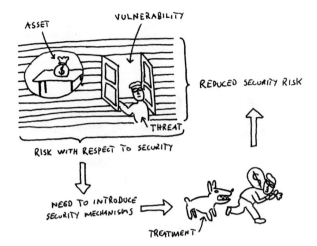

Fig. 6.2 Illustration of the relationship between the concepts of a risk analysis

In order to exemplify the terminology, the analysis leader shows the illustration depicted in Fig. 6.2. The figure presents a scenario in which a burglar breaks into a house. It is suitable as an illustration since it can be easily related to by people of all backgrounds.

The analysis leader explains that without the three elements of asset, vulnerability and threat there cannot be any risk; it is only when there is something of value that there is something to harm. Moreover, for harm to happen there must exist a threat towards the asset, as well as a vulnerability that the threat may exploit.

In the given example the risk is the potential for the theft of valuables from the house. If the risk level is intolerable, there is a need for measures to mitigate the risk, that is, treatments must be implemented in order to reduce the risk level. In this case, the treatment is directed towards the threat; a guard dog is placed outside the house with the intention of removing the threat towards the asset.

The analysis leader finally links the terminology to the CORAS risk modelling language by showing the diagram depicted in Fig. 6.3. The diagram illustrates both how the various terms are related and how the terms are manifested as building blocks in the risk modelling language. The diagram also shows that the treatments may be directed towards threats as well as vulnerabilities with the aim of reducing the probability of the unwanted incident. The analysis leader moreover explains that treatments can be implemented with the purpose of reducing the damage of the incident, for example, by depositing some of the valuables in a safe-deposit box.

The presentation of the CORAS method should be brief and only serve as a preparation of what lies ahead; a bulleted list over one or two slides with a short description of each step suffices. The results of the risk analysis rely on the engagement and involvement of the target team, and it is therefore crucial that we motivate the tasks, explain their goals and place each of them in the overall context of the analysis process.

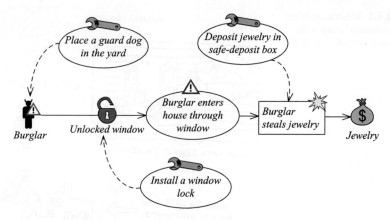

Fig. 6.3 Example CORAS diagram

Example 6.2 The analysis leader presents the method by showing the overview depicted in Table 6.3 and briefly describing the main activities, goals and results for each step. The analysis leader focuses particularly on Step 2 since it describes the agenda and goals for the current step. The activities of Step 3 are also emphasised in order to prepare the target team for the subsequent tasks.

6.2.2 Presentation of the Goals and Target of the Analysis

The presentation of the target of the analysis as well as the overall goals, an overview of which is given in Table 6.4, is given by one or more representatives of the parties. Since the success of this task relies on the target team, we must ensure that the parties are well informed about the responsibility of performing this task and that the personnel assigned to the task have good knowledge about the target.

It is during this task and the discussions of the subsequent task that we gather most of the information about the target of analysis, so the presentation must be as precise and informative as possible. The target team should be prepared to present the system or organisation it wishes to have analysed, the relevant parties that have an interest in the target in the context of the analysis, the assets of main concern within the target, and what kind of incidents the parties are most worried about.

> **Definition 6.1** The *target of analysis* is the system, organisation, enterprise, or the like that is the subject of the analysis.

If there are other parties than the customer that need to be taken into account, we must make sure that their viewpoints are represented. In some cases, the interests and concerns of other parties are well known, for example, the interests of hospital patients in an analysis of risks towards hospital health records and services. In other cases, we may need to include representatives of these parties in the target team.

6.2 Conducting the Tasks of Step 2

Table 6.3 Overview of the CORAS method

The eight steps of the CORAS method

1	**Preparations for the analysis**
	The customer briefly informs the analysis team about the target it wishes to have analysed, and the analysis team prepares for the analysis
2	**Customer presentation of the target**
	The customer presents the system or organisation it wishes to have analysed; the focus and scope of the analysis is identified and an analysis plan is set up
3	**Refining the target description using asset diagrams**
	The analysis team presents its understanding of the target of analysis; the assets are identified, as well as the most important related threats and vulnerabilities
4	**Approval of the target description**
	The analysis team presents the documentation of the target of analysis for finalisation and approval by the customer; values are assigned to the identified assets, and the risk evaluation criteria are established
5	**Risk identification using threat diagrams**
	Risks are identified through a structured brainstorming
6	**Risk estimation using threat diagrams**
	The likelihoods and consequences for the identified risks are estimated
7	**Risk evaluation using risk diagrams**
	The risks are evaluated against the risk evaluation criteria
8	**Risk treatment using treatment diagrams**
	Treatments for the mitigation of unacceptable risks are identified and evaluated

Table 6.4 Overview of Step 2b

Presentation of the goals and target of the analysis

- **Objective:** The analysis team understands what the parties wish to have analysed and what the goals of the analysis are
- **How conducted:** Presentation given by representative(s) of the target team explaining what the parties wishes to have analysed and what kind of incidents they are most worried about occurring; the target presentation typically includes business goals, work processes, users and roles, contracts and policies, hardware and software specifications and network layout
- **Input documentation:** Information about the target that may have been provided by the parties in advance, as well as information that is gathered by the analysis team about the kind of target that is addressed
- **Output documentation:** Informal drawings and sketches describing the target, as well as other relevant target documentation provided to the analysis team

The purpose of the presentation of the intended target is for us to achieve knowledge about what the parties wish to have analysed. Depending on who the customer is and what the parties wish to achieve through the analysis, this presentation may cover aspects such as business goals and processes, users and roles, work processes, contracts and policies, hardware and software specifications and network layout. To complement the information provided through this presentation, we should also ask for all the relevant documentation the parties may have about the target, such as system specifications, policies and reports from previous risk analyses.

In addition to getting the first grip of what is to be the target of analysis, we need to achieve a precise understanding of what are to be the perspectives of the analysis, in other words, who the parties are. The concept of risk is inherently related to the concept of party; without a stakeholder with interest in the target in question there is no one that assigns value to the target in the meaning that there are no assets at stake and therefore nothing to risk. We must therefore ensure that the presentation of the target addresses the issues of party, asset and risk.

Very often it is the customer itself that is the party of the analysis, but in some cases it may be important to also take the perspective of other parties into account. We should therefore question the target team on this issue and have the participants to discuss it. Once it has been clarified who are the parties of the analysis, the discussion should move on to the issue of assets. For each of the identified parties, what is it that the analysis should address? What are the entities or aspects of the target that have value and that the parties need to protect?

The final issue that should be addressed during the presentation of the target are the threats and incidents of most worry for the parties. This will increase our understanding of the goals of the analysis and which type of risks the parties need to have identified and analysed. The focus on these threats and incidents may also help in increasing our understanding of what should be the focus of the analysis and which assets that are most important. This discussion should be related to the identified parties and their interest in the target of analysis.

Example 6.3 A representative of AutoParts presents the target of analysis by displaying the drawing depicted in Fig. 6.4. The main concern is the online store and the databases keeping track of the inventory and the customer information. He explains that employees have access to the system from their workstations, but also remotely via the Internet. Both customers and sub-suppliers interact with the online store over the Internet via a web application.

AutoParts' objective with the analysis is to identify and evaluate the risks towards the online store and the databases. The most important issue with respect to the online store is the availability of its services to the customers and to the sub-suppliers. This relies on the availability of the databases.

The confidentiality of the customer database is important to ensure compliance with data protection laws. They believe that their routines for processing and protecting personal information about AutoParts' customers are good, but want a proper analysis and evaluation of potential risks in case the current situation is not satisfactory.

The target team explains that in addition to evaluating risks that may arise from threats and vulnerabilities within the target, potential attacks via the web application or via the remote access used by the employees should be analysed.

The target team finally emphasises that the analysis should take into account the effect incidents may have on the satisfaction of the customers of AutoParts, as well

6.2 Conducting the Tasks of Step 2

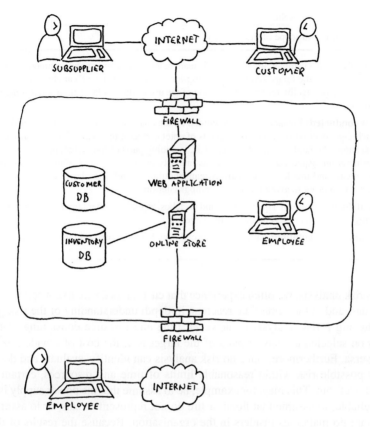

Fig. 6.4 Presentation of target

as AutoParts' reputation in general and the sub-suppliers' trust, since these issues have direct impact on the success of their business.

AutoParts wants to conduct the analysis solely from their own perspective, which means that there will be only one party of the analysis. The analysis will take into account potential concerns of the customers and the sub-suppliers, but only the assets of AutoParts.

6.2.3 Setting the Focus and Scope of the Analysis

The tasks of setting the focus and scope of the analysis, as well as establishing the assumptions of the analysis, an overview of which is given in Table 6.5, are conducted as a plenary discussion that we lead. The analysis leader explains the task and its motivation to the target team and guides the discussion, while the analysis secretary documents the results.

Table 6.5 Overview of Step 2c

Setting the focus and scope of the analysis

- **Objective:** The customer and the analysis team reach an agreement and common understanding of what is to be the focus, scope and assumptions of the analysis; a clear characterisation of the focus and scope facilitates the optimisation of the use of the available time and resources for the analysis
- **How conducted:** Plenary discussion guided by the analysis leader with the aim of determining which aspects or parts of the target that is of chief concern; the discussion must also determine what should be held outside of the analysis, including parts of the system or organisation under consideration, aspects of security such as integrity or confidentiality of information, or particular threats and incidents; the analysis team should be careful to clarify uncertainties or open questions in order to avoid misunderstandings later on
- **Input documentation:** The drawings and sketches, as well as other target information, gathered during the customer presentation of the target
- **Output documentation:** Notes and informal sketches produced on-the-fly during the discussion

As risk analysts, we often experience that customers desire to cover a wide range of issues, and at the same time reach an in-depth understanding of the risk picture. Conducting a risk analysis is, however, time and resource consuming. Particular focus on selected aspects or parts of the target is at the cost of a wide scope, and vice versa. Furthermore, since no risk analysis can identify, evaluate and document every possible risk within reasonable limits of time and resources, certain issues must be left out. This may, for example, be to assume that the power supply is stable and reliable, to assume that flood or fire do not represent threats, or to assume that there are no malicious insiders in the organisation. Because the results of the risk analysis depend on these assumptions, they must be explicitly documented.

Definition 6.2 The *scope of the analysis* is the extent or range of the analysis; the scope defines the border of the analysis, that is, what is held inside of and what is held outside of the analysis. The *focus of the analysis* is the main issue or central area of attention in the risk analysis; the focus is within the scope of the analysis. The *environment of the target* is the surrounding things of relevance that may affect or interact with the target; in the most general case, the rest of the world. The *assumptions of the analysis* are what we take as granted or accept as true (although they may not be so); the assumptions may be about the target and about the environment; the results of the analysis are valid only under these assumptions.

Example 6.4 AutoParts has already in advance decided that the main objective of the analysis is to identify and evaluate risks towards the inventory database and the customer database. With respect to the latter, the management have decided that they need to better understand potential risks related to breach of compliance with laws and regulations for data protection since the database stores some personal

6.2 Conducting the Tasks of Step 2   91

data. Additionally, AutoParts wants to identify risks towards the availability of the services provided to its customers, in particular the availability of the online store.

At a more general level, AutoParts is concerned about customer satisfaction, the company's reputation, and the trust of the sub-suppliers in AutoParts as a business partner.

Following the presentation of the main objectives of AutoParts for the analysis, the analysis leader guides the further discussion with the aim of making the focus and scope more precise. As to the databases, the analysis leader asks the target team whether all the aspects of confidentiality, integrity and availability should be considered. The target team members believe that the backup solution is satisfactory, but since the potential damage of loss of data is large they decide to include integrity. Taking into account confidentiality of the data in the analysis is held as vital. The target team agrees that availability is not a main worry and should be held outside the scope, unless loss of availability of the databases affects the availability of the online store.

The analysis leader further explains that the range of threats and incidents that generally may affect reputation, trust and customer satisfaction may be very wide and too much to consider given the main objectives of the analysis. The customer is therefore advised to consider risks towards these in the context of the threats, vulnerabilities and incidents in relation to the main objectives of the analysis.

Evaluating compliance to data protection laws and regulations, as well as the severity of compliance breach, requires expert knowledge within this domain. As neither the analysis team nor the customer possess such expertise, the customer decides to bring in a consultant to attend at the relevant meetings and workshops.

Having reached a satisfactory description of the focus and scope of the analysis, the analysis leader brings the discussion over to establishing possible analysis assumptions. The target team soon agrees that environmental threats such as flood, fire or power failure should be held outside the analysis. Since the automated online store has yet to be launched, they cannot disregard failure of their own system.

The analysis leader asks the target team whether there are any human threats that can be disregarded, in particular malicious insiders. The participants firmly believe that all employees are trustworthy, but since the risks that may result from a deliberate human threat within the company may be very severe they insist on considering insiders in the analysis.

After the completion of this task, the analysis secretary sums up what has been agreed and gets approval from the target team.

6.2.4 Determining the Meeting Plan

The objectives of this task, an overview of which is given in Table 6.6, is to decide who should participate at the meetings and workshops of the analysis, to set up a meeting plan, and to determine a schedule for finalisation and delivery of documentation and the analysis report.

Table 6.6 Overview of Step 2d

Determining the Meeting Plan

- **Objective:** The target team and the analysis team agree on meeting dates, meeting participants and date of delivery of the analysis report
- **How conducted:** Plenary discussion lead by the analysis team; the plan must schedule sufficient time between the analysis steps for the analysis team to process and analyse the result from the previous step and to prepare for the next step; the analysis team must also ensure that personnel with adequate background are present at the various meetings and workshops
- **Input documentation:** The tentative time schedule that were put up during Step 1
- **Output documentation:** List of analysis participants and their roles; meeting plan with dates, tasks and participants; list of deadlines for delivery of documentation, including the final analysis report

Table 6.7 Risk analysis roles

Role	Name	Organisation	Background/Expertise
Analysis leader	Alice	CORAS Risk Consulting	Risk analysis
Analysis secretary	Bob	CORAS Risk Consulting	Risk analysis
Decision maker	David	AutoParts	Corporate management
Domain expert (security)	John	AutoParts	Data security officer
Domain expert (development)	Peter	AutoParts	System development
Domain expert (user)	Irene	AutoParts	Sales and marketing
Domain expert (law)	Claire	Berger & Jones	Lawyer

The recommended participants for each of the analysis steps are presented in the respective chapters of this book. In general, the people involved throughout a risk analysis are the analysis leader and the analysis secretary, at least one decision maker, personnel with expert knowledge on relevant domains or aspects of the target, and system users. It is only the analysis team that is required to participate in all steps of the analysis, but we recommend that one representative of the customer is also attending all meetings and workshop. The latter participant is preferably the appointed contact person.

Example 6.5 Given the objectives, scope and focus of the analysis, the analysis team and AutoParts agree to involve four persons in the target team. In addition to the chief executive officer as the decision maker representing AutoParts, the target team will involve one system developer and one system user, as well as AutoParts' data security officer. AutoParts will also contact a law firm and bring in a lawyer as consultant. The documentation of the risk analysis roles is given in Table 6.7.

When we decide on the meeting plan for the risk analysis, we must allow for the time we need between the analysis steps for processing and analysing the result

6.2 Conducting the Tasks of Step 2

Table 6.8 Risk analysis plan

Date	Tasks	Participants
March 4	Identification of target, focus and scope of the analysis	Analysis team, decision maker, system developer, data security officer
March 25	Asset identification; high-level analysis	Analysis team, decision maker, system developer, system user, data security officer
April 8	Setting of risk evaluation criteria; approval	Analysis team, decision maker, system developer, system user, data security officer
April 15	Risk identification	Analysis team, system developer, system user, lawyer, data security officer
April 23	Risk estimation	Analysis team, decision maker, system developer, system user, lawyer, data security officer
May 6	Risk evaluation	Analysis team, decision maker, system developer, data security officer
May 14	Risk treatment; clean up and finalisation of results	Analysis team, decision maker, system developer, system user, data security officer

Table 6.9 Schedule for delivery of documentation

Date	Document
April 4	Draft of target description
April 10	Finalised and approved target description
April 17	Finalised threat diagrams
April 30	Finalised risk diagrams
May 19	Finalised treatment diagrams
May 31	Risk analysis report

from the previous step, and to prepare for the next step. For each of the meetings and workshops, we must specify who should attend. It is the responsibility of the customer, commonly the appointed contact person, for calling the meetings and setting them up.

Example 6.6 The agreed meeting plan is presented in Table 6.8. It is decided that the data security officer and the system developer should attend all meetings since they have key competence with respect to all the activities that are to be conducted. The planned time span of the whole analysis from the initial meeting to the delivery of the analysis report is about three months.

The agreed schedule for delivery of documentation delivered by the analysis team is presented in Table 6.9.

6.3 Summary of Step 2

Step 2 of the CORAS risk analysis method is summarised by the overview presented in Table 6.10.

Table 6.10 Summary of Step 2

Customer presentation of the target	

Analysis tasks:
• The risk analysis method is introduced
• The target team presents the goals and the target of the analysis
• The focus and the scope of the analysis are set
• The meetings and the workshops are planned
	People that should participate:
	• Analysis leader (required)
	• Analysis secretary (required)
	• Representatives of the customer:
	  – Decision makers (required)
	  – Technical expertise (optional)
	  – Users (optional)

Modelling guideline: 1. At this early stage of the analysis, it can be useful to describe the target with informal drawings, pictures or sketches on a blackboard 2. The presentation can later be supplemented with more formal modelling techniques such as the UML or data flow diagrams	

Chapter 7
Refining the Target Description Using Asset Diagrams

A correct and successful risk analysis requires that the target of analysis, as well as its focus, scope and assumptions, are correctly and commonly understood by all the participants of the analysis. The success of the analysis furthermore requires that we spend the available time and resources efficiently and focused. The purpose of Step 3 of the analysis is to establish a common understanding of what is to be analysed, what are the main issues, and what is the main direction of the analysis. In particular, Step 3 is used as a means to correct and improve the understanding of the analysis team about the target and about the objectives of the customer.

As with the previous step, we commonly conduct the tasks of Step 3 as a face-to-face meeting with the target team. Alternatively we may, for example, organise it as a video or phone conference, or through the exchange of written documentation.

7.1 Overview of Step 3

Our main objective of Step 3 is to achieve a refined and more precise characterisation of the target, including its scope and focus and its assets, as well as getting an overview of the potential risks at the enterprise level, for example, the kind of risks the decision makers are aware of and that motivated the risk analysis in the first place. Such a characterisation will subsequently direct the analysis towards the most important issues, thus ensuring that we spend the time and resources in the most efficient way.

An overview of Step 3 is given in Table 7.1. The first task is for us to present the target of analysis as we have understood it based on the previous steps and the information provided to us by the customer. The presentation should be a first step towards a model of the target. At the same time, it should serve as a means to ensure that we have correctly understood the input from the customer and other parties. Typically, the outcome is that our understanding to a large extent is correct, but that certain issues need to be corrected, and more details are required for other aspects. We prepare the task by documenting the target using a suitable formal or

Table 7.1 Overview of Step 3

Refining the target description using asset diagrams

- **Objective:** Ensure a common and more precise understanding of the target of analysis, including its scope, focus and main assets
- **How conducted:** The analysis team presents their understanding of the target of analysis; asset identification and high-level risk analysis are conducted as interactions between analysis team and representatives from the customer, preferable in a face-to-face meeting
- **Input documentation:** Target models in a suitable formal or semi-formal language prepared by the analysis team based on information gathered during the previous step and other target information provided to the analysis team
- **Output documentation:** Updated and corrected models of the target of analysis, including CORAS asset diagrams; high-level risk analysis table documenting the results of the high-level analysis
- **Subtasks:**
  a. Presentation of the target by the analysis team
  b. Asset identification
  c. High-level analysis

semi-formal language such as the UML, or an in-house notation that the customer is accustomed to.

The second task of Step 3 is the asset identification. Based on the discussions and the information gathered during Step 1 and Step 2, we explicitly specify the main assets of the relevant parties. We furthermore classify the assets according to whether they are direct or indirect, and we specify how the assets are related. The result of the asset identification is documented using CORAS asset diagrams.

The third and final task of Step 3 is to conduct a high-level risk analysis. This is conducted as a brainstorming session guided by the analysis leader. During the high-level analysis, we identify the most important threats, vulnerabilities and risks with respect to the identified assets and as seen from the perspective of a decision maker at the enterprise level. The purpose of the task is to contribute to a precise characterisation of the focus of the analysis and the most important issues to address in more details later on. We document the results of the high-level analysis using a high-level risk analysis table.

The three tasks together are essential in establishing the target description that gives direction to and serves as basis for the subsequent steps of the risk analysis.

Definition 7.1 The *target description* is the documentation of all the information that serves as the input to and the basis for the risk analysis. This includes the documentation of the target of analysis, the focus and scope of the analysis, the environment of the target, the assumptions of the analysis, the parties and assets of the analysis, and the context of the analysis.

Table 7.2 Overview of Step 3a

Presentation of the target by the analysis team

- **Objective:** Ensure that the analysis team has correctly understood the target and the objectives of the analysis; establish a common understanding of the target among the analysis team and the target team; refine the target description and bring it closer to finalisation
- **How conducted:** Presentation given by the analysis team of both static and dynamic features of the target; static features may be hardware configurations, network design and organisational structure, whereas dynamic features may be work processes and information flow; corrections and comments from the target team derived through a plenary discussion are implemented
- **Input documentation:** Models in a suitable formal or semi-formal language such as the UML prepared by the analysis team; the models are based on the information gathered during the previous steps and on target documentation provided to the analysis team; UML class diagrams and UML collaboration diagrams, for example, are suited to model static system features, whereas UML activity diagrams and UML sequence diagrams, for example, are suited to capture dynamic features
- **Output documentation:** Updated and corrected models describing the target of analysis

7.2 Conducting the Tasks of Step 3

We now go through the tasks of Step 3 in more detail and describe how we conduct each of them. The accomplishment of the tasks of this step requires the participation of decision makers and technical expertise. System users or others with first-hand expertise may also be brought in if necessary.

7.2.1 Presentation of the Target by the Analysis Team

The main reason why we, the analysis team, present the target is that we want to ensure that we have correctly understood what has been presented to us. We therefore encourage the target team to make comments and corrections and to fill in missing pieces during our presentation. Conducting this task, an overview of which is given in Table 7.2, also contributes to ensure that the target of analysis is commonly understood by the participants of the analysis, which is crucial for the efficiency and the very correctness of the risk analysis and its results.

Our presentation of the target furthermore intends to refine the description of the target and to make it more precise. Based on the results of and the documentation from the previous two steps of the risk analysis, we prepare the target models using a standardised, (semi-)formal language such as the UML or the like. In some cases, the target team is accustomed to a particular language or notation that is used in-house. We should therefore ask whether our choice of language is suitable, or whether there are preferable alternatives.

The target models should include descriptions of both static and dynamic features. Static features may be network design, physical and logical organisational structures, and conceptual relationships, whereas dynamic features may be work processes, data processing and information flow.

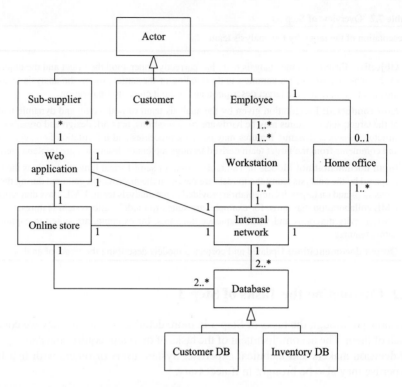

Fig. 7.1 UML class diagram showing a conceptual view of the target

The analysis secretary is responsible for documenting changes and amendments on-the-fly while the presentation is given and the target team makes corrections and comments. After this step and before Step 4 of the analysis, we finalise and polish all the target models.

Example 7.1 The analysis team from CORAS Risk Consulting has prepared diagrams using the UML describing the target from different viewpoints. The diagrams focus on the parts and aspects of the target of most relevance for the objectives of AutoParts regarding the analysis. The diagrams include more details than the sketches previously presented by the customer, and during a walk-through of the diagram given by the analysis leader the members of the target team make comments and corrections. The following is a description of the updated and corrected diagrams that make up the output of this task.

The UML class diagram of Fig. 7.1 depicts the most relevant concepts of the target, and shows how they are related. The relevant actors are the sub-suppliers, the customers and AutoParts employees. The former two are related to AutoParts' system via the web application only. The multiplicities show that there is one web application that is related to arbitrary numbers of sub-suppliers and customers. Employees, on the other hand, are via workstations or their home offices related to all relevant parts of the system through the internal network of AutoParts.

7.2 Conducting the Tasks of Step 3

Fig. 7.2 Collaboration diagram showing the physical communication lines relevant for customers and sub-suppliers

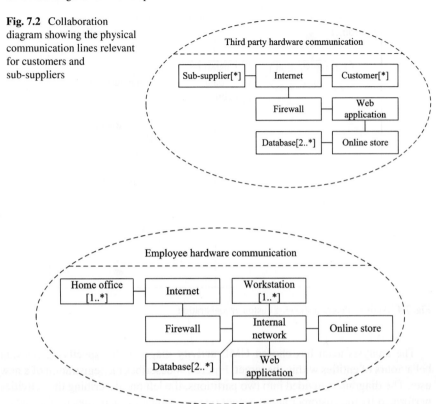

Fig. 7.3 Collaboration diagram showing the physical communication lines relevant for employees

The internal network is modelled as one entity that is related to the web application, the online store and the databases. The multiplicity of the databases shows that there may be more than two of them that are related to the internal network and the online store, but it is only the customer database and the inventory database that are explicitly modelled since they are of main concern in the analysis.

> *Remark UML collaboration diagrams* specify relationships between interacting entities. In Fig. 7.2 and Fig. 7.3 collaboration diagrams are used to show the relevant physical communication lines between entities within the target of analysis. The multiplicities show how many instances there are of each entity. When the multiplicity is omitted the default is one.

Figure 7.2 shows the communication lines of relevance for the sub-suppliers and the customers of AutoParts. They interact with the web application via Internet and the firewall. Figure 7.3 shows the communication lines of relevance for AutoParts employees who interact with the internal network either via workstations or via their home offices.

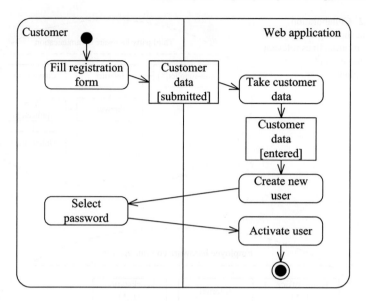

Fig. 7.4 Activity diagram showing customer registration

The analysis team has chosen UML activity diagrams for specifying relevant behaviours of entities within the target. Figure 7.4 describes the registration of a new user. The diagram is divided into two partitions, the left one describing the activities performed by the customer and the right one describing the activities performed by the web application. The customer data is shown as an object that flows through the network of activities.

> *Remark UML activity diagrams* specify how a certain behaviour is conducted through carrying out several activities. The separate activities are represented by rounded rectangles, and activities are conducted sequentially or concurrently. The arrows from one activity to another shows flow of control, and one activity cannot be initiated until all the activities from the incoming arrows are finalised. Activities are concurrent when there is a branching of control flow from an activity. A solid black circle represents the starting point of the control flow and the activities, whereas a circle with a smaller black dot inside represents the termination. Activity diagrams can also specify object flow. An object is depicted by a square with the object name. The current state of an object is specified in square brackets under the object name. The activity diagram can be divided vertically into so-called partitions that show the entity or actor that is conducting the activities within the partition. The name of the entity or actor is placed at the top of the partition.

The activity diagram in Fig. 7.5 shows customer usage of services offered through AutoParts' web application. In particular, the diagram shows the placing of an order and the state of the order as it is processed. Since there are two outgoing

7.2 Conducting the Tasks of Step 3

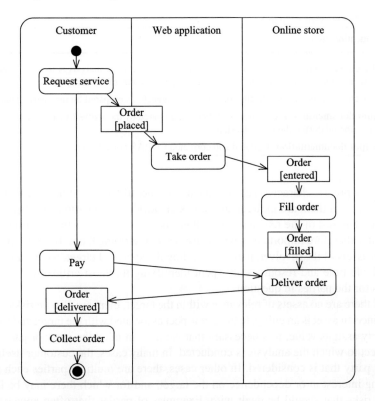

Fig. 7.5 Activity diagram showing customer usage of services

flows of control from service request, the following activities are concurrent. However, the two ingoing arrows to the deliver order activity mean that both payment and filling of order must have finalised before the activity of delivering the order is conducted.

Since data protection is one of the main issues of the analysis, the analysis leader asks the target team which personal data the customer is required to submit to the web application during the various activities. The following list is then produced.

- Full name
- Home address
- Billing address
- Phone number
- Email address
- Credit card number

7.2.2 Asset Identification

The asset identification, an overview of which is given in Table 7.3, was initiated during the previous step of the analysis. The purpose at this point in the analysis is

Table 7.3 Overview of Step 3b

Asset identification

- **Objective:** Identify precisely with respect to which assets the analysis is to be conducted; increase the understanding of the appropriate scope and focus of the analysis
- **How conducted:** Plenary discussion based on suggestions prepared by the analysis team
- **Input documentation:** CORAS asset diagram prepared by the analysis team based on information gathered during the previous step
- **Output documentation:** Updated and corrected CORAS asset diagram

to give a precise characterisation of the assets, identify which of the assets are direct and which of them are indirect, as well as identifying the relationships between the assets. The rest of the risk analysis will be specialised towards the identified assets, and it is therefore important to be precise when describing them. By focusing on the main assets of the involved parties regarding the specified target, we ensure that the available time and resources are spent identifying the most critical and important risks for the parties in question.

If there are no assets of relevance within the target, there are also no risks towards it. Since an asset is an entity, feature or aspect associated with the target and to which a party assigns value, it is necessary that we have in mind the party or parties with respect to which the analysis is conducted. In many cases, the customer itself is the only party that is considered. In other cases, there are multiple parties each having strong interest in or dependence on the target, and for which there may be intolerable risks that should be dealt with. Examples of parties that often appear in risk analysis in addition to the customer are a governmental inspectorate concerned with privacy or data protection issues, end-users concerned with availability of services, or shareholders of the customer concerned with the reputation and bottom line of the enterprise.

Remark A *party* is an organisation, company, person, group or other body on whose behalf the risk analysis is conducted.

The target team may be inclined to think of assets as something concrete and tangible such as computers and communication lines, or something they use on a regular basis such as information and services. The notion of asset is, however, defined quite generally as something to which a party assigns value. In addition to physical things such as computers and communication lines, assets may be more abstract and conceptual such as trust and reputation, or compliance with laws and regulations. An asset may also be an aspect of an entity, such as availability of a service or confidentiality of information. We must therefore make sure that the target team thoroughly discusses what are the most important parts, features or aspects of the target and that should be the focus of the analysis. For example, if a party seeks protection of certain information, is it because the information itself is valuable, for example because it contains trade secrets? Is it because the information is needed for the provision of certain services? In the latter case, is it the information itself

7.2 Conducting the Tasks of Step 3

Table 7.4 Examples of assets

Asset types	Examples
Information assets	Databases and data files, voice records, image files, system documentation, policies, user manuals, training material, business strategies, trade secrets, customer information
Paper documents	Contracts, guidelines, company documentation, business results
Software assets	Application software, system software, development tools
Physical assets	Computers and networks, communication equipment, storage media, power supply equipment, accommodation, inventory
Services	Computing services, communication services, power services
Human assets	Human health and life, know-how and expertise
Marketing assets	Company image and reputation, customer and partner trust, public relations
Business assets	Market share, stock price, revenue, bottom line, turnover
Organisational assets	Internal regulations and routines, policies and procedures
Compliance	Compliance with legal and regulatory requirements, adherence to ethical standards

that is the main concern or is it the availability of the services in question? And are there aspects that are more important than others, such as availability and integrity?

> **Remark** An *asset* is something to which a party assigns value and hence for which the party requires protection.

Table 7.4 gives an overview of some typical kinds of asset as well as examples of each of the kinds. The list is not exhaustive, neither by the kinds nor by the examples, as the purpose is to give an idea of what can be held as valuable for a party and with respect to which a risk analysis thus may be conducted. The table may nevertheless serve as a checklist to support the asset identification task.

It should be noted that some of the asset kinds may overlap. Paper documents, physical assets and human assets may, for example, all serve the same purpose as information carriers. Sometimes we may also have to decide whether a specific asset should be specified as one kind or another, or whether one and the same entity should be represented as several assets. Should a computer, for example, be protected for its value by the information it stores, by the services it provides, or for its value as a piece of hardware? Should humans be protected for being assets of their own, for example, human health from the perspective of a hospital, or should they be protected for the resources they represent by their expertise?

Once we have identified the assets, we turn to the issue of identifying possible relationships between the assets. The purpose of this task is to describe how harm to one asset may lead to harm to other assets. Such harm relations between assets are important to understand and document since the complete risk picture can only be understood by investigating all of the possible impacts of identified risks. Certain sensitive information may, for example, be one of the identified assets and we need to identify and evaluate risks against it. Another asset may be the availability of a critical service. Harm to the information asset may then imply harm to the service asset if the delivery of the latter depends on the availability of the former. A risk with respect to a given asset can be properly understood and evaluated only by taking into account the other assets to which it is related.

The final issue of the identification and documentation of the assets is to determine which of the assets are direct and which of the assets are indirect with respect to the target of analysis. An indirect asset is an asset that, with respect to the target and scope of the analysis, is only harmed via harm to other assets. Leakage of sensitive information caused by a hacker attack may, for example, lead to the reputation being damaged. The information asset may therefore in this case be classified as direct and the reputation asset as indirect. Importantly, however, this depends on the target of analysis and its scope. If public relations is a core issue, we may choose to include media coverage and exposure of the corporate management in the target of analysis. With such a perspective there may be incidents that directly affects reputation, and reputation must be treated as a direct asset.

> **Remark** An *indirect asset* is an asset that, with respect to the target and scope of the analysis, is harmed only via harm to other assets. A *direct asset* is an asset that is not indirect.

In practice, we may always treat all assets in the same way and omit the distinction between direct and indirect assets. The purpose of the distinction is to simplify parts of the analysis by focusing on the immediate damage of the unwanted incidents. Since the risks towards the indirect assets can be determined by considering their relationships to the direct assets, and the unwanted incidents associated with the direct assets, we normally conduct Step 5 and Step 6 by considering the direct assets only. During the risk evaluation in Step 7 and the treatment identification of Step 8, however, we need to take into account also the indirect assets in order to obtain and address the complete risk picture.

In principle, the target team should provide the information required by the analysis team, since it is the target team that has first-hand knowledge and expertise about the target of analysis. We must nevertheless be prepared to stimulate the discussion. A potential pitfall of any risk analysis that we must avoid is that the members of the target team for various reasons may be reluctant to throw themselves into discussions and to account for their views and opinions. A useful strategy is that we prepare suggestions for discussion; it is often easier for the members of the target team to respond to and comment on something that we present to them than coming up with something from scratch on their own. Based on the results of the previous step of the analysis, we therefore prepare asset diagrams that can serve as a starting point for discussion. As a rule of thumb, the analysis should focus on a selection of four or five direct assets for each party in question.

The result of the asset identification is documented using CORAS asset diagrams, one diagram for each party with respect to which the risk analysis is conducted. By definition, an asset is something to which a party assigns value, so without any parties there are also no assets. The reason for identifying one set of assets for

7.2 Conducting the Tasks of Step 3

each party is twofold. First, as different parties may have different interest in and dependence on the target of analysis, the parts or features of the target to which they assign value also differs. Second, even though two or more parties assign value to and require protection of the same parts of the target, the level of importance or value of the same part of the target may vary from one party to another party. A hospital patient will, for example, have interest in the protection of health records since his or her health and privacy depends on it, whereas a hospital have interest in the protection of health records in order to maintain public trust and to comply with data protection laws, in addition to safeguard the interests of the patients.

Example 7.2 The task is to identify and specify the assets of AutoParts, since AutoParts is the only party in this risk analysis. The analysis team has, based on the previous meetings, obtained quite a clear picture of the most important assets. The analysis leader triggers the discussion by displaying the following list of assets: *Online store, Inventory database, Customer database, Compliance with data protection laws and regulations, Customer satisfaction, Sub-supplier's trust,* and *Reputation*.

The target team agrees that the presented list summarises the assets with respect to which the analysis is to be conducted, but stresses that the main concern is the online store and the services provided to the customers. Since the online store relies on the availability of customer information and the inventory database, the availability of the databases are also of important concern.

Confidentiality and integrity are often relevant aspects that must be taken into account when considering information and databases. The analysis leader has the target team to discuss these aspects. Obviously, integrity of the data is important to ensure quality and correctness of the services. The target team furthermore soon agrees that confidentiality of the customer database is crucial for both data protection laws and AutoParts' reputation. Although less critical, the target team also argues that confidentiality of the inventory database is important from a business oriented perspective.

The discussion has already covered parts of the next issue to which the analysis leader turns, namely identifying the relationships between the assets. Since the online store depends on the databases, harm to the latter two may lead to harm to the former. Loss of confidentiality of the customer database may furthermore imply a compliance breach. While the target team discusses the asset relations for each asset in turn, the analysis leader creates the CORAS asset diagram depicted in Fig. 7.6. An arrow from one asset to another means that harm to the former may lead to harm to the latter, and is referred to as the harm relation.

Initially the analysis leader represents all the assets in the diagram as direct assets. After the identification of the relationships between the assets is completed, she moves the discussion to the final issue of identifying the indirect assets. With respect to the target and focus of the analysis, it is clear that the three assets of customer satisfaction, reputation and sub-supplier's trust can only be harmed indirectly via one or more of the other assets. The target team therefore concludes that these three assets are indirect in this analysis, which is documented in the diagram by using the white indirect asset symbol.

Fig. 7.6 Asset diagram

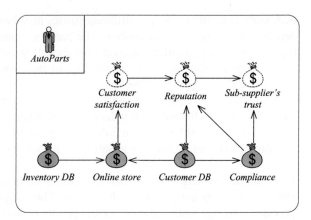

Table 7.5 Overview of Step 3c

High-level analysis

- **Objective:** Identify what the parties are most worried about happening, thus increasing the understanding of the correct and appropriate focus and scope for the analysis; establish an initial risk picture at enterprise level and from a decision maker perspective
- **How conducted:** Threats, vulnerabilities, threat scenarios and unwanted incidents are identified through plenary discussion or brainstorming lead by the analysis team
- **Input documentation:** Asset diagram and target description
- **Output documentation:** High-level risk analysis table

7.2.3 High-level Analysis

The purpose of the high-level risk analysis, an overview of which is given in Table 7.5, is to complement the previous two tasks in the objective of characterising the scope and focus of the target of analysis. Sometimes it may be difficult to decide exactly what we should and should not include in a risk analysis, and what should be the issues of main focus. Identifying the most important or most relevant assets, for example, may be hard without also looking at the most relevant risks at the same time. The parties may furthermore be tempted or inclined to include as much as possible in the analysis, which could make thorough, in-depth risk identification and evaluation impossible given the available time and resources.

When conducting the high-level analysis we aim at establishing an initial, high-level overview of risks at an enterprise level. In particular, we aim at identifying and documenting the main threats, vulnerabilities and unwanted incidents from the perspective of decision makers. Involving one or more decision makers at this initial risk analysis ensures that we conduct the in-depth analysis of the subsequent analysis steps by focusing on the kinds of risks that the decision makers are aware of or most worried about, and that motivated the risk analysis in the first place.

The results of the high-level analysis, in addition to refining the focus and scope of the analysis, serve as a starting point for the risk identification activity of Step 5

7.2 Conducting the Tasks of Step 3

Table 7.6 High-level risk analysis table

Who/what causes it?	How? What is the scenario or incident? What is harmed?	What makes it possible?
Accidental and deliberate human threats, as well as non-human threats	Things that can go wrong; incidents and scenarios leading up to them that are caused by the threats and that harm assets	Weaknesses or lack of safeguards that make it possible for threats to cause harm to assets

of the analysis. We may therefore see the high-level risk analysis as a first iteration of the overall risk analysis.

The technique we recommend for conducting the high-level analysis is a so-called structured brainstorming. This means that the target team freely discusses the risk picture by suggesting what may go wrong, what the possible causes may be, and how it is possible. The brainstorming is structured in the sense that the discussions are guided by the analysis leader. We must ensure that the discussion is kept relevant and that the necessary issues are covered during the brainstorming.

The basis that we use for the brainstorming is the table format shown in Table 7.6. The table is divided into three columns, and facilitates both the structuring of the brainstorming and the documentation of the results.

The table should be displayed to the target team while being filled in one-the-fly by the analysis secretary. Depending on the discussions and how the we choose to guide the brainstorming, the table may be filled in column by column or row by row. In practice, it is however likely that the cells will be filled in and updated in a somewhat arbitrary order. What is important is that we give each of the members of the target team the chance to express his or her opinions and main concerns.

A good starting point is that we first ask the target team what the main worries are. What scenarios and incidents can occur in relation to the target of analysis, and which assets would be harmed? These issues are addressed by the second column of the high-level analysis table, which is filled in by the analysis secretary as the discussion guided by the analysis leader proceeds.

Once we have identified the most important scenarios and incidents, we switch the focus to the identification of the threats. Who or what is the initial cause of the scenarios? It is important to stress that the threats may be both human and non-human, where non-human threats can be things like computer virus, system failure or power failure. The human threats may for example be attackers or malicious insiders, but they may also be trusted personnel that cause risks, for example, by accident or because of lack of competence or training.

This part of the brainstorming is addressed by the first column of the high-level risk analysis table. When we conduct this threat identification, we must ensure that we identify threats for each of the scenarios and incidents already plotted into the second column. We must furthermore seek to identify further threats that may be relevant.

Table 7.7 Documenting the high-level analysis

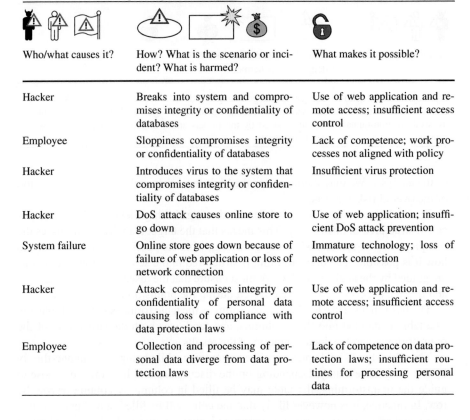

Who/what causes it?	How? What is the scenario or incident? What is harmed?	What makes it possible?
Hacker	Breaks into system and compromises integrity or confidentiality of databases	Use of web application and remote access; insufficient access control
Employee	Sloppiness compromises integrity or confidentiality of databases	Lack of competence; work processes not aligned with policy
Hacker	Introduces virus to the system that compromises integrity or confidentiality of databases	Insufficient virus protection
Hacker	DoS attack causes online store to go down	Use of web application; insufficient DoS attack prevention
System failure	Online store goes down because of failure of web application or loss of network connection	Immature technology; loss of network connection
Hacker	Attack compromises integrity or confidentiality of personal data causing loss of compliance with data protection laws	Use of web application and remote access; insufficient access control
Employee	Collection and processing of personal data diverge from data protection laws	Lack of competence on data protection laws; insufficient routines for processing personal data

The high-level analysis should finally identify the vulnerabilities, which are the system weaknesses or lack of safeguards that make it possible for the threats to cause the scenarios and unwanted incidents. The result of the vulnerability identification is plotted into the relevant rows of the third column of the high-level analysis table.

Since the identification of threats and vulnerabilities may lead to other scenarios and incidents to be thought of, we should do some iterations and let the discussions go back and forth to ensure that all the relevant issues are covered.

Example 7.3 The analysis secretary displays the high-level analysis table shown in Table 7.7 with all the cells empty. He is ready to document the results on-the-fly while the analysis leader explains the task to the target team.

The analysis leader begins by asking the target team what scenarios or incidents they are mostly concerned about. The general agreement is that the main concern is the availability of AutoParts' services to the customers. This relies on the databases and the online store. As documented in the second column of Table 7.7, the target team focuses on loss of integrity and confidentiality of the databases, as well as incidents that cause the online store to go down. They are also worried about compliance with data protection laws in relation to confidentiality of the databases and their internal processing of personal data.

When the analysis leader brings the discussion to the causes of these scenarios to occur, the target team brings up attacks by hackers and virus infection as possible threats. The analysis leader reminds the target team members that they should also consider things may have inside causes. The target team agrees that they should look into their own work routines and processes in order to investigate the potential of incidents that are caused by system failures or their own employees. The various threats are documented in the first column of the high-level analysis table.

Finally, the analysis leader turns the attention to the vulnerabilities, which are documented in the third column of the table. The use of the web application and the possibility for remote access, both via the Internet, can be utilised in attacks. The target team also believes that the competence and training of their own employees with respect to data security and data protection laws and regulations may be insufficient, and thereby open for incidents to occur.

7.3 Summary of Step 3

Step 3 of the CORAS risk analysis method is summarised by the overview presented in Table 7.8.

Table 7.8 Summary of Step 3

Refining the target description using asset diagrams	
Analysis tasks:	People that should participate:

• The target as understood by the analysis team is presented
• The assets are identified
• A high-level analysis is conducted
	• Analysis leader (required)
	• Analysis secretary (required)
	• Representatives of the customer:
	  – Decision makers (required)
	  – Technical expertise (required)
	  – Users (optional)

Modelling guideline:
- Asset diagrams:
  1. Draw the diagram frame and place the name of the party in the compartment in the upper left corner
  2. Place the assets within the frame
  3. Indicate with arrows which assets may be harmed via other assets
  4. Assets that are found to be indirect are modelled using the white indirect asset symbol; indirect assets are with respect to the target of analysis harmed only as a consequence of another asset being harmed first
  5. Repeat the above steps for each party of the analysis

Table 7.8 (Continued)

Refining the target description using asset diagrams

- Target descriptions:
    1. Use a formal or semi-formal notation such as the UML, but ensure that the notation is explained thoroughly so that the target team understands it; if the customer has specific preferences or an in-house notation, such notations should then be used
    2. Create models of both the static and the dynamic features of the target; static features may be hardware configurations, network design, and so forth, while dynamic features may be work processes, information flow, and so forth
    3. For the static part of the description, UML class diagrams and UML collaboration diagrams (or similar notations) are suitable
    4. For the dynamic parts of the description, UML activity diagrams and UML sequence diagrams (or similar notations) are suitable

Chapter 8
Approval of the Target Description

As the name indicates, our main objective with Step 4 is to obtain the approval from the involved parties of the target description with all assumptions and preconditions. This means that after this meeting the documentation of the target should be complete and correct. This includes the assets, the focus and the scope of the analysis, as well as the assumptions of the analysis. After this step the description of the target, and therefore the basis for the risk analysis tasks to follow, is in principle fixed.

During this step, we furthermore define the risk function for calculating risks levels, and the target team establishes the criteria for evaluating the risks. Step 4 finalises the preparatory phases of the analysis. After this step, we should therefore have established the necessary vocabulary and means for identifying, evaluating and categorising risks.

The step is typically held as a face-to-face meeting, but may be organised as a video or phone conference or by the exchange of written documentation. People that will be involved in one or more of the remaining steps of the analysis may be brought in at this step in order to introduce them to the target of analysis.

8.1 Overview of Step 4

The results of any risk analysis depend on the correctness and completeness of the description of the target of the analysis. We therefore do not initialise Step 5 until all the involved parties have explicitly approved the target description. This includes confirming that the focus and scope of the analysis match their main objectives of the analysis.

An important motivation for obtaining the explicit approval of the target description from all parties is to make it clear that it is the customer and its identified parties that decide on the overall direction and setting of the analysis. This makes it difficult for the customer to later claim that the analysis results are off target because the chosen focus, scope or assumptions are inadequate since they have themselves approved it.

Table 8.1 Overview of Step 4

Approval of the target description

- **Objective:** Ensure that the background documentation for the rest of the analysis, including the target, focus and scope, is correct and complete as seen by the customer; decide a ranking of the assets according to importance in order to prioritise the time and resources during the analysis; establish scales for estimating risk and criteria for evaluating risks
- **How conducted:** The step is conducted as a structured walk-through of the target description, preferably at a face-to-face meeting; any errors or omissions in the documentation are identified and corrected; based on plenary discussions and guidance of the analysts, the target team decides on a ranking of the assets, a risk scale and the risk evaluation criteria
- **Input documentation:** Full target description prepared by the analysts based on information and models derived through the previous steps
- **Output documentation:** Correct and complete target description, as well as risk evaluation criteria with definitions of all scales; all the documentation must be approved by the parties before the next step of the analysis
- **Subtasks:**
  a. Approval of the target description
  b. Ranking of assets
  c. Setting the consequence scales
  d. Setting the likelihood scale
  e. Defining the risk function
  f. Deciding the risk evaluation criteria

Table 8.1 gives an overview of the tasks of Step 4. Following the first subtask of obtaining the approval of the target description, we rank the identified assets according to importance or value to the associated party. Since the available time and resources for conducting a risk analysis are always limited, we may have to do some prioritising. Such a prioritising were already done when we decided on the focus and scope of the analysis, but this may be refined by the ranking of the assets. Notice that the set of identified assets with respect to which the analysis is conducted is the most important ones, and the number of assets should not exceed what we can cover during the analysis. When we rank the assets, we therefore do not rule out any of them from the analysis, but rather let the ranking serve as a guidance for planning and structuring the analysis, for focusing on the most important issues, and for prioritising among scenarios, for example, during risk identification.

In order to evaluate the risks that we identify during the analysis tasks to come, we must establish scales for estimating risk levels. A risk is the potential of unwanted incidents to occur, where an unwanted incident is an event that harms or reduces the value of assets. The risk level, or risk value, is derived from the likelihood of the unwanted incident to occur and the consequence of the unwanted incident in terms of asset damage. One of the subtasks of Step 4 is therefore to define a risk function that takes a likelihood value and a consequence value and yields the risk level. Since defining the risk function requires well-defined notions of likelihood and consequence, we must first conduct the separate tasks of establishing consequence and likelihood scales.

Both the likelihood scales and the consequence scales may be quantitative or qualitative. What is important for the analysis is that we operate with a granularity or level of abstraction that is sufficiently low to model the relevant risks and distinguish between levels of risks where the difference has significance.

Likelihoods of unwanted incidents are given as frequencies or probabilities. Consequence values describe the level of damage of unwanted incidents on assets. In some cases, we can describe asset damage in the same terms for all of the assets, for example, by loss of monetary value. However, the various assets in one and the same risk analysis are often not immediately comparable, which requires separate consequence scales to be established for separate assets. In the case of a hospital, for example, both uptime of critical services and human health may be important assets. These two assets are not easily comparable, and we should allow for this in a risk analysis by, for example, describing damage to the former asset in terms of downtime and damage to the latter in terms of reduction of human health condition.

The final subtask of Step 4 is to establish the risk evaluation criteria. The objective of this task is to determine what level of risk the parties are willing to accept or tolerate with respect to the identified assets. We specify these criteria in terms of what losses that can be tolerated over a given period of time, or what losses that can be tolerated for a given probability. For each combination of a likelihood value and a consequence value, the resulting risk level is either accepted or must be further evaluated for possible treatment. It should be noticed that the risk evaluation criteria are not absolute in the sense that we always identify and implement treatments for unacceptable risks, and never do so for acceptable risks. The risk evaluation criteria are a means to prioritise the identified risks and determine what should be of main concern with respect to risk mitigation. Risk treatment is a question of cost and benefit, and if the cost of reducing a risk is higher than the obtained benefit, the treatment should not be implemented. Likewise, if a small and in the first place acceptable risk can be easily and cheaply mitigated, it should be treated.

8.2 Conducting the Tasks of Step 4

In this section, we present each of the subtasks of the approval step in more detail, focusing on how we conduct each of them. The approval of the target description and the establishment of the risk evaluation criteria requires the participation of decision makers representing the involved parties. Additionally, personnel with expert knowledge about the target of analysis are required to participate to ensure that the finalised and approved target description is correct and complete.

Table 8.2 Overview of Step 4a

Approval of the target description

- **Objective:** Finalise the documentation and characterisation of the target of analysis, including the assets, the focus and scope of the analysis as well as the assumptions made for the analysis
- **How conducted:** The analysis team gives a plenary walk-though of the documentation; errors or omissions that are identified are recorded for corrections; the task is terminated once the full documentation has been approved by the customer
- **Input documentation:** The target description consisting of models and other documentation of the target of analysis and the assets prepared by the analysis team on the basis of the previous steps
- **Output documentation:** Correct and complete target description; the models and other documentation are cleaned up and properly prepared by the analysis team and serve as the finalised target description to be used during the remaining analysis

8.2.1 Approval of the Target Description

After the three first steps of the risk analysis, we and the target team should have reached a common understanding of what is the target of analysis, what is the focus and scope of the analysis, as well as which are the assets with respect to which the analysis is to be conducted. The objective of the approval, an overview of which is given in Table 8.2, is to ensure that the target description is correct and complete, and that the chosen target of analysis matches the goals of the analysis and the parties' main concerns with respect to the risk picture.

In principle, we fix the target description once it has been approved. The risk analysis tasks to be conducted, including risk identification, risk evaluation and identification of risk treatments, are all based on the target description; conducting the analysis tasks and obtaining correct results rely on the correctness of the target description and its persistence throughout the analysis. We may of course, based on explicit agreement with the customer, backtrack and implement changes to the target description at later points in time, but this is time and resource consuming and may require us to conduct parts of the analysis over again.

During our preparations for this task we clean up and polish the documentation of the target description that resulted from the previous steps of the analysis. The task is preferably held as a face-to-face meeting with the target team where we obtain the approval of the target description by giving a walk-through of all the documentation. While the various parts of the target description are displayed, the target team either approves the description or points out errors and omissions that must be corrected. The corrections may be implemented on-the-fly or recorded by the analysis secretary for being implemented later.

The approved documentation of the target of analysis is polished and finalised between this and the next step of the analysis. Since the finalisation may include implementation of corrections, we should send the documentation to the customer for inspection and formal approval before we proceed to Step 5 of the analysis.

8.2 Conducting the Tasks of Step 4

Table 8.3 Overview of Step 4b

Ranking of assets

- **Objective:** Facilitate identification of the most important assets to support prioritising of risks and thereby also prioritising the time and resources for risk analysis and treatment identification
- **How conducted:** Plenary discussion lead by the analysis team
- **Input documentation:** CORAS asset diagrams from previous step
- **Output documentation:** A list of assets ordered by importance; each asset can be given a value and/or a number from a scale defined for ranking assets

8.2.2 Ranking of Assets

Each of the assets that we have identified is within the focus and scope of the analysis, and we must consider them all during the actual risk analysis for the identification of risks towards them. Due to time and resource constraints, however, it is often necessary to do prioritising, that is, that we focus particularly on selected aspects or parts of the target and that we prioritise between scenarios during risk identification. In order to guide the prioritising, we may rank the assets according to importance or value, such that we consider the most important assets first and foremost. The purpose is not to designate some of the assets as unimportant, but rather describe their relative importance.

The task, an overview of which is given in Table 8.3, is conducted as a plenary discussion guided by the analysis leader. We display in turn each of the CORAS asset diagrams from the target description, and the target team agrees on an appropriate ranking. It may be that individual members of the target team are most concerned about issues or aspects of the target that are most relevant for their area of responsibility or work. In case of disagreements, it is important to involve the decision makers to clarify the main objectives of the analysis and what are the main concerns.

Example 8.1 The target team is asked to rank the assets according to their importance using a scale from 1 (very high importance) to 5 (very low importance). The participants agree that the online store itself is the most important one and should have rank 1 since it is crucial for the business and since the other assets are almost irrelevant without the online store. The databases are also crucial since the online store depends on them and since they store sensitive information. They both are therefore given rank 2. The customer satisfaction is important since the business and the turnover relies on it, and is given rank 2.

The compliance asset causes some discussion since the participants do not have deep knowledge about the data protection laws and regulations. They believe, however, that the asset should have a high priority since breaches with compliance may have very negative effect. They choose to give it rank 2, but will clarify the issue when the lawyer is brought in at later stages.

The overall reputation and the sub-supplier's trust are obviously important, but they agree that the five aforementioned assets are more important and therefore give the latter two the rank 3. The resulting ranking of the assets is documented in the asset table shown in Table 8.4.

Table 8.4 Asset table

Asset	Importance	Type
Online store	1	Direct asset
Customer DB	2	Direct asset
Inventory DB	2	Direct asset
Customer satisfaction	2	Indirect asset
Compliance	2	Direct asset
Reputation	3	Indirect asset
Sub-supplier's trust	3	Indirect asset

Table 8.5 Overview of Step 4c

Setting the consequence scales

- **Objective:** Establish consequence scales for describing the harm of identified unwanted incidents on direct assets; to be used when estimating risk values
- **How conducted:** Determined by the target team through plenary discussion possibly based on suggestions provided by the analysis team
- **Input documentation:** Consequence scales suggested by the analysts
- **Output documentation:** A consequence scale defined for each direct asset

8.2.3 Setting the Consequence Scales

In order to estimate and evaluate risks, we need to be able to describe and talk about the potential harm that may be caused by the risks. The consequence of a risk describes the level of damage the associated unwanted incident inflicts on an asset when the incident occurs. When we are setting the consequence scale, an overview of which is given in Table 8.5, we define the set of values that we use for describing the possible consequences.

> **Remark** A *consequence* is the impact of an unwanted incident on an asset in terms of harm or reduced asset value.

The consequence scale may be quantitative or qualitative. A qualitative scale allows a precise characterisation of the harm of unwanted incidents on assets, but requires that we have the expertise and data that are needed for the consequence estimations. Using qualitative values such as *insignificant* or *major* may often be more appropriate. A precise description of what the various values of the scale mean should nevertheless be given.

In some cases, we may operate with one consequence scale for all assets. The kinds of assets in one analysis often varies, however, and in practice it may be difficult to describe the damage inflicted on all of them using the same terms. One asset may, for example, be sensitive information, another yearly revenue and a third reputation. Conveniently, we can choose to use the same notions, such as *insignificant*

8.2 Conducting the Tasks of Step 4

and *major* to describe consequences for all assets, which facilitates the discussions. The specific meaning of each such notion must nevertheless be defined in a separate consequence scale for each asset.

The consequence scales we choose must be appropriate for the assets, and appropriate for the desired abstraction level of the analysis. The latter means that we must define the scales such that it is possible to differ between consequences where the difference is significant for the analysis and for obtaining an adequate risk picture.

In case we choose to operate with scales of intervals of consequence values, we should ensure that the relative difference between to adjacent intervals are more or less the same all over the scale. The absolute difference between two intervals at the lower end of the scale is then smaller than the absolute difference between two intervals at the higher end of the scale, but the differences are comparable in relative terms.

If it later in the analysis process turns out that the scales we have chosen are not appropriate they may be adjusted, although this should be avoided if practical since it may be at the cost of time and resources.

In the same way as the value of an asset is determined by the party associated with the asset, so is the consequence scale with respect to an asset. It is therefore the target team that is supposed to come up with the consequence scales, but we must ensure that the scales are appropriate for the analysis. Since the target team cannot be expected to have experience is risk analysis, we should come up with suggestions for discussion.

The risk identification of Step 5 is conducted with respect to the direct assets only; the issue of determining the full risk picture when also the indirect assets are taken into account is addressed during the risk evaluation of Step 7. Whereas the consequence scales for the direct assets must be defined before, we proceed to the next step, we may leave setting the consequence scales for the indirect assets to the risk evaluation.

Example 8.2 There are four direct assets in the target of analysis and with respect to which the full analysis will be conducted. These are the online store, the customer database, the inventory database, and compliance. The databases are of the same type, so a common consequence scale can be defined for them.

The analysis team has prepared the task by making suggestions for consequence scales that are presented to the target team. The suggestions are scales of five levels ranging from *insignificant* to *catastrophic*. The lowest level, as the name suggests, means that an incident with such a consequence can be ignored if it does not occur with a very high likelihood. The highest level refers to harm that is so severe that it may put enterprises out of business.

The consequence scale for the assets of the inventory database and the customer database are given in Table 8.6. The analysis team suggests defining these scales in terms of how many database records are affected, that is, records that are deleted, disclosed, corrupted, and so forth. After the plenary discussion, the target team agrees to operate with the description given in the right hand column of Table 8.6.

The description of harm to databases in terms of the number of records that are affected may not be adequate for all unwanted incidents that affect the databases.

Table 8.6 Consequence scale for databases

Consequence	Description
Catastrophic	Range of [50%, 100%] of records are affected
Major	Range of [20%, 50%) of records are affected
Moderate	Range of [10%, 20%) of records are affected
Minor	Range of [1%, 10%) of records are affected
Insignificant	Range of [0%, 1%) of records are affected

Table 8.7 Consequence scale for online store

Consequence	Description
Catastrophic	Downtime in range [1 week, ∞)
Major	Downtime in range [1 day, 1 week)
Moderate	Downtime in range [1 hour, 1 day)
Minor	Downtime in range [1 minute, 1 hour)
Insignificant	Downtime in range [0, 1 minute)

The disclosure of only a very few records, for example, may be catastrophic if this particular data is very sensitive. The description of the various consequences therefore intends to serve only as a point of reference in order to explain and understand the meaning of the values.

The consequence scale for the online store is given in Table 8.7. Damage to this asset is given in terms of downtime as described in the right-hand column. Again, the description of the consequences is intended to capture the severity. Therefore, if there are incidents that harm the online store in other ways than causing downtime, for example, that only a few services are malfunctioning or slow, the values ranging from *insignificant* to *catastrophic* are still used.

The fourth asset, namely compliance with data protection laws and regulations, is less concrete than the other three in the sense that a breach with compliance may not have any immediate consequence. In reality, it may not have any consequence at all since it may not be discovered and reported. The analysis team therefore suggests that the consequences are defined in terms of sanctions in case of infringement, irrespective of whether incidents are reported. Given an incident that represents a compliance breach, the consequence will then be given by the relevant sanction assuming that the incident is reported and that AutoParts is prosecuted. The consequence scale for the compliance asset is given in Table 8.8.

8.2.4 Setting the Likelihood Scale

The purpose of the task of setting the likelihood scale, an overview of which is given in Table 8.9, is to establish means for specifying the probability for or frequency of occurrences of unwanted incidents. The scale is to be used when we estimate risk

8.2 Conducting the Tasks of Step 4

Table 8.8 Consequence scale for compliance

Consequence	Description
Catastrophic	Chief executive officer is sentenced to jail for more than 1 year
Major	Chief executive officer is sentenced to jail for up to 1 year
Moderate	Claim for indemnification or compensation
Minor	Fine
Insignificant	Illegal data processing is ordered to cease

Table 8.9 Overview of Step 4d

Setting the likelihood scale

- **Objective:** Establish a scale for describing the likelihood of unwanted incidents to occur; the scale is defined in terms of frequencies or probabilities; to be used when estimating risk values
- **How conducted:** Determined by the target team through plenary discussion possibly based on suggestions provided by the analysis team
- **Input documentation:** A likelihood scale suggested by the analysis team
- **Output documentation:** A likelihood scale

levels, since it is the combinations of likelihoods and consequences that yield risk values. From a practical point of view, it does not make sense to operate with more than one likelihood scale.

> *Remark* A *likelihood* is the frequency or probability of something to occur.

As for the consequence scale, the likelihood scale may be quantitative or qualitative. A quantitative scale of exact values may be suitable when we have access to data about the precise probabilities or frequencies of unwanted incidents to occur. Operating with qualitative values such as *unlikely* and *certain* is suitable for cases in which it is difficult or infeasible to come up with exact likelihood estimates. As for the consequence scale, if we choose to operate with intervals the relative difference between two adjacent likelihoods should be more or less the same all over the scale.

If the customer has conducted risk analyses previously, there may be likelihood scales that can be reused. If not, we should present a suggestion that the target team can use as a starting point for discussion. The likelihood scale is agreed on through plenary discussion guided by the analysis leader. As a rule of thumb, the lowest likelihood should correspond to approximately one occurrence during the lifetime of the target of analysis.

If necessary, we can adjust the likelihood scale during later phases of the analysis process. In principle, however, the likelihood values should be fixed throughout the analysis.

Example 8.3 The analysis leader triggers the discussion by presenting a likelihood scale of intervals ranging from *rare* to *certain* as depicted in Table 8.10. The re-

Table 8.10 Likelihood scale

Likelihood	Description	
Certain	Five times or more per year	$[50, \infty) : 10y$
Likely	Two to five times per year	$[20, 49] : 10y$
Possible	Once a year	$[6, 19] : 10y$
Unlikely	Less than once per year	$[2, 5] : 10y$
Rare	Less than once per ten years	$[0, 1] : 10y$

maining lifetime of AutoParts is expected to exceed one decade, but the target team agrees that the new online store system that is to be launched should be considered in a ten year perspective.

The description of the likelihood scale is given in the right hand column. A rough description is given in natural language followed by a precise description of the intervals. The precise description specifies the interval of occurrences over a ten year period. The likelihood *unlikely*, for example, is described by $[2, 5] : 10y$ which means at least two and at most five occurrences per ten years.

8.2.5 Defining the Risk Function

The risk function yields for each combination of a likelihood and a consequence the resulting risk level. As with the value of an asset, it is the associated party that determines the severity of a given risk. This may depend on the type or importance of the asset in question, the degree to which the party is risk aversive, the potential gain of accepting risks, and so forth.

> *Remark* A *risk* is the likelihood of an unwanted incident and its consequence for a specific asset. The *risk level* is the level or value of a risk as derived from its likelihood and consequence.

Since the risk function is a mapping from likelihoods and consequences to risk values, we must define a separate risk function for each of the consequence scales. Each direct asset is thereby associated with one risk function.

Before defining the risk function, an overview of which is given in Table 8.11, we must establish an appropriate scale for risk levels. This scale may be quantitative or qualitative, depending on the type of values we used for specifying likelihoods and consequences.

If the likelihood scale and the consequence scales are continuous or very fine-grained, for example, probabilities ranging from 0 to 1 and monetary consequences ranging from 0 to infinite, the risk level scale may correspondingly fine grained. The risk function may in that case, for example, be defined by multiplication of likelihood and consequence.

If we use more coarse-grained scales of intervals, we can use risk matrices to represent risk functions. A risk matrix is a simple and efficient way of representing

8.2 Conducting the Tasks of Step 4

Table 8.11 Overview of Step 4e

Defining the risk function

- **Objective:** Determine the level of risk an unwanted incident represents as a function of the likelihood and consequence of its occurrence
- **How conducted:** The analysis team will typically draw a matrix for each of the consequence scales with the likelihood scale horizontally and the consequence scale vertically; based on suggestions provided by the analysis team and on plenary discussions, the target team decides the risk value for each entry in the matrix; a risk value scale is set with an appropriate granularity for the analysis; if the risk matrix representation is not appropriate, an alternative risk function must be defined
- **Input documentation:** The consequence scales and the likelihood scale set during the previous tasks
- **Output documentation:** The decided risk functions

the risk function, where the likelihood scale is represented vertically and the consequence scale is represented horizontally, as exemplified in Table 8.12. Each entry in the matrix then represents a risk value. The risk scale can then have at most as many values as there are entries. In the given example, there are 25 matrix entries, but in practice we need only a few risk values, typically two to five. The adequate number of risk values depends on the extent to which we need to distinguish between risk levels. Moreover, various combinations of a likelihood and a consequence should yield the same risk value; if, for example, the risk function is given by multiplication, the risk value of a probability of 0.2 and a consequence of 1,000,000 equals the risk value of a probability 0.8 and a consequence of 250,000. Operating with risk matrices thus roughly means that each risk level is distributed along a diagonal line upwards from left to right.

The task of defining the risk function for each of the assets, or each of the consequence scales, is also conducted through a plenary discussion. We should prepare by setting up matrices that we present and fill in on-the-fly. It is usually straightforward for the target team to agree on the extremes; the combination of *certain* and *catastrophic* is definitely a very high risk, whereas the combination of *rare* and *insignificant* is a very low risk. The challenge is therefore to determine between which entries in a row and between which entries in a column the risk value changes from one level to another.

Example 8.4 The analysis leader starts by having the target team agree on the scale for specifying risk levels. She displays the matrix showing the combinations of frequencies and consequences and explains how each entry maps to a risk level. They decide to operate with the four risk values of *very low*, *low*, *high* and *very high*. Operating with colour codes such that green, yellow, orange and red represent the respective risk values, the risk functions are defined by filling in each cell with the agreed colour. The task is conducted for each of the three consequence scales, resulting in three risk functions.

The resulting risk function for the databases is depicted in Table 8.12. Initially, the analysis leader displays the matrix with blank entries. She then triggers the discussion by filling the bottom right cell with the red colour. The members of the target

Table 8.12 Risk function for databases

		Consequence				
		Insignificant	Minor	Moderate	Major	Catastrophic
Frequency	Rare					
	Unlikely					
	Possible					
	Likely					
	Certain					

Table 8.13 Risk function for online store

		Consequence				
		Insignificant	Minor	Moderate	Major	Catastrophic
Frequency	Rare					
	Unlikely					
	Possible					
	Likely					
	Certain					

Table 8.14 Risk function for compliance

		Consequence				
		Insignificant	Minor	Moderate	Major	Catastrophic
Frequency	Rare					
	Unlikely					
	Possible					
	Likely					
	Certain					

team discuss how many cells they can move to the left until the risk value should be lowered. This exercise is repeated for each row until the entire matrix is filled in. Subsequently, the analysis leader has the target team to discuss the result, particularly focusing on the borders where the risk level shifts. After some discussion and adjustments, the risk function for the database assets is then fixed.

The same exercise is repeated for the online store asset and the compliance asset. The resulting matrices defining the risk functions are depicted in Table 8.13 and Table 8.14, respectively.

8.2.6 Deciding the Risk Evaluation Criteria

Having defined the risk function for each of the assets, the target team must determine the risk evaluation criteria, an overview of which is given in Table 8.15. These criteria are a specification of the level of risk the parties of the analysis are willing to accept. When we have identified the risks in later phases of the analysis process, we use the criteria to determine which of the risks must be further evaluated for possible treatment. In principle, all risks that are unacceptably high should be mitigated. However, since risk treatment may be costly, the mitigation of risks is always a question of cost and benefit. The risk evaluation criteria are therefore a means to guide the analysis and the evaluation process, and to determine which risks should be of main concern.

8.2 Conducting the Tasks of Step 4

Table 8.15 Overview of Step 4f

Deciding the risk evaluation criteria

- **Objective:** Determine what level of risk the customer is willing to accept; will later be used to determine which risks that must be evaluated for possible treatment
- **How conducted:** The customer determines for which combinations of consequence and likelihood the resulting risk is acceptable and for which combinations the risk must be evaluated for possible treatment
- **Input documentation:** The risk function defined in the previous task
- **Output documentation:** Risk evaluation criteria for each asset

The risk evaluation criteria for a given asset is basically a function from a risk level to one of the statements *acceptable* and *unacceptable* (i.e., *must be evaluated*). This means that the target team must determine for each combination of a likelihood and a consequence whether the risk should be accepted or require further evaluation for possible treatment.

For convenience, we should define the risk functions such that a given set of risk levels represents acceptable risks and the complementary set of risk levels represents unacceptable risks. Given the matrix in Table 8.12, for example, if the target team decides that the combination of *likely* and *moderate* is unacceptable, whereas the combination of *possible* and *moderate* is acceptable, the risk function should be reconsidered since both combinations map to the same risk level.

The task of determining the risk evaluation criteria is conducted through a plenary discussion where the analysis leader presents the risk function for each of the assets in turn. If the risk function is represented by a matrix, the task is for the target team to agree on which half of the matrix represents acceptable risks and which half represents risks that require further evaluation.

Since there are several assets there are also several sets of risk evaluation criteria. We may, however, in some cases conveniently combine the criteria for different assets into one representation since the criteria basically are a mapping from risk levels. In other words, if the mapping from risk levels to risk evaluation criteria is uniform for different assets, we may combine the risk evaluation criteria for these assets into one representation.

The target team may find it difficult to determine exactly what should be the risk evaluation criteria, in particular to decide at which points the risk level shifts from acceptable. As the understanding of and knowledge about the risk picture increases throughout the following phases of the analysis, it may be that the target team finds that the criteria should be reconsidered and adjusted. The criteria as established during this task are therefore not definitive. The adjustment and final confirmation of the risk evaluation criteria is a separate task of a later step in the analysis process.

Example 8.5 The analysis leader displays the risk matrix shown in Table 8.12 in order to have the target team to determine the risk evaluation criteria for the two database assets. She asks the participants to decide which part of the matrix represents risks that are intolerably high and that should be evaluated for possible treatment. The target team soon agrees that the risks of value *high* and *very high*, that

Table 8.16 Risk evaluation criteria

Risk level	Criterion
Very high	Evaluate for possible treatment
High	Evaluate for possible treatment
Low	Accept
Very low	Accept

is, the red and orange areas of the matrix, require further analysis and evaluation, whereas all other risks are acceptable.

When moving on to the online store asset and the matrix of Table 8.13, there is some discussion as to whether the combination of *certain* and *insignificant* should be acceptable. The risk function maps these values to the risk level *low*, so the analysis leader explains that if this risk is held as unacceptable they should modify the risk function such that it maps to *high*. This will facilitate the specification of the risk evaluation criteria and it also harmonises the risk evaluation criteria for the database assets with the online store asset. After some discussion, the target team concludes that the risk function should be left unmodified and that the matrix entry in question should be defined as acceptable. As for the database assets, the risk evaluation criteria are then specified such that *high* and *very high* risk are unacceptable, while *low* and *very low* risks must be evaluated further.

The exercise is repeated for the compliance asset and the matrix of Table 8.14. The result of the task of deciding the risk evaluation criteria is documented in Table 8.16.

8.3 Summary of Step 4

Step 4 of the CORAS risk analysis method is summarised by the overview presented in Table 8.17.

Table 8.17 Summary of Step 4

Approval of the target description

Analysis tasks:	People that should participate:

• The target team approves the target description
• The assets may be ranked according to importance (optional)
• Consequence scales must be set for each direct asset within the scope of the analysis
• A likelihood scale must be defined
• A risk scale and a risk function must be defined
• The target team must decide risk evaluation criteria for each direct asset within the scope of the analysis
	• Analysis leader (required)
	• Analysis secretary (required)
	• Representatives of the customer:
	  – Decision makers (required)
	  – Technical expertise (required)
	  – Users (optional)

Modelling guideline:
• Modelling is not part of this step

Chapter 9
Risk Identification Using Threat Diagrams

At this point, we begin the detailed, target specific and asset-driven risk identification. The objective of Step 5 is to identify and document the risks of relevance for the target of analysis, given the chosen focus and scope. Our method is asset-driven in the sense that the assets direct the risk identification process; we view the target description with the sole aim of protecting the identified assets, and use the enterprise level risk picture that we documented during the high-level analysis to guide us in identifying the most important scenarios, incidents and risks.

We typically organise the risk identification as a full day workshop of about six hours where we conduct a structured brainstorming. The target team should consist of personnel with various backgrounds and different insights into the problem in order to elicit as much relevant information about potential risks as possible.

When we bring personnel of various backgrounds together, we face challenges with respect to communication and common understanding of problems and opinions. The CORAS language helps us to mitigate this challenge by providing support for representing risk pictures in a structured and intuitive way. We furthermore use CORAS threat diagrams to document the results of the risk identification.

9.1 Overview of Step 5

The risk identification process is organised as a structured brainstorming, that is to say as a structured walk-through of the target of analysis. As input to the brainstorming, we use the target models that we developed during the previous steps of the analysis. These models describe the relevant parts of the target from different viewpoints and at an adequate level of abstraction for the analysis. By bringing in personnel of various background, expertise and interest, we also get various viewpoints on the risk picture. This is to ensure that we identify as many risks as possible, since personnel with the same background are likely to focus on only a few issues which could result in an incomplete or biased risk picture.

In order to identify risks, we try to find as many potential unwanted incidents as possible since a risk is always associated with an unwanted incident; we get one risk for each asset that is harmed by an unwanted incident. Moreover, in order to understand how unwanted incidents—and thereby risks—arise, what causes them and how they are caused, we also identify threats, vulnerabilities and threat scenarios. To the extent that likelihoods and consequences are provided, for example through the brainstorming or through other sources, these are also documented. However, it is not until Step 6 that we really address the issue of estimating likelihoods and consequences for the identified risks.

By our choice of focus and scope of the analysis, as well as an appropriate level of abstraction, we have already determined what should be held outside of the analysis. This is important since we need to ensure that we spend the limited time and resources on the most relevant issues. In the same way as we made choices about what to include and not in the target description, we must also make choices about which threat scenarios and unwanted incidents to include during the risk identification. The risks that we identify should of course all be within the target of analysis, but we must ensure that we focus on the risks that really matter and that we need to understand. The risk identification is therefore asset-driven.

During the preceding analysis steps leading up to the current step of risk identification, we have basically produced three types of documentation that each serve as input to and guidance of the risk identification task. First, the CORAS asset diagrams that we used to document the asset identification of Step 3 explicitly specify what the parties of the analysis seek to protect, and therefore with respect to which we do the risk identification. Second, the high-level analysis table that we used to document the results of the high-level analysis of Step 3 shows the main worries of the decision makers and therefore their objectives with the analysis. Third, the target description that was finalised and approved during Step 4 describes the architecture, behaviour, processes, and so forth that we are to consider when we are identifying risks. Whereas the assets served as a starting point for the high-level analysis, we use the scenarios of the high-level analysis as a starting point for the risk identification. In particular, we use the results of the high-level analysis as a guidance for selecting parts from the target description that serve as input to the risk identification.

Table 9.1 gives an overview of Step 5. Our first task is to decide a suitable categorisation of the threat diagrams that we will make during the risk identification. This should be set before the workshop begins, in other words it should be part of our preparations for the workshop. The purpose of the categorisation is to facilitate the structuring of the brainstorming session and to address separate issues separately. The categorisation furthermore yields a categorisation of the risk that we identify and document. We choose the categorisation that we believe is most convenient for the analysis, for example, by type of threats or by assets.

The brainstorming session itself begins with the identification of threats and unwanted incidents with respect to the direct assets. While the discussion proceeds and the target team comes up with potential unwanted incidents, these are plotted into fragments of CORAS threat diagrams with relations to the assets they harm. The threats that cause these incidents are plotted into the diagrams with arrows to the incidents.

Once we have identified and documented the threats and the unwanted incidents, we turn to the questions of how the threats may cause unwanted incidents to occur

9.1 Overview of Step 5

Table 9.1 Overview of Step 5

Risk identification using threat diagrams

- **Objective:** Identify the risks that must be managed; determine where, when, why and how they may occur
- **How conducted:** Conducted as a brainstorming session involving a target team consisting of people of different backgrounds with different insight into the problem at hand; using the assets and the high-level analysis as starting point the risks are gradually identified by identifying unwanted incidents, threats, threat scenarios and vulnerabilities; the risk identification is conducted with respect to the target description, and the results are documented on-the-fly by drawing CORAS threat diagrams as the information is gathered
- **Input documentation:** Target description including CORAS asset diagrams; high-level risk analysis table
- **Output documentation:** CORAS threat diagrams and notes made by analysis secretary during the risk identification; the diagrams are cleaned up and completed by the analysis team offline before the next step of the analysis
- **Subtasks:**
  a. Categorising threat diagrams
  b. Identification of threats and unwanted incidents
  c. Identification of threat scenarios
  d. Identification of vulnerabilities

and what makes it possible. This leads to the identification of threat scenarios and the identification of vulnerabilities. We continue with the initial, incomplete threat diagrams, expanding them by filling in the identified threat scenarios between threats and unwanted incidents. The identified vulnerabilities are placed as annotations on the relations.

The general structure of the brainstorming session is to address each category of threat diagram in turn, and gradually establish the complete risk picture by conducting the subtasks sequentially. In practice, we often experience that it is a constant challenge to keep all the participants on the same track, and to keep them focused on the same issue simultaneously; the members of the target team have different roles and interests, and may be inclined to front their own causes and worries, and to pursue their own agenda. It is therefore important that we firmly guide the discussion and maintain the structure that we have chosen. At the same time, we should try not to make unnecessary obstacles for the discussion; the very purpose of a brainstorming session is that ideas and views can be quite freely proposed and thrown back and forth. The structure we propose in the following sections is also not to be very strictly understood; in practice, the risk identification process is often somewhat iterative, and the subtasks may also be somewhat interleaved.

One of the main objectives with the development of the CORAS language was to provide support for the brainstorming session that we conduct in order to identify risks. First, the language is intended to mitigate the challenge of having personnel of different backgrounds and interests to communicate and to establish a common understanding between them. Second, the language is intended to facilitate the risk identification process itself by offering expressiveness and language constructs that match the core concepts of risk analysis, such as threat, unwanted incident and vulnerability. Third, the models that we make serve as input to the subsequent tasks of estimating, evaluating and treating risks. Fourth, the documentation of the risks using CORAS diagrams can be used in the final risk analysis report that documents the results of the analysis. The CORAS language is therefore a corner stone of our model-driven approach to risk analysis.

9.2 Conducting the Tasks of Step 5

In this section, we look more closely at each of the subtasks of Step 5 and describe how they are conducted. As already mentioned, the ordering of Step 5b through Step 5d is not very rigid. We may go through them sequentially for each category of threat diagram in turn, we may do several iterations, and we may choose to do some interleaving. It all depends on how the discussions flow and what we find suitable during a given risk analysis. Whatever specific structuring we choose, we must, however, always try to keep all the participants focused on the same issue simultaneously. We must furthermore make sure that we cover all parts and aspects of the target of analysis.

9.2.1 Categorising Threat Diagrams

The purpose of the categorisation of the threat diagrams, an overview of which is given in Table 9.2, is to provide structure to the brainstorming session, and to categorise the risks that we identify. The structure we choose is for our own convenience and what we see as most suitable. One strategy is to look at the target description and structure these models into categories that we address one at the time. We may, for example, divide the target into various physical, logical or organisational domains and identify risks with respect to each of them in turn. Or we can look at individual work processes, information flow processes, and so forth separately. Another strategy is to look at the results of the high-level analysis and categorise diagrams according to, for example, assets or kinds of threats.

It is crucial for the success of the brainstorming session that we plan and prepare well in advance. In addition to decide a categorisation of the risk diagrams that we make throughout the workshop, we must decide how the target models we have made should serve as input to the brainstorming. For each of the categories, we pick the relevant target models that we present before the risk identification starts.

Example 9.1 During the preparations for the risk identification workshop, the analysis team from CORAS Risk Consulting does the last polishing and finalisation of the approved target description for the AutoParts risk analysis. While going through the documentation the analysts plan for how to structure the brainstorming and how to make sure that they cover all parts of the target.

9.2 Conducting the Tasks of Step 5

Table 9.2 Overview of Step 5a

Categorising threat diagrams
• **Objective:** Facilitate structuring of the risk identification brainstorming and a categorisation of the identified risks   • **How conducted:** Decision of analysis team based on the description of the target of analysis and the high-level analysis; threat diagrams may be categorised according to for example assets, threats, work processes, organisational domains, and so forth   • **Input documentation:** Target description and high-level risk analysis table   • **Output documentation:** A set of categories used by the analysis team to guide and structure the risk identification brainstorming session; for each category, the analysis team selects relevant target models such as UML class, sequence and activity diagrams describing features and aspects of the target of analysis at an appropriate level of detail for the analysis

One possible distinction is to look at the system from an internal perspective on the one hand and an external perspective on the other hand. From the internal perspective, issues such as work process, policies, security routines, software, and so forth are relevant. From the external perspective they must focus on the various access pathways to the system, both from customers and sub-suppliers, as well as employees with remote access. Based on the high-level analysis, they also consider categorising threat diagrams with respect to threats, and address various types of threats separately during the brainstorming.

The third option they consider, and which they choose, is to structure the brainstorming according to assets. This will yield separate diagrams for the separate assets, and perhaps some overlap between the diagrams. Overlapping diagrams should not imply that they have to do the same risk identification over again, but rather that they can reuse parts of one diagram in another. Moreover, overlapping diagrams allow them to check consistencies between diagrams and perhaps discover mistakes or things that should be reconsidered.

While going through each asset in turn during the brainstorming, they must make sure that they cover all aspects of the target with respect to the given asset. According to the plan there will be a system developer, a system user and a data security officer participating, in addition to a lawyer that the customer has brought in. For each of the members of the target team, the analysis team prepares checklists and questions that the analysis leader will use in case there is a need to bring the discussions forward or to ensure that all the participants are involved. The analysts also prepare some initial, preliminary CORAS threat diagrams based on the high-level analysis that they plan to use as proposals to trigger discussions in case it is needed.

9.2.2 Identification of Threats and Unwanted Incidents

A good starting point for the risk identification is to go straight to business: What can go wrong? We start the workshop by explaining to the target team how the risk

Table 9.3 Overview of Step 5b

Identification of threats and unwanted incidents

- **Objective:** Identify how assets may be harmed and the threats that cause the harm; facilitate the subsequent task of identifying threat scenarios
- **How conducted:** The analysis secretary presents on a whiteboard or some visual display a set of assets adjusted to the very right; the analysis leader has the target team to discuss and suggest how these may be harmed and which threats causes the harm by referring to the models of the target of analysis; the suggestions are added on-the-fly by the analysis secretary as unwanted incidents and threats, respectively; the threats are adjusted to the very left and the unwanted incidents are placed next to the assets; relations are added between threats and unwanted incidents and between unwanted incidents and assets; the analysis leader constantly probes the target team and, if necessary, stimulates the discussion by giving suggestions and by leading the attention to relevant parts of the target models; the activity is repeated for each category of threat diagram; assets for which unwanted incidents are not identified are removed from the diagrams
- **Input documentation:** Target description, high-level risk analysis table and sets of initial, preliminary threat diagram fragments structured according to the decided categorisation
- **Output documentation:** CORAS threat diagrams depicting which threats cause which unwanted incidents, and which unwanted incidents harm which assets

identification will be conducted, what we expect from the participants, and what we want to achieve. In order to trigger the discussions and to start identifying the risk picture from the very beginning, we start by presenting an asset and asking the target team how this asset can be harmed. In other words, we begin by identifying unwanted incidents.

> *Remark* An *unwanted incident* is an event that harms or reduces the value of an asset.

We always identify risks towards an asset by considering the target of analysis with its focus, scope and level of abstraction. For a given asset, we therefore also present to the target team a selected part of the target description and discuss how risks may arise given this selection. Our strategy for selecting relevant parts of the target description is to use the scenarios that we identified during the high-level analysis as starting point combined with our planned categorisation of threat diagrams. For example, if virus attack was identified as a potential threat scenario during the high-level analysis, we identify the parts of the target description in which such an attack may be relevant. By the structured walk-through of these parts of the target description during the brainstorming, we aim at describing more specifically how and why a relevant scenario such as a virus attack may occur, as well as the unwanted incidents that may result from it.

We start building the CORAS threat diagrams by capturing the identified elements one by one. One possibility is that we use a whiteboard or the like to draw the diagrams by hand as the discussions proceed. The analysis secretary is then responsible for saving the results by documenting everything on-the-fly. More conveniently, we may use the CORAS tool and project the diagrams on a screen. The analysis secretary then continuously creates and expands the diagrams while they are displayed to the target team.

Usually, the members of the target team come up with typical incidents that they know of, and incidents that they are particularly worried about. We document the findings continuously by placing the unwanted incidents to the left of the assets. If the discussion comes to a halt, we must be ready to bring it forward and stimulate it by asking questions or making suggestions. We can do this by pointing at other parts or aspects of the target, and by making sure that each of the participants is involved and contributes to the discussion. Another strategy we may use is to make suggestions based on the high-level analysis.

While we identify unwanted incidents and plot them into the diagrams, we also identify and document the related threats. For each of the unwanted incidents, we therefore ask the target team who or what is the initial cause of the incident. The participants often tend to think of people, for example hackers, when they think of threats. We must therefore make sure that non-human threats such as viruses, environmental threats such as power failure or heat, and accidental threats such as deletion of data are also discussed.

> *Remark* A *threat* is a potential cause of an unwanted incident.

We place the identified threats to the very left in the diagrams, and use arrows to relate threats to the unwanted incidents they may cause. We also place arrows from unwanted incidents to the assets they harm when they occur. An overview of the identification of threats and unwanted incidents is given in Table 9.3.

While the CORAS threat diagrams evolve, we can use them as a basis for eliciting more information from the target team. For a particular threat, for example, we ask if the participants can think of other unwanted incidents that this threat can cause. For a particular unwanted incident, we ask if further assets can be harmed by it, possibly via other scenarios. As always, we must firmly keep the discussion relevant by referring to the target description and the identified assets.

The structure we have chosen for the brainstorming is important in order to keep the discussions relevant and to ensure that we cover all parts of the target of analysis. The structure should, however, not obstruct or hinder the participants. For example, if we are currently addressing one particular asset with respect to one particular part of the target, and the target team identifies important threats and incidents that are relevant for other assets, we document also this and follow it up either immediately and in the same diagram, or later on in a separate diagram.

After having addressed all assets and covered all parts of the target description from the different viewpoints of the target team, we are ready to move to the identification of threat scenarios that will be described in Sect. 9.2.3. The initial CORAS diagrams we have made now show the unwanted incidents we have identified, along with the threats that cause them and the assets they harm. During the subsequent activities, we identify and fill in the missing pieces until we eventually have the full risk picture.

As mentioned above, the sequence of the subtasks of the risk identification is not rigid. In practice, we may choose to complete the tasks for some diagrams, and then begin over for other diagrams. It may also be that we identify some threat scenarios and vulnerabilities before the first subtask is completed, and that we identify more incidents and threats during the identification of threat scenarios and vulnerabilities. We may even anticipate the events of Step 6 by documenting likelihood and consequence values if such values are proposed. We should, however, try to maintain the structure we have chosen for the brainstorming and to follow the sequential procedure of Step 5. It is challenging to keep the discussions focused and directed, and we need to avoid the pitfall of a brainstorming that flows aimlessly in all directions.

In the examples of this chapter, we show only one diagram and describe how it evolves from the initial phases to a full description of some risks. At the end of the chapter, we present other finalised threat diagrams that describe further results of the risk identification.

Example 9.2 The analysis team uses the CORAS tool to display the diagrams and to document the results of the risk identification as the brainstorming session proceeds. The analysis secretary does the editing on-the-fly while the analysis leader guides the brainstorming and leads the discussions.

Each direct asset is in turn placed to the very right and serves as a starting point for the risk identification and modelling. As to the database storing the customer information, the target team is mostly concerned about confidentiality, worrying that sensitive information leaks to third parties.

One option for representing risks for loss of confidentiality of customer information is to model *Confidentiality breach* as an unwanted incident. This representation is, however, abstract in the sense that it does not say what kind of information that is lost. The analysis leader therefore asks the target team to be more precise, and explains that this will make it easier to later evaluate the severity of the incident and the likelihood of its occurrence. One kind of information that is particularly sensitive is payment card data, so they decide to model this as a separate incident. The lawyer of the target team also says that information that can be used to identify customers should be considered separately, since it is relevant not only for confidentiality in isolation, but also for data protection and the reputation of AutoParts.

Following the identification of the unwanted incidents for the various assets and the various parts of the target description, the analysis leader moves the discussion to the identification of threats. The most obvious threats for the target team are hackers that break into the system and steal or cause leakage of sensitive information. But they also know that AutoParts employees may themselves cause sensitive information to leak, for example by carelessness or by not strictly adhering to security policies and routines. They therefore want employees to be considered as threats in the analysis in order to identify and evaluate risks that may be caused by them.

Figure 9.1 shows the initial threat model that documents the threats causing harm to confidentiality of the customer database. In a similar way, the brainstorming session results in identification and documentation of unwanted incidents with respect to the inventory database, to the online store and to compliance.

9.2 Conducting the Tasks of Step 5

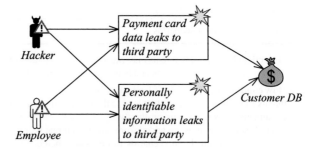

Fig. 9.1 Threats to confidentiality of customer DB

9.2.3 Identification of Threat Scenarios

Once we have identified the unwanted incidents, as well as the threats that may cause them to occur, we turn to the identification of threat scenarios, an overview of which is given in Table 9.4. Threat scenarios are scenarios that may result in one or more unwanted incidents. They are initiated by threats, and explain and describe how a threat can be the initial cause of the occurrence of unwanted incidents.

> Remark A *threat scenario* is a chain or series of events that is initiated by a threat and that may lead to an unwanted incident.

We use the CORAS threat scenario construct for the modelling and documentation of the results. For each of the initial threat diagrams resulting from the previous subtask, we use the unwanted incidents as starting point for the further discussion. The goal is to have the target team bridging the gap from the incidents to the right to the threats to the left. As before, we use the results of the high-level analysis to guide the selection of relevant parts of the target description that we present to the target team. The task is then to have the target team to identify the threat scenarios that may occur within a given part of the target of analysis.

Threat scenarios are described by an explanatory text and it is for us to decide the level of detail in each scenario, and to decide the number of threat scenarios in each diagram. Often, we need to decide whether a given scenario should be made more precise by splitting it up into several subscenarios that describe special cases. Other times, it may be suitable to combine two or more scenarios into one scenario if there is no need to understand and represent them separately.

The guiding principle in our decisions with respect to granularity and abstraction is that the models should support the risk analysis method. On the one hand, the models should support the brainstorming session and capture what the target team wants to communicate. On the other hand, the models should explain and describe the risk picture at a level of detail that is necessary for us to properly understand and evaluate the risks, and to understand how the risks can be treated.

The CORAS language allows the modelling of threats that initiate threat scenarios and unwanted incidents, threat scenarios that lead to other threat scenarios, threat scenarios that lead to unwanted incidents, unwanted incidents that lead to

Table 9.4 Overview of Step 5c

Identification of threat scenarios

- **Objective:** Explain how threats may cause unwanted incidents by identifying the threat scenarios that are initiated by the threats and that may lead to unwanted incidents
- **How conducted:** The analysis leader has the target team to discuss and explain how the identified threats can initiate scenarios that lead to the identified unwanted incidents; the analysis secretary documents the suggestions on-the-fly by adding threat scenarios ordered by time from left to right; directed relations are added from threats to threat scenarios, from threat scenarios to threat scenarios and from threat scenarios to unwanted incidents; relations may also be added from unwanted incidents to unwanted incidents and to threat scenarios if relevant; the analysis leader constantly probes the target team and stimulates the discussion by referring to relevant parts of the target models and, if necessary, giving suggestions
- **Input documentation:** Target description, high-level risk analysis table and the CORAS threat diagrams from Step 5b showing threats, unwanted incidents and assets
- **Output documentation:** The CORAS threat diagrams from Step 5b extended with the relevant threat scenarios and relations

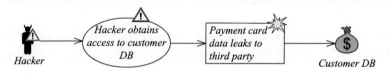

Fig. 9.2 Threat scenario that leads to harm to customer DB

threat scenarios, and unwanted incidents that lead to other unwanted incidents. We are therefore provided much flexibility and choices with respect to the identification and modelling of threat scenarios. Since the objective of the identification of threat scenarios is to explain how unwanted incidents may occur, we utilise the flexibility of the language to distinguish between scenarios and between chains of scenarios whenever such distinctions are relevant for explaining sources of risks.

Example 9.3 In order to have the AutoParts target team to discuss and explain how the unwanted incidents affecting the confidentiality may occur, the initial threat diagram of Fig. 9.1 is presented. The analysis leader begins with the unwanted incident *Payment card data leaks to third party* and the question of how the threat *Hacker* may be the cause of this incident.

One possibility that the target team suggests is that the hacker retrieves this data by obtaining access to the customer database. The analysis secretary uses the CORAS tool to continuously document and display the suggestions, and plots the threat scenario *Hacker obtains access to customer DB* between the threat and the unwanted incident. The relevant part of the diagram is shown in Fig. 9.2. The analysis leader explains that in order to fully understand this scenario they need to be more explicit about how a hacker may go by to obtain this access. She shows relevant parts of the target description to the target team, whereupon the data security

9.2 Conducting the Tasks of Step 5

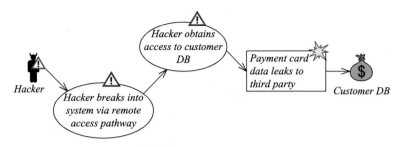

Fig. 9.3 Threat scenarios that lead to harm to customer DB

officer explains that the hacker can try to break in via one of the remote access pathways.

One choice of modelling at this point is to make the previous threat scenario more precise and replace *Hacker obtains access to customer DB* with *Hacker obtains access to customer DB by breaking in via remote access pathway*. A better choice, however, is to introduce *Hacker breaks into system via remote access pathway* as a separate threat scenario and relate it to *Hacker obtains access to customer DB*. This allows the two scenarios to be addressed separately in the sequel, for example, to discuss and investigate whether any of the two may lead to other threat scenarios or unwanted incidents in isolation. The representation of the scenarios separately may also facilitate likelihood estimation and treatment identification. The analysis secretary plots the threat scenario into the diagram, the relevant part of which is shown in Fig. 9.3.

Figure 9.4 shows a selection of the identified threat scenarios leading up to incidents that harm the confidentiality of the customer database. In addition to attacks via remote access pathways, the hacker may cause the two identified unwanted incidents to occur by introducing malicious code (malcode) such as viruses, worms or Trojan horses to the system. As shown by the diagram, the target team believes that this could happen via the web application or via email to employees.

The target team furthermore explains that if the information that is stored on the databases is stored, transmitted or processed irregularly there is a potential for security breaches. Employees can therefore be threats that initiate threat scenarios that lead to loss of confidentiality of the customer database, as captured by the threat diagram of Fig. 9.4.

The analysis leader uses this threat diagram to focus on the asset of the customer database only, particularly on confidentiality. The findings are, however, relevant for further assets, and the analysis leader will use parts of the diagram to lead the discussion in further directions. Most of the diagram is, for example, relevant for the inventory database also. In case the identified threat scenarios affect the integrity and availability of the databases, the threat scenarios may also lead to harm to the online store asset. Both of the unwanted incidents in Fig. 9.4 may furthermore be relevant for data protection laws and regulations, and therefore the compliance asset.

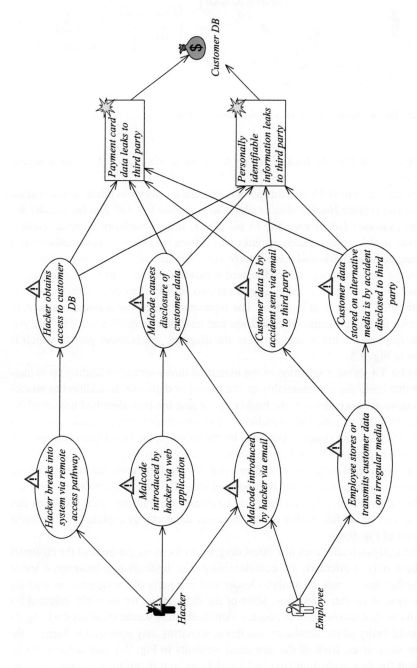

Fig. 9.4 Threat scenarios leading to loss of confidentiality of customer DB

9.2 Conducting the Tasks of Step 5

Table 9.5 Overview of Step 5d

Identification of vulnerabilities

- **Objective:** Explain and identify the vulnerabilities, which are the flaws or weaknesses of the target of analysis that opens for unwanted incidents to occur; this completes the risk identification step
- **How conducted:** The analysis leader has the target team to discuss and explain why threats can initiate threat scenarios, why one threat scenario can lead to another and why threat scenarios can lead to an unwanted incident by identifying the relevant system vulnerabilities; the analysis secretary documents the suggestions on-the-fly by inserting vulnerabilities on the relations leading to the unwanted incidents or to the threat scenarios; if necessary, the analysis leader again probes the target team and stimulates the discussion by giving suggestions
- **Input documentation:** Target description, high-level risk analysis table and the CORAS threat diagram from Step 5c showing threats, threat scenarios, unwanted incidents and assets
- **Output documentation:** The CORAS threat diagrams documenting the results of the risk identification brainstorming session; the diagrams are cleaned up by the analysis team offline before the next step of the analysis

9.2.4 Identification of Vulnerabilities

During the risk identification so far, we have focused on what can go wrong (unwanted incidents), who or what causes it (threats), and how it is caused (threat scenarios). The final task of the risk identification is to explain and describe how all this is possible, which we do by the identification and modelling of vulnerabilities, an overview of which is given in Table 9.5. A vulnerability may, for example, be used deliberately by a threat to cause harm, or it may be lack of mechanisms that would prevent one threat scenario to lead to unwanted incidents or to other threat scenarios.

> **Remark** A *vulnerability* is a weakness, flaw or deficiency that opens for, or may be exploited by, a threat to cause harm to or reduce the value of an asset.

Vulnerabilities are an inherent aspect of risk, since without vulnerabilities the target is unassailable and cannot be harmed. In order to properly understand risks and how to deal with them, we therefore need to identify and understand the vulnerabilities. It is easy to think of vulnerabilities as something that is deliberately utilised or exploited by a malicious threat. But vulnerabilities are also what make accidents happen or what prevents us from recovering from accidents or attacks. Lack of backup solutions, for example, is a vulnerability that opens for accidental deletion or loss of data to become a risk. We must therefore be careful to have a broad view on the risk picture when we are identifying the vulnerabilities. For each of the unwanted incidents and threat scenarios, as well as the relations between them, we must try to determine what makes it possible.

We can think of vulnerabilities as control mechanisms that ideally should be in place, but for some reason are missing or are not sufficiently robust. Vulnerabilities can also be exceptional circumstances that have not been planned for or that nullify the effect of existing, otherwise satisfactory, controls. Vulnerabilities can furthermore be organisational or system characteristics that cannot be removed since they represent necessary or required parts or features of the target, for example, an Internet connection that is crucial for a system.

During the brainstorming session, we use the CORAS threat diagrams that we made during the previous subtask as input to the identification of the vulnerabilities. Vulnerabilities are a part of the explanation of how unwanted incidents and threat scenarios can occur. For each of the threat diagrams we have made, we can structure the discussion by directing the attention of the target team to each of these diagram elements in turn and ask what makes it possible. It is useful to look at the incoming relations and have the target team to explain how it is possible for a threat to initiate a threat scenario, how one threat scenario can lead to another threat scenario, and so forth. We also use the results of the high-level analysis as well as the relevant parts of the target description to support the task. Each of the vulnerabilities we identify are documented by placing them as annotations on the relations between diagram elements. Vulnerabilities can be placed on all relations except on the harms relations from unwanted incidents to assets, since these relations represent the harm on an asset and not events or scenarios propagating to other events and scenarios.

Example 9.4 The analysis leader shows the threat diagram of Fig. 9.4 and explains and motivates the task of identifying vulnerabilities. Each path of the diagram starting from a threat and ending in an unwanted incident describes a possible way in which a risk may arise. The analysis leader chooses to focus on each path in turn and have the target team to discuss what makes the path, and therefore the risk, possible.

For a hacker breaking into the system via a remote access pathway, the very use of remote access is a vulnerability; if it was not for the use of remote access, this type of attack would obviously be impossible. The target team argues, however, that a successful break-in in itself need not lead to leakage of customer data if access to the customer database can be prevented. The possibility of a hacker obtaining access

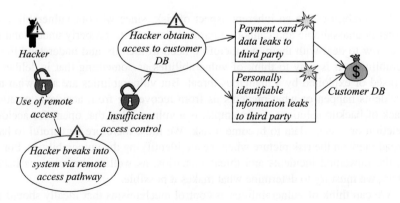

Fig. 9.5 Vulnerabilities that are exploited by hacker

9.2 Conducting the Tasks of Step 5

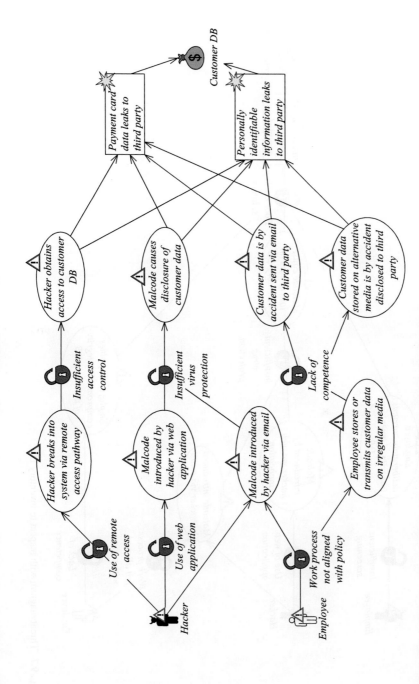

Fig. 9.6 Vulnerabilities that are exploited in causing harm to confidentiality of customer DB

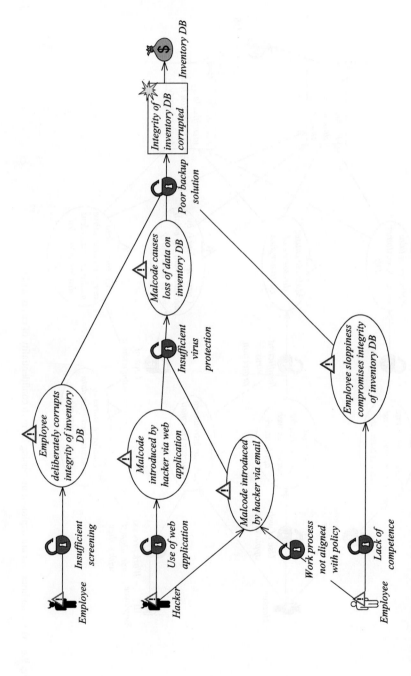

Fig. 9.7 Threat diagram with respect to inventory DB

9.2 Conducting the Tasks of Step 5

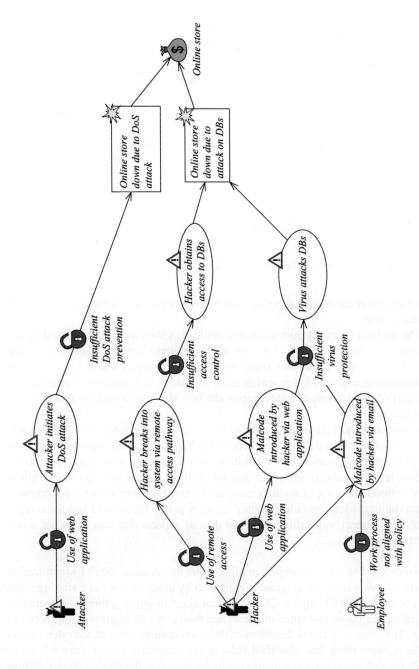

Fig. 9.8 Threat diagram with respect to online store

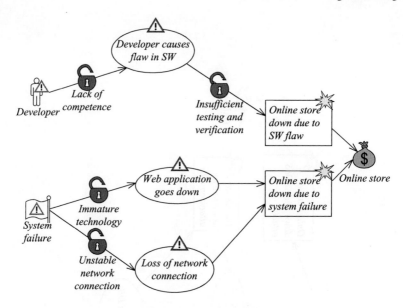

Fig. 9.9 Threat diagram with respect to online store

to the customer database is therefore due to insufficient or lacking mechanisms for access control.

The analysis secretary documents the findings as the discussions proceed by annotating the diagram of Fig. 9.4 with vulnerabilities. The documentation is displayed to the target team, thus supporting both the analysis leader in guiding the brainstorming, and the participants in communicating their opinions to each others. The part of the diagram that documents the two aforementioned vulnerabilities is shown in Fig. 9.5.

Further results of the identification of vulnerabilities that open for harm to confidentiality of customer database are depicted in Fig. 9.6. The diagram shows that the use of the web application in combination with insufficient virus protection makes it possible for a hacker to introduce malcode to the system. The target team furthermore believes that lack of security awareness among the employees may represent vulnerabilities. More specifically, they mention policy breach and lack of competence with respect to handling sensitive data as aspects that may open for threat scenarios to occur.

Example 9.5 The above examples in this chapter demonstrate the risk identification step by showing how one diagram is expanded by using an asset as a starting point and ending up with a complete CORAS threat diagram showing threats, vulnerabilities, threat scenarios, unwanted incidents and assets, as well as the relations between them. The remaining threat diagrams of this section show some of the other results of the brainstorming that identified risks in the AutoParts case. Figure 9.7 documents the risk identification with respect to the inventory database. The risk picture with respect to the customer database is obviously similar to the risk picture with

9.2 Conducting the Tasks of Step 5

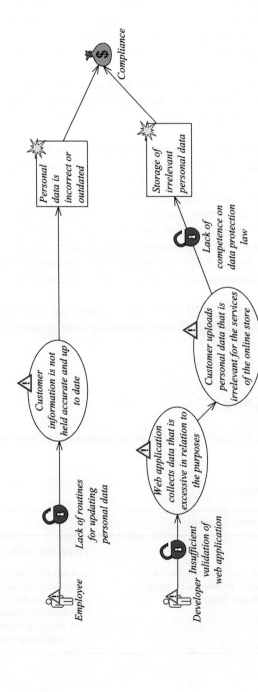

Fig. 9.10 Threat diagram with respect to compliance

respect to the inventory database. Elements of the diagram in Fig. 9.6 are therefore reused.

The threat diagram of Fig. 9.7 shows that it may be relevant to take into account the possibility of malicious insiders, which are represented by the deliberate threat *Employee*. Although the likelihood of employees deliberately causing damage or harm may be small, the level of damage may be high and therefore represent a significant risk.

The threat diagrams in Figs. 9.8 and 9.9 show some of the risks that were identified with respect to the online store. More specifically, the diagrams focus on incidents that cause the online store to go down, since that is the main concern for AutoParts.

Since the online store relies on the databases, some of the threats, threat scenarios and vulnerabilities that were documented in the threat diagrams of Figs. 9.6 and 9.7 are reused as shown in Fig. 9.8. Additionally, the possibility of a denial of service (DoS) attack was considered as a risk towards the availability of the online store.

The online store has currently yet to be launched, and the target team is concerned about possible flaws in the software or the technology that can lead to the online store to go down. This is documented in the threat diagram of Fig. 9.9 where both *Developer* and *System failure* are represented as potential threats, the latter demonstrating how non-human threats may be relevant to take into account when considering all risks.

The threat diagram of Fig. 9.10 finally shows some of the issues of relevance for understanding and documenting the risks towards the asset of compliance with data protection laws.

9.3 Summary of Step 5

Step 5 of the CORAS risk analysis method is summarised by the overview presented in Table 9.6.

Table 9.6 Summary of Step 5

Risk identification using threat diagrams	
Analysis tasks:	People that should participate:

• Identify and document risks through the identification and documentation of unwanted incidents, threats, threat scenarios and vulnerabilities	• Analysis leader (required)
	• Analysis secretary (required)
	• Representatives of the customer:
	  – Decision makers (optional*)
	  – Technical expertise (required)
	  – Users (required)

9.3 Summary of Step 5

Table 9.6 (Continued)

Risk identification using threat diagrams

Modelling guideline:

- Threat diagrams:
  1. Decide how to structure the threat diagrams; a diagram may either focus on one asset at the time, a particular aspect of the target, or a specific kind of threat; for instance, deliberate sabotage in one diagram, mistakes in another, environmental in a third, and so forth; this makes it easier to generalise over the risks, for example, "these risks all harm asset X", "these risks are caused by human errors", or "these risks are related to the network"
  2. Assets are listed to the very right
  3. Relevant threats are placed to the very left
  4. Unwanted incidents are placed in between with relations from the threats that may cause them, and relations to the assets they impact
  5. Assets that are not harmed by any incidents can be removed from the diagram
  6. Add threat scenarios between the threats and the unwanted incidents in the same order as they occur in real time (in other words, in a logical consequence)
  7. Insert the vulnerabilities before the threat scenario or unwanted incident to which they lead; for example, a vulnerability called "poor backup solution" is typically placed before the threat scenario "the backup solution fails to run the application database correctly"

*This workshop usually has a technical focus; the competence of decision makers is more relevant in the next step

Chapter 10
Risk Estimation Using Threat Diagrams

After having completed the identification and documentation of risks, we are ready to estimate the risks. Risk estimation is to estimate the risk level of the identified risks. The objective is to determine the severity of the risks which allows us to subsequently prioritise and evaluate the risks, as well as determining which of the risks should be evaluated for possible treatment.

A risk is the likelihood of an unwanted incident and its consequence for an asset, and the risk level is derived from the combination of the likelihood and the consequence. We use the threat diagrams from Step 5 as input and estimate risk levels by estimating likelihoods and consequences of the identified unwanted incidents. The risk estimation is conducted as a brainstorming involving a target team consisting of personnel with various backgrounds. It is important to include people with the competence, knowledge and expertise needed to estimate realistic likelihoods and consequences, meaning that technical expertise, users and decision makers must be included.

The risk estimation is typically conducted as a separate workshop. Often it may be convenient to conduct the risk estimation of Step 6 together with the risk evaluation of Step 7 in a full day workshop of about six hours.

10.1 Overview of Step 6

At this point in the analysis process, we have reached a description of the relevant risks by our identification and documentation of the possible ways the unwanted incidents may harm the assets. We have furthermore established a thorough explanation and description of how the risks may arise by documenting the threats, threat scenarios and vulnerabilities in CORAS threat diagrams. We may also have gathered some information about the likelihood of threat scenarios and unwanted incidents to occur, as well as the consequence of unwanted incidents, in case such information has already been provided, for example, during the risk identification workshop. The objective of Step 6 is to complete the estimation of these values, and particularly to determine the severity and significance of the identified risks.

Table 10.1 Overview of Step 6

Risk estimation using threat diagrams
• **Objective:** Determine the risk level of the identified risks
• **How conducted:** Conducted as a workshop involving a target team representing various backgrounds, including technical expertise and decision making; the threat scenarios and the unwanted incidents are annotated with likelihoods based on input from the target team; each relation between an unwanted incident and an asset is annotated with the consequence describing the impact of the incident on the asset; risk levels are documented using CORAS risk diagrams modelling each of the identified risks and their risk values as calculated from the estimated likelihoods and consequences
• **Input documentation:** CORAS threat diagrams from the risk identification step; the likelihood scale, the consequence scales and the risk functions defined during Step 4
• **Output documentation:** CORAS threat diagrams completed with a likelihood assigned to each unwanted incident and a consequence assigned to each relation between an unwanted incident and an asset; CORAS risk diagrams modelling the risks and their estimated risk levels
• **Subtasks:** a. Likelihood estimation b. Consequence estimation c. Risk estimation

Since one unwanted incident may harm several assets, each unwanted incident may represent several risks, one for each of the assets it harms. When we are estimating the risk levels, we therefore focus on the unwanted incidents that we identified and documented during Step 5, along with each of the assets an incident is related to. For a given unwanted incident and an asset it harms, we derive the risk level from the likelihood for the incident to occur and its consequence in terms of the damage it inflicts on the asset.

Coming up with good and precise likelihood and consequence estimates may be challenging. We must therefore make sure that the target team includes personnel with adequate competence and good knowledge about the target of analysis at the risk estimation workshop. Additionally, we should prepare by, for example, gathering statistical and historical data if available, consulting domain experts, and retrieving any data that the customer might have available from previous risk analyses or the like. To the extent that we gather such background information, we prepare suggestions that we are ready to present to the target team in case it is needed.

We use the CORAS threat diagrams that we made during the risk identification as input to the risk estimation task. The threat diagrams that we made on-the-fly during the brainstorming session of the risk identification process are most likely not well-structured and polished. Sometimes they are also not really finalised since the analysis secretary often does some of the documentation of the findings by writing down and sketching supplementary information beside the diagrams we make. As part of the preparations to the risk estimation of Step 6, we therefore finalise the threat diagrams and make them ready for the workshop.

Table 10.1 gives an overview of Step 6 and its subtasks. Since risk estimation amounts to estimating likelihoods and consequences, we conduct these two activities as separate subtasks before we do the actual risk estimation in the third and final subtask.

The objective of the likelihood estimation is to assign likelihoods to the unwanted incidents. In principle, we can look at the unwanted incidents in isolation and assign likelihoods to them directly. This may, however, be challenging, and we therefore use the information in the full threat diagrams to support us; knowing and considering the threats, threat scenarios and vulnerabilities are therefore important in order to determine the risk level. In particular, our method is to assign likelihoods to the threat scenarios and then use this information to support the estimation of likelihoods of the unwanted incidents to which the threat scenarios may lead. This method is moreover facilitated by the expressiveness in the CORAS language to assign conditional likelihoods to the leads-to relations that express the likelihood of a threat scenario or an unwanted incident to lead to another threat scenario or unwanted incident in case the former occurs.

Estimating and documenting the likelihood of threat scenarios furthermore give useful information about the most important sources of risks, which is important in order to fully understand the risk picture and in order to identify adequate treatments for unacceptable risks.

The objective of the consequence estimation is to assign consequences to the impacts relations from unwanted incidents to assets. Whereas a likelihood is independent of the parties of the risk analysis, this is not so for the consequences; it is only the party of an asset that can decide the degree to which a given unwanted incident is harmful. We therefore rely on the participation of decision makers or others that can speak on behalf of the parties when we are conducting the consequence estimation.

The likelihoods and consequences we use can be quantitative or qualitative, depending on what is suitable for the analysis. This decision was made during Step 4, and we use the scales that we defined at that point.

In the final subtask of risks estimation, we use the results of the likelihood and consequence estimation. The risk estimation is then straightforward since we already have defined the risk functions during Step 4. In addition to the threat diagrams annotated with the estimated likelihoods and consequences, we document the risk estimation using CORAS risk diagrams. The risk diagrams give an overview of all the risks with their risk values.

10.2 Conducting the Tasks of Step 6

In this section, we go through each of the subtasks of Step 6 and describe in more detail how we conduct them. To the extent that likelihoods and consequences have already been gathered, we bring these values in as input to the tasks and present them to the target team for verification or adjustment.

Table 10.2 Overview of Step 6a

Likelihood estimation

- **Objective:** Estimate the likelihood of the unwanted incidents documented under the risk identification step to occur; to be used when estimating risk levels
- **How conducted:** The CORAS threat diagrams resulting from the risk identification step are taken as starting point; based on discussions between the target team members guided by the analysis leader, each of the identified unwanted incidents is assigned a likelihood estimate; the likelihood scale used is the one that were set during Step 4 of the analysis; if likelihoods cannot be assigned directly to the unwanted incidents, these are estimated for the threat scenarios leading up to them, providing data for calculating the resulting likelihood for the unwanted incident; likelihoods can also be estimated for the initiates relations and conditional likelihoods for the leads-to relations for further input data; support for the estimation should be gathered by the analysis team in advance, for example by consulting domain experts, historical data and statistical data; when a likelihood estimate is arrived at for a threat scenario, an unwanted incident or a relation, the analysis secretary annotates the threat diagram accordingly
- **Input documentation:** The CORAS threat diagrams resulting from the risk identification step; the likelihood scale defined during Step 4
- **Output documentation:** CORAS threat diagrams annotated with a likelihood estimate for each of the unwanted incidents and any number of the identified threat scenarios, initiates relations and leads-to relations

10.2.1 Likelihood Estimation

A likelihood is a description of the chance for something to happen. We may specify likelihoods as frequencies or probabilities. A frequency specifies a number of occurrences within a given period of time, whereas a probability is a number ranging from 0 to 1 specifying the possibility for a scenario or an incident to occur.

> Remark A *likelihood* is the frequency or probability of something to occur.

The likelihood values we use are those we defined during Step 4 when we set the likelihood scale. These values are either quantitative or qualitative. Quantitative values allow the specification of exact likelihoods, such as a frequency of 3 times per year or a probability of 0.2. Using qualitative values, we may describe likelihoods using terms such as *unlikely* and *possible*. Such values can also be used to denote intervals of a likelihood scale.

The objective of the likelihood estimation, an overview of which is given in Table 10.2, is to estimate the likelihood for the unwanted incidents to occur. The unwanted incidents are those we identified during the risk identification of Step 5, and that we documented using CORAS threat diagrams. To conduct the task of likelihood estimation, we present to the target team each of these threat diagrams. Through a brainstorming session that we lead, the target team is then invited to come up with likelihood estimates that we assign to each of the unwanted incidents.

The task of assigning likelihoods to the unwanted incidents can be challenging, and we therefore need to prepare well and to use methods to support the estima-

10.2 Conducting the Tasks of Step 6

tion. In principle, it is the target team of the workshop that is supposed to come up with the likelihood estimates. We must therefore make sure that the required expertise is present. The type of expertise that is needed varies depending on the target of analysis. Personnel responsible for system design, maintenance and support may have good knowledge about how the system is used, what kind of failures that may occur, what kind of problems that are run into, and so forth. Data security officers may have good knowledge about how often various types of security incidents and attacks may occur. System users know how the system is used, and perhaps misused, and have valuable knowledge about which problems that typically are encountered.

In addition to prepare by gathering adequate expertise, we should gather as much information and data to support us as possible. This can be data that the customer already has got, for example from system monitoring or from previous analyses. We can also gather data by consulting domain experts or statistics and reports about typical incidents and attacks. During the brainstorming we can use this data as starting points for discussions, or as suggestions to bring the discussions forward.

Given our preparations and the expertise present at the workshop, it may still be a challenge to estimate directly the likelihoods for the unwanted incidents to occur. Our method is then to use the information that we gathered during the risk identification in the form of the threats that cause incidents to occur and the threat scenarios that lead to the incidents. By assigning likelihoods to the threat scenarios, we use these values to support the estimation of the unwanted incidents that the treat scenarios may lead to. We may also assign likelihoods to the relations between the elements of the treat diagram. A likelihood on an initiates relation, for example, specifies the likelihood for a threat to initiate a threat scenario. A likelihood on a leads-to shows the conditional likelihood of some scenario/incident to lead to another scenario/incident.

In many risk analyses, assigning likelihoods to leads-to relations is infeasible because they are too hard to estimate, or because the complexity becomes too high for the target team. On some occasions, however, this is a very helpful tool in the risk estimation process. Our experience is that probabilities or probability intervals are best suited from a practical point of view to capture conditional likelihoods of leads-to relations. Throughout this book we therefore restrict ourselves to that.

Example 10.1 One of the unwanted incidents that were identified with respect to the online store asset of AutoParts was *Online store down due to system failure* as documented in the threat diagram of Fig. 9.9. The target team members are reluctant to assign a likelihood directly to the unwanted incident; after all, the new system with the online store has yet to be launched, so they do not have much experience with the stability of the system apart from the testing they have conducted.

The analysis leader therefore brings the attention to the two threat scenarios that may lead to the unwanted incident, namely *Web application goes down* and *Loss of network connection*. If the target team can assign likelihoods to these two scenarios, they can combine these values and use the result to support the estimation of the likelihood for the unwanted incident.

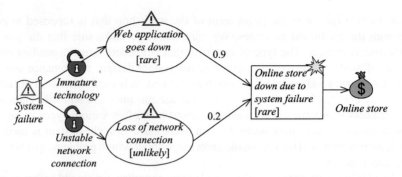

Fig. 10.1 Combining likelihood values

The target team agrees that loss of network connection is not very likely, and the system developer also explains that they have good redundancy and backup solutions. The analysis leader refers to the likelihood scale defined in Table 8.10 and asks which of the values are most suitable. The target team agrees on the value *unlikely*, and the analysis secretary annotates the threat scenario accordingly as shown in Fig. 10.1.

As to the threat scenario *Web application goes down*, the system developer explains that they have done thorough testing, and that they use state of the art technology that is reliable. He is therefore confident that *rare* is the correct likelihood estimation for the threat scenario in question.

The next issue is to use the likelihood estimates of the threat scenarios *Web application goes down* and *Loss of network connection* in the estimation of the likelihood of the unwanted incident *Online store down due to system failure*. In some cases, the likelihood of an unwanted incident equals the aggregate of the likelihoods of the threat scenarios that leads to it. But it may also be higher in case there are other causes for the incident to occur that are not documented. If these other causes are really significant, it may be that the risk identification and documentation is incomplete and should be reconsidered. Often we also find that the likelihood of the unwanted incident is less than the aggregate of the threat scenarios in question. This is because the occurrence of a given threat scenario not necessarily leads to an unwanted incident.

The analysis leader has the target team to consider the relations from the threat scenarios in question and to discuss the probability of the threat scenarios to lead to the unwanted incident. The participants explain that if the web application goes down, it is very likely that the online store becomes unavailable and agree that the leads-to relation should be annotated with the probability 0.9. Loss of network connection should usually not be a problem due to redundancy, and they therefore estimate the probability to 0.2. The probabilities are annotated on the leads-to relation of the threat diagram shown in Fig. 10.1.

Assuming that there are no other threat scenarios that contribute significantly to the likelihood of the unwanted incidents, the results of the estimations can be roughly combined and aggregated as shown in Table 10.3. The likelihood of the

10.2 Conducting the Tasks of Step 6

Table 10.3 Aggregating likelihood values

Threat scenario	$Rare = [0, 1] : 10y$	$Unlikely = [2, 5] : 10y$
Leads to	0.9	0.2
Combined	$([0, 1] : 10y) \cdot 0.9 = [0, 0.9] : 10y$	$([2, 5] : 10y) \cdot 0.2 = [0.4, 1] : 10y$
Aggregate	$[0, 0.9] : 10y + [0.4, 1] : 10y = [0.4, 1.9] : 10y$	
Approximation	*Rare*	

threat scenario *Web application goes down* is *rare*, which is defined as the interval $[0, 1] : 10y$. Since the probability of this threat scenario leading to the unwanted incident is 0.9 the contribution of the threat scenario can be derived by multiplying the likelihood with the probability, as shown in the third row of the second column of Table 10.3. The resulting likelihood interval is $[0, 0.9] : 10y$.

The same exercise is conducted in the third column for the threat scenario *Loss of network connection*. The combined contribution of the two threat scenarios is derived by adding the two resulting intervals, as shown by the aggregate in the fourth row. The resulting interval is $[0.4, 1.9] : 10y$. Since this is less than 2 times per 10 years, the best approximation is *rare*.

The combined contribution of the two threat scenarios therefore yields *rare* as the estimate of the likelihood of the unwanted incident *Online store down due to system failure*. The analysis secretary inserts the results into the threat diagram as shown in Fig. 10.1.

Notice, importantly, that the correct calculation of likelihoods using CORAS threat diagrams depends on several factors. An important issue is whether the diagram in question is complete in the sense that it documents all the possible causes for unwanted incidents and threat scenarios to occur. The statistical dependencies between scenarios or incidents are also a factor that must be taken into account. Rough aggregations may be adequate for more course-grained calculations, where we only need to determine to which likelihood interval a given threat scenario or unwanted incident belongs, rather than estimating an exact likelihood. For more fine-grained and exact calculations, we need to be correspondingly careful in the calculations. More thorough likelihood analysis is supported in CORAS, an issue that is presented in detail in Chap. 13 of Part III.

In summary, the main objective of the likelihood estimation is to assign likelihoods to the unwanted incidents we have identified such that we later can use these values in the estimation of risk values. As explained above, the purpose of assigning likelihoods to the threat scenarios and to the relations is to support our estimation of likelihoods of the unwanted incidents. However, knowing and documenting the likelihoods of threat scenarios is important also because it contributes to our understanding of the full risk picture. It gives us a better foundation for understanding how risks arise and what the main causes of unwanted incidents are. The documentation of the likelihoods furthermore constitute valuable input to the identification of adequate and effective treatments of unacceptable risks.

Table 10.4 Overview of Step 6b

Consequence estimation

- **Objective:** Estimate the consequence of the unwanted incidents on each of the assets they affect; to be used when estimating risk levels
- **How conducted:** The CORAS threat diagrams annotated with the likelihood estimates resulting from the previous subtask are taken as starting point; based on discussions between the target team members guided by the analysis leader, each relation between an unwanted incident and an asset is assigned a consequence estimate; the consequence scale used is the one that were set for the given asset during Step 4 of the analysis; when a consequence has been decided upon, the analysis secretary annotates the threat diagram accordingly
- **Input documentation:** The CORAS threat diagrams annotated with likelihood estimates resulting from the previous subtask; the consequence scales defined during Step 4
- **Output documentation:** CORAS threat diagrams annotated with a likelihood estimate for each unwanted incident and a consequence estimate for each relation between an unwanted incident and an asset

10.2.2 Consequence Estimation

A consequence is a description of the level of damage an unwanted incident inflicts on an asset when the incident occurs. For some analyses, we may talk about and specify consequences using the same terms for all assets, for example, in terms of loss of monetary value. As explained and exemplified in Chap. 8, we often operate with separate consequence scales for separate assets, since assets are often incomparable.

> *Remark* A *consequence* is the impact of an unwanted incident on an asset in terms of harm or reduced asset value.

The consequence scales we use are those we defined during Step 4. As for the likelihoods, the consequences are either quantitative or qualitative. Quantitative values allow the specification of exact consequences, such as loss of 12,000 Euros or 30 minutes downtime of services. With qualitative values we describe consequences using terms such as *insignificant* or *major*. Such terms can also be used to denote intervals of a consequence scale, or they can be described by illustrative examples of harm that serve as reference points.

Our objective with the consequence estimation, an overview of which is given in Table 10.4, is to assign consequences to each of the impacts relations from unwanted incidents to assets that we identified and documented during Step 5. As with the previous task of likelihood estimation, we conduct the consequence estimation as a brainstorming session. We present the threat diagrams annotated with likelihood estimates that resulted from the previous subtask, and invite the target team to come up with consequences through discussions. To the extent that we have gathered consequences that are already known, we insert these immediately and have the target team to verify or adjust them.

The likelihood of a scenario or an incident to occur does not depend on the parties of the analysis, and we can therefore use background information, statistical data

10.2 Conducting the Tasks of Step 6

and expert opinions to support us. This is not as straightforward for consequences, since it is the parties that must determine how bad an incident is. It is therefore crucial that all the involved parties are represented during the consequence estimation. Alternatively, decision makers or other personnel that can make judgements and speak on behalf of the parties may attend. It is, however, important that we involve other personnel also, such as users, developers, lawyers, and so forth since they have good insight into the problems and can explain the significance and impact of scenarios and incidents.

The full threat diagrams give useful support for the estimation of likelihoods, since we can use information about the likelihood of the threat scenarios to estimate the likelihood of the unwanted incidents. This is not the case for the consequence estimation, since it is only unwanted incidents that harm assets. The threat diagrams may nevertheless provide useful information since they describe why and how unwanted incidents occur. By considering the threats that initiate the unwanted incidents, as well as the threat scenarios that lead to them, we can make better judgements about the severity of an unwanted incident than by considering the incidents in isolation only. We therefore present the full diagrams to the target team in order to give support to the discussions and estimations.

Example 10.2 Figure 9.6 shows a CORAS threat diagram that documents some of the risks that were identified during Step 5 with respect to the confidentiality of the customer database of AutoParts. The threat diagram annotated with likelihoods and consequences is shown in Fig. 10.2.

There are two unwanted incidents regarding which likelihood and consequence estimations must be assigned, namely *Payment card data leaks to third party* and *Personally identifiable information leaks to third party*. There are two threats and eight threat scenarios in the diagram, all of which may lead to the unwanted incidents. By means of a walk-through, the analysis leader have the target team to assign likelihoods to the threat scenarios. She chooses to go through the diagram from left to right and thereby use the likelihoods of the preceding scenarios to support the estimation of the likelihoods of the subsequent scenarios as explained in Example 10.1.

Although the two unwanted incidents may be caused by exactly the same threats and through the same threat scenarios, the target team explains that the likelihood of *Payment card data leaks to third party* is less than the likelihood of *Personally identifiable information leaks to third party*. This is because the protection of the payment card data is strong, and that the employees generally are much more security aware with respect to highly sensitive data.

As to the consequence estimation, they all agree that leakage of payment card data is very severe; the data is highly sensitive, and the customers of AutoParts expect nothing but complete security of this information. The consequence is therefore *catastrophic*, as annotated on the impacts relation from the unwanted incident to the asset in the threat diagram of Fig. 10.2.

Leakage of personally identifiable information is less critical, and the severity of course depends on how much information that is compromised. The target team

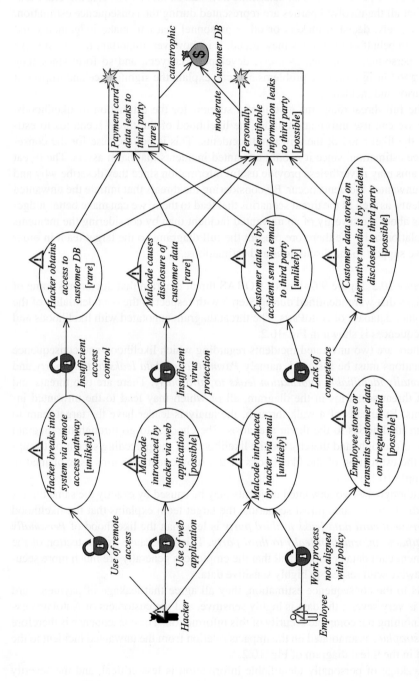

Fig. 10.2 Likelihood and consequence estimations

10.2 Conducting the Tasks of Step 6

Table 10.5 Overview of Step 6c

Risk estimation
• **Objective:** Calculate the risk level for each of the identified risks
• **How conducted:** For each relation between an unwanted incident and an asset, the risk level is calculated based on the estimated likelihood and consequence; the risks and their risk levels are documented using CORAS risk diagrams
• **Input documentation:** The CORAS threat diagrams with likelihood and consequence estimates resulting from the previous subtask; the risk functions defined during Step 4
• **Output documentation:** CORAS risk diagrams annotated with the estimated risk level of each of the identified risks

members all acknowledge that leakage of such information should never occur, but if only small individual pieces of information are compromised, such as a name or an email address, it is less critical. To allow for incidents in which larger amounts of information is compromised, they agree to assign the consequence *moderate* to the impacts relation in question, as depicted in Fig. 10.2.

Recall that the consequence scale for the AutoParts case that were defined in Table 8.6 of Chap. 8 is in terms of how many database entries are affected by the occurrence of unwanted incidents. In Fig. 10.2 in the previous example, however, the unwanted incidents refer to the type of data that is compromised. For the unwanted incident *Personally identifiable information leaks to third party*, for example, the consequence *moderate* is described by *Range of* [10%, 20%) *of records are affected*. This description of the consequence is really not adequate for the unwanted incident in question. However, in this case the description is used only as a reference point in order to understand more precisely what is meant by notions such as *moderate* and *catastrophic*. In other words, the severity of the unwanted incident *Personally identifiable information leaks to third party* is held as equal to the severity of an unwanted incident that result in 10% to 20% of the database records to be corrupted, compromised or otherwise harmed.

Example 10.3 Some of the other results of the likelihood and consequence estimation are shown in the threat diagrams of Figs. 10.3 and 10.4. In particular, these diagrams document unwanted incidents that represent risks against the online store asset.

10.2.3 Risk Estimation

An unwanted incident may harm several assets, and it therefore may represent several risks, namely one risk for each of the assets it harms. The chance of loosing a laptop, for example, represents at least two risks for the owner. One is the risk of loosing the laptop itself as a valuable piece of hardware. Another is the risk of loosing the information stored on the laptop.

158　　10 Risk Estimation Using Threat Diagrams

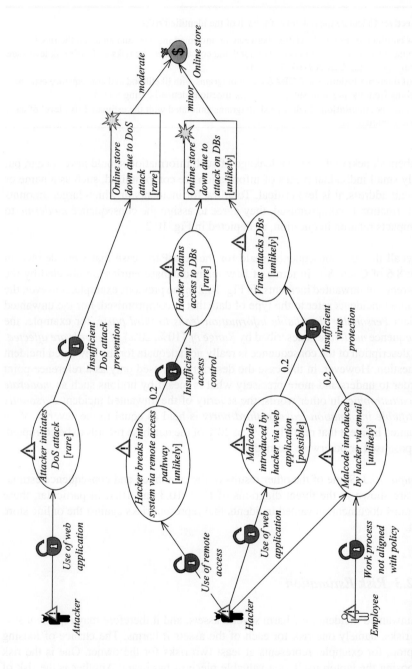

Fig. 10.3 Likelihood and consequence estimation of online store to go down

10.2 Conducting the Tasks of Step 6

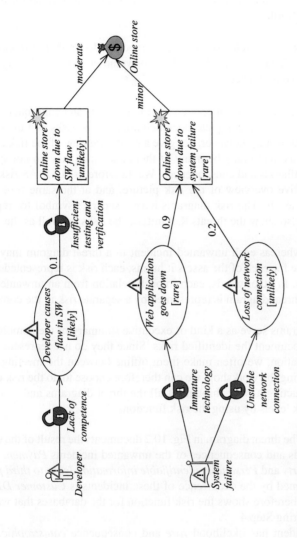

Fig. 10.4 Likelihood and consequence estimation of online store to go down

When we are estimating risks, an overview of which is given in Table 10.5, we therefore consider each of the impacts relations of the threat diagrams that resulted from the likelihood and consequence estimations. The risk level is derived from the likelihood and the consequence of an unwanted incident. During Step 4, we defined a risk function for each of the direct assets that takes a likelihood and a consequence and yields the risk level. Given these risk functions and the completed estimations of the likelihoods and consequences, the risk estimation is therefore quite straightforward.

> *Remark* A *risk* is the likelihood of an unwanted incident and its consequence for a specific asset. The *risk level* is the level or value of a risk as derived from its likelihood and consequence.

Given the threat diagrams annotated with likelihoods and consequences and the risk functions, we have a complete documentation of the risks and the risk levels. The threat diagrams may, however, not give a good overview of the risk picture and the risk level since there may be many of them and since they do not specify risk levels, only likelihoods and consequences. We therefore use CORAS risk diagrams to give an intuitive overview of the risk picture, and at the same time explicitly specify the risk levels. The risk diagrams have a separate symbol for representing risks, and they also show the threats that initiate the risks, as well as the assets that are harmed.

Notice that whereas each unwanted incident in a threat diagram may represent several risks, one for each of the assets it harms, each risk is represented separately in risk diagrams. In other words, each impacts relation from an unwanted incident to an asset in a threat diagram is represented by a separate risk in the corresponding risk diagram.

The risk diagrams serve as a kind of executive summary of the completed threat diagrams that document the identified risks. Since they do not represent or require any new information, we often make them offline between the meetings with the target team. During the workshop, we can therefore choose to do the risk estimation by simply conducting a walk-through of all the threat diagrams and consecutively calculate the risk levels by using the risk functions.

Example 10.4 The threat diagram in Fig. 10.2 documents the result of the estimation of the likelihoods and consequences of the unwanted incidents *Payment card data leaks to third party* and *Personally identifiable information leaks to third party*. The asset that is harmed by the occurrence of these incidents is *Customer DB*, and the analysis leader therefore shows the risk function for the databases that was defined in Table 8.12 during Step 4.

The first incident has likelihood *rare* and consequence *catastrophic*. The risk function is represented by a matrix that is divided into four sections, each section representing one of the risk levels *very low*, *low*, *high* and *very high*. The matrix entry in the *rare* row and the *catastrophic* column is orange, which represents the risk level *high*. The second incident has likelihood *possible* and consequence *moderate*, which also yields the risk level *high*.

Fig. 10.5 Estimation of risks towards customer DB

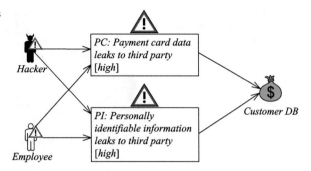

Fig. 10.6 Unwanted incident harming several assets

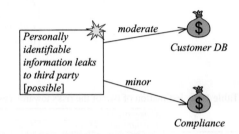

The risk diagram that represents these two risks is shown in Fig. 10.5, where each risk is annotated with the estimated risk level. The description of the risk is the same as the description of the corresponding threat diagram. It is often useful to have a short name or an id for each risk, for example, for reference or overviews. The analysis leader therefore places an abbreviation in front of the description. The unwanted incident *Payment card data leaks to third party* is therefore described by *PC: Payment card data leaks to third party*, where *PC* is the risk id.

In this case, the given unwanted incident harms only one asset, and therefore represents only one risk. If several assets are harmed, there are also several risks that correspond to the same unwanted incident. In order to operate with a unique risk id for each of the risks, we may, for example, give them unique numbers.

Example 10.5 As an example of an unwanted incident that represents several risks, consider again the unwanted incident *Personally identifiable information leaks to third party*. As explained by the lawyer representing AutoParts in the risk analysis, this incident is relevant for data protection, and therefore represents a risk towards the compliance asset.

Figure 10.6 shows the relevant fragment of the threat diagram of Fig. 10.2. The consequence of the unwanted incident with respect to the asset *Compliance* is estimated to *minor*. By the risk function for *Compliance* as defined in Table 8.14, the derived risk level is *low*.

Fig. 10.7 Estimation of risks towards customer DB and compliance

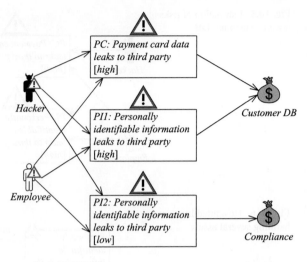

Table 10.6 Estimation of two of the risks towards customer DB

		Consequence				
		Insignificant	Minor	Moderate	Major	Catastrophic
Frequency	Rare					PC
	Unlikely					
	Possible			PI1		
	Likely					
	Certain					

Updating the risk diagram of Fig. 10.5 to take into account the risk towards compliance yields the risk diagram of Fig. 10.7. The unwanted incident *Personally identifiable information leaks to third party* is now represented by two risks, one for each of the affected assets. In order to distinguish between the two separate risks that the unwanted incident represents they are labelled with the unique short names *PI1* and *PI2*.

If we used the matrix format to define the risk function during Step 4, we may choose to plot all the identified risks into the matrices in order to provide a simple overview of the risks and their risk level. This is a convenient way of estimating risk levels, and at the same time we get an overview of all the identified risks.

Example 10.6 While the analysis leader is going through all the completed threat diagrams in order to conduct the risk estimation, the analysis secretary plots the results into the risk function matrices. When the risk estimation is complete, the filled in matrices are presented to the target team for overview. The risks towards the customer database shown in Fig. 10.5, for example, are plotted into the matrix as shown in Table 10.6.

10.3 Summary of Step 6

Step 6 of the CORAS risk analysis method is summarised by the overview presented in Table 10.7.

Table 10.7 Summary of Step 6

Risk estimation using threat diagrams	
Analysis tasks:	People that should participate:

• Provide likelihood estimates for the unwanted incidents
• Do an estimate of the consequences of the unwanted incidents
• Do an estimate of the risk levels on basis of the likelihood and consequence estimates
	• Analysis leader (required)
	• Analysis secretary (required)
	• Representatives of the customer:
	  – Decision makers (required)
	  – Technical expertise regarding the target (required)
	  – Users (required)

Modelling guideline:

- Risk estimation on threat diagrams:
  1. Add likelihood estimates to the threat scenarios if this kind of information is available
  2. Add likelihood estimates to each unwanted incident, either directly or based on the threat scenarios or unwanted incidents that lead up to it
  3. Annotate each impacts relation from an unwanted incident to an asset with a consequence from the respective asset's consequence scale

- Risk diagrams:
  1. Use the threat diagrams and replace each unwanted incident with one risk for each impacts relation, where each risk shows the risk description, the risk id, and the risk value
  2. Remove threat scenarios and vulnerabilities, but keep the relations between the threats and the risks

Chapter 11
Risk Evaluation Using Risk Diagrams

At this point in the analysis, we have finalised the identification of risks and the estimation of the risk levels. The main objective with Step 7 is to evaluate the risks such that we can prioritise between them, and determine which of them should be evaluated for treatment.

The evaluation of risks basically amounts to comparing the estimated risk levels against the risk evaluation criteria that we established during Step 4. The criteria determine for each risk whether it is acceptable or needs to be evaluated further. However, although individual risks may be acceptable when considered in isolation, they may not be so if each of them contributes to the same overall risk. During Step 7, we therefore also group individual risks that each can be understood as a special case of the same, more general risk. In order to estimate the level of the combined risks, we accumulate their separate risk values.

We furthermore take the indirect assets into account and estimate their contribution to the overall risk picture. It is only by considering the potential of risks to propagate from direct assets to indirect assets that we can fully understand the full risk picture, evaluate the risks and determine which risks that need to be considered for treatment.

The evaluation of risks requires the participation of decision makers and personnel that can speak on behalf of the parties of the analysis. Additionally, domain experts such as technical expertise and users should be brought in. The risk evaluation is typically held as a meeting of about three hours. For convenience, the risk evaluation may be conducted as a full day workshop combined with either the risk estimation of the previous step or the treatment identification of the next step.

11.1 Overview of Step 7

In principle, we have at this point all the information and documentation we need for evaluating the risks that we have identified with respect to the direct assets. That is, during the previous steps of risk identification and risk estimation we have documented all the relevant risks and their risk values. Additionally, we have already

Table 11.1 Overview of Step 7

Risk evaluation using risk diagrams
• **Objective:** Decide which of the identified risks are acceptable and which of the risks that must be further evaluated for possible treatment
• **How conducted:** Likelihood and consequence estimates are confirmed or adjusted; adjustments of the risk evaluation criteria are made if needed; risks that each contributes to the same overall risk are grouped and their accumulated risk value calculated; using the threat diagrams with respect to the direct assets combined with the asset diagrams, the risks with respect to the indirect assets are identified and estimated; each of the identified risks is evaluated by comparing against the risk evaluation criteria
• **Input documentation:** CORAS threat diagrams with estimated likelihoods and consequences; CORAS risk diagrams; CORAS asset diagrams; the risk evaluation criteria
• **Output documentation:** CORAS threat diagrams for both direct and indirect assets; CORAS risk diagrams with evaluation results
• **Subtasks:** a. Confirming the risk estimates b. Confirming the risk evaluation criteria c. Providing a risk overview d. Accumulating risks e. Estimating risks with respect to indirect assets f. Evaluating the risks

established the criteria for evaluating the identified risks. However, during the various tasks that we conduct throughout the analysis, both we and the target team gain new knowledge about and new insight into the risk picture. We therefore need to reconsider the risk evaluation criteria, and also conduct a quality assurance of the results of the risk estimation.

Table 11.1 gives an overview of Step 7 and its subtasks. The first subtask is to confirm or adjust the estimated risk levels as documented in the CORAS threat diagrams and risk diagrams. Subsequently, in the second subtask, we revisit the risk evaluation criteria that were established during Step 4 and invite the parties of the analysis to confirm or adjust them. The third subtask is to give an overview of all the identified risks with their final and confirmed risk levels so as to prepare for the risk evaluation.

The three first subtasks are basically a recapitulation and a quality assurance of what we have achieved and documented so far. Through the various analysis tasks up to now, we have gained good knowledge about the general risk picture with respect to the direct assets, and we have established a thorough documentation of the relevant unwanted incidents and the risks that these incidents represent. What still remains before we can evaluate the identified risks, however, is to investigate their potential impact in a wider setting.

On the one hand, we may have several individual risks that can only be properly understood and evaluated by considering them in combination. This is addressed in the fourth subtask where we identify risks that can be grouped and accumulated so as to represent a more general risk. The accumulation of such risks shows the full potential of several smaller, individual risks, and it is only by considering this potential that we can determine whether these risks are acceptable.

On the other hand, we must take into account the indirect assets and identify risks with respect to them. The fifth subtask is therefore to investigate the extent to which the identified risks with respect to the direct assets propagate to risks with respect to the indirect assets. This reveals the full impact of the identified risks with respect to the target of analysis, and is a requirement for evaluating the general risk picture and determining which of the risks must be evaluated for mitigation through risk treatment.

Once we have completed the fifth subtask of Step 7, we have finalised the identification and documentation of the full risk picture with respect to both the direct and the indirect assets. The final subtask is then to evaluate the risks and determine which of them should be considered for possible treatment.

11.2 Conducting the Tasks of Step 7

In this section, we present each of the subtasks in more detail, focusing on how we conduct them. The former three subtasks are a kind of approval and quality assurance of the results so far, whereas the latter three finalise the risk identification and evaluation.

11.2.1 Confirming the Risk Estimates

The activities we conduct during a risk analysis involve the processing of large amounts of information. The brainstorming sessions of Step 5 and Step 6 where we identify and estimate risks, respectively, may be particularly demanding. It may therefore be a challenge for both us as risk analysts and for the target team to maintain a full and good overview of the risk picture. After having completed Step 6, we therefore need to recapitulate and assure the quality of the current documentation. An overview of this task of confirming the risk estimates is given in Table 11.2.

The CORAS diagrams that we make on-the-fly to support the analysis tasks and to document the results usually need to be cleaned up and polished. Furthermore, we often gather additional and supplementary information on the side that we do not bother to properly insert into the diagrams while the workshop discussions proceed. Between Step 6 and Step 7, we therefore go through all the documentation, polish and complete the diagrams, and make them ready for inspection and quality assurance.

When we have polished the diagrams we send them to the customer for review and inspection. If necessary, we accompany the diagrams with complementary and explanatory information. Such information may, for example, be additional explanations of the diagrams in prose, or explanations of the relationships between the CORAS diagrams and the target description.

Table 11.2 Overview of Step 7a

Confirming the risk estimates

- **Objective:** Ensure that the estimates conducted during Step 6 are appropriate
- **How conducted:** The analysis team cleans up and quality checks the documentation resulting from the risk estimation step; before the risk evaluation step is conducted, the analysis team sends the documentation to the customer for internal review and quality check; any adjustments are implemented by correcting the threat diagrams and risk diagrams from the previous analysis step
- **Input documentation:** The CORAS threat diagrams with likelihood and consequence estimates from Step 6
- **Output documentation:** Finalised and confirmed CORAS threat diagrams and risk diagrams

The contact person of the customer is responsible for distributing the diagrams to the decision makers, to the parties of the analysis and to other relevant personnel. Typically, the diagrams should be sent to all or most of the personnel that have participated throughout the risk analysis so far as target team members. After the inspection of the diagrams by the relevant personnel, the contact person must report back to us. In case there are any adjustments of the likelihood and consequence estimates, we implement these by correcting the diagrams. The corrected and finalised diagrams should be returned such that the target team can prepare for the meeting of Step 7.

It is important that we allow for sufficient time between Step 6 and Step 7 to properly do the quality assurance and finalisation of the documentation. In order to ensure that we have enough time to do our preparations, we should agree with the contact person on a deadline for the feedback. If we choose to conduct the risk estimation and the risk evaluation in a combined workshop, we need to do the quality assurance before the final step of risk treatment instead.

11.2.2 Confirming the Risk Evaluation Criteria

The customer and the parties of a risk analysis usually only have superficial insight into the risk picture at the initial phase of the analysis. Decision makers, for example, are often aware of risks at a business or enterprise level, but lack insight into the deeper causes of risks. At this point in the analysis, however, the decision makers and other personnel that have been involved in the analysis have gained much new knowledge about the risk picture. As a consequence, the basis for evaluating risks is more solid at this point as compared to the initial phases.

Before we evaluate the identified risks, we therefore revisit the risk evaluation criteria that we established during Step 4. For each party, we present to the target team the associated risk evaluation criteria, and have the participants to judge whether the criteria are appropriate. For a given asset, it is only the associated party that can decide how risks towards this asset should be evaluated. We must therefore make

11.2 Conducting the Tasks of Step 7

Table 11.3 Overview of Step 7b

Confirming the risk evaluation criteria

- **Objective:** Ensure that the risk evaluation criteria defined during Step 4 are appropriate
- **How conducted:** The target team, and especially the decision makers, decides whether the risk evaluation criteria are appropriate or whether they should be adjusted; new knowledge and insight that are gathered during the analysis process may result in adjustments
- **Input documentation:** Risk evaluation criteria from Step 4 of the analysis
- **Output documentation:** Finalised and confirmed risk evaluation criteria

Table 11.4 Overview of Step 7c

Providing a risk overview

- **Objective:** Establish an overview of the general risk picture
- **How conducted:** The risk overview is presented as a walk-through of the CORAS risk diagrams
- **Input documentation:** The finalised CORAS threat diagrams with confirmed likelihoods and consequences; the finalised and confirmed CORAS risk diagrams
- **Output documentation:** None

sure that decision makers that represent the parties of the analysis are present, or alternatively personnel that can make judgements and speak on behalf of the parties. In case the target team demands any adjustments, we update the risk evaluation criteria accordingly. An overview of this task of confirming the risk evaluation criteria is given in Table 11.3.

11.2.3 Providing a Risk Overview

In order to summarise what we have achieved so far, we present to the target team the identified risks and their confirmed, quality assured risk levels. An overview of this task is given in Table 11.4. We may choose to conduct a walk-through of the threat diagrams with the finalised likelihood and consequence estimates. However, because we have sent all the diagrams to the customer for inspection and review before this workshop, it should not be necessary to do a detailed presentation of all the threat diagrams. Therefore, if we do not find it necessary with a detailed walk-through and this is also not requested by the target team, we only give a presentation of the CORAS risk diagrams that show an overview of all the risks and their risk levels.

We made the risk diagrams to capture the outcome of the risk estimation of Step 6. As a result of the inspection of the threat diagrams with possible adjustments of likelihoods and consequences, we may need to make possible adjustments of the risk levels that annotate the risk diagrams before we present them.

Example 11.1 As part of the preparations for Step 7, the analysis team from CORAS Risk Consulting has cleaned up and polished all the threat diagrams. The AutoParts

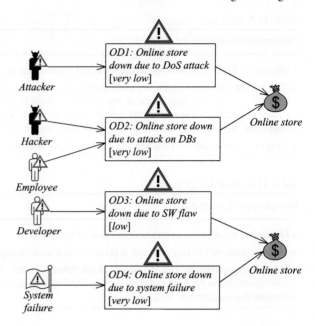

Fig. 11.1 Risks for online store to go down

target team has furthermore assured the quality of and approved the results before the risk evaluation workshop. After having presented the risk evaluation criteria to the target team for the final approval, the analysis leader summarises the results of the risk identification and risk estimation by conducting a walk-through of the finalised CORAS risk diagrams.

The risk diagrams of Fig. 11.1 give an overview of the risks towards the online store that were documented in the threat diagrams of Figs. 10.3 and 10.4. The risk values annotated on the risks are derived from the consequence and likelihood estimates of the threat diagrams and the risk function that is documented in Table 8.13.

The analysis leader similarly presents the risk diagrams derived from the other finalised threat diagrams, thus conducting a walk-through of all the risks with respect to each of the direct assets.

11.2.4 Accumulating Risks

The CORAS risk analysis method is asset-driven, and a core issue during a risk analysis is to identify potential unwanted incidents that harm assets. When we were estimating risks during Step 6, we addressed unwanted incidents individually with their estimated likelihood and consequences. While this estimation yields the risk value represented by a pair of an unwanted incident and an asset in isolation, it does not take into account that several risks together may pull in the same direction. In other words, although we find that individual risks may be acceptable when they are considered in isolation, they may not be so when considered together.

11.2 Conducting the Tasks of Step 7

Table 11.5 Overview of Step 7d

Accumulating risks

- **Objective:** Calculate and document the accumulated value of several risks that contribute to the same overall risk
- **How conducted:** The individual risks that contribute to the same general risk are modelled in a combined CORAS risk diagram; the accumulated likelihoods and consequences are calculated on the basis of the respective CORAS threat diagrams
- **Input documentation:** The finalised CORAS threat diagrams with confirmed likelihoods and consequences; the finalised CORAS risk diagrams
- **Output documentation:** CORAS risk diagrams documenting the accumulated risks

Assume, for example, a risk analysis for a hospital where *Availability of health records* is one of the assets. If the health records are stored and accessed electronically, we may have unwanted incidents such as *Loss of availability of health records due to power failure* and *Loss of availability of health records due to loss of network connection*. When these two unwanted incidents are considered separately and in isolation, and their likelihoods and consequences are estimated, we may find that the risk levels are low and acceptable. However, these two incidents are actually special cases of the more general risk *Loss of availability of health records*; we may understand the two former risks as subrisks of the latter. In order to determine whether the two subrisks are acceptable or not, we need to accumulate their risk levels to derive the level of the general risk that they represent. An overview of this task is given in Table 11.5.

For a set of unwanted incidents with separate likelihoods and consequences, we accumulate their risk values into one risk value by accumulating their likelihoods on the one hand and their consequences on the other hand.

When we are accumulating likelihoods, we aggregate the likelihoods of the unwanted incidents in question. If we are operating with coarse-grained likelihood scales with intervals, it may suffice to do rough aggregations in order to determine to which likelihood interval a combined unwanted incident belongs. For a more fine-grained likelihood analysis with more precise likelihoods, we need to be more careful and also take into account possible statistical relationships between the unwanted incidents that are combined. The reader is referred to Chap. 13 for rules and guidelines for how to analyse likelihood using CORAS diagrams.

When we are accumulating the various consequences, we can obviously not add up the values. This is because the occurrence of the general unwanted incident, such as *Loss of availability of health records*, is an occurrence of one of the special cases of unwanted incidents, such as *Loss of availability of health records due to power failure* and *Loss of availability of health records due to loss of network connection*. For example, if the consequence of both of the special cases is *moderate*, then also the consequence of every occurrence of the general unwanted incident should be *moderate*. Furthermore, when combining unwanted incidents, we may also have to take into account the case in which several incidents occur simultaneously and thereby cause greater harm than individual incidents alone. The modelling and analysis of these cases are addressed in Sect. 13.3.

Fig. 11.2 Grouping of incidents that cause online store to go down

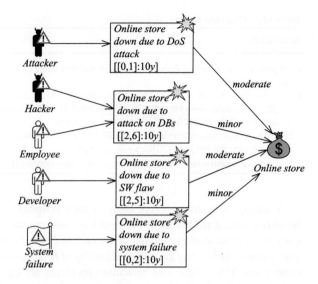

In practice, however, a rough rule of thumb is to accumulate the likelihoods and keep the consequences when we are accumulating risk values. If the consequences of the various unwanted incidents vary, we need to calculate their average or agree on an adequate general value by consulting the decision makers of the customer or representatives of the party associated with the asset in question. In each case, we nevertheless need to carefully consider the result of each accumulation to see if the combined likelihood and the combined consequence are adequate.

Example 11.2 Some of the risks that were identified with respect to the online store of AutoParts are risks for the online store to go down, as documented in the risk diagrams of Fig. 11.1. Separately, the four identified risks are all of value *low* or *very low* which represent acceptable risk levels. However, since the four risks contribute to the same overall risk, namely *Online store down*, their accumulated impact must be calculated in order to determine whether the risk for the online store to go down is acceptable or not.

The analysis leader shows the finalised threat diagrams describing the risks for the online store do go down that were documented as Figs. 10.3 and 10.4 of the risk estimation step. For the purpose of accumulating the values of the four risks, each of the likelihoods are described using as precise intervals as possible. The result is documented on-the-fly resulting in the diagram of Fig. 11.2. This diagram fragment extracts only the unwanted incidents with the threats that cause them and the assets they harm. Alternatively, the complete threat diagrams can be presented, but once the precise likelihood intervals have been estimated it suffices to focus on the unwanted incidents.

The precise likelihood interval for each of the four unwanted incidents in Fig. 11.2 is determined on the basis of the respective likelihoods of the threat scenarios leading up to them as documented in Figs. 10.3 and 10.4. By summarising the results, the accumulated likelihood is that incidents that cause the online store

11.2 Conducting the Tasks of Step 7

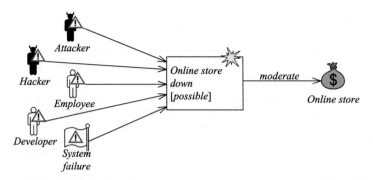

Fig. 11.3 Accumulated likelihoods and consequences for online store to go down

to go down occurs between 4 and 14 times over a period of 10 years. This interval lies in the intersection of the likelihoods *unlikely* and *possible* that were defined in Table 8.10 during Step 4. The target team agrees that the accumulated likelihood interval fits the likelihood *possible* best, which is documented accordingly as shown in Fig. 11.3.

The unwanted incident *Online store down* shown in Fig. 11.3 refers to any of the four unwanted incidents of Fig. 11.2. The consequences of these four incidents vary between the adjacent consequences *minor* and *moderate* as defined in Table 8.7. The consequence *minor* is described by the downtime interval [1 minute, 1 hour), whereas *moderate* is described by the downtime interval [1 hour, 1 day). The target team suggests using the *moderate* to represent the accumulated consequence since the average of the consequences gravitates to *moderate* rather than *minor*. The impacts relation of the diagram in Fig. 11.3 is annotated accordingly.

The result of the accumulated risk is documented in a CORAS risk diagram with the risk *OD: Online store down* annotated with the risk value *high*. This risk value is derived from the likelihood *possible* and the consequence *moderate* using the risk function defined in Table 8.13.

11.2.5 Estimating Risks with Respect to Indirect Assets

The indirect assets are assets that, with respect to the target and scope of the analysis, can be harmed only via harm to the other assets. At this point in the analysis, when we have completed the identification and estimation of risks with respect to the direct assets, we can therefore determine how these risks propagate to the indirect assets.

Through the risk analysis we have conducted so far with respect to the direct assets, we have gathered basically all the information and documentation we need in order to address the indirect assets. What remains to do is to determine the extent to which the indirect assets are harmed by the unwanted incidents via harm to the direct assets, an overview of which is given in Table 11.6.

Table 11.6 Overview of Step 7e

Estimating risks with respect to indirect assets

- **Objective:** Document and estimate risks with respect to the indirect assets
- **How conducted:** Risks towards the indirect assets are identified by considering the identified risks towards the direct assets; using the asset diagrams, determine the extent to which the indirect assets are harmed via harm to the direct assets and estimate the resulting risk levels; conducting the task requires establishing consequence scales, risk functions and risk evaluation criteria for each of the indirect assets
- **Input documentation:** The finalised CORAS threat diagrams for the direct assets; the CORAS asset diagrams
- **Output documentation:** Consequence scales, risk functions and risk evaluation criteria for indirect assets; CORAS threat diagrams and CORAS risk diagrams documenting risks towards indirect assets

In a sense, we can consider the identification and estimation of risks with respect to the indirect assets as a condensed version of the full risk analysis with respect to the direct assets. What we need to do with respect to the indirect assets, in addition to mapping the risks of direct assets to risks of the relevant indirect assets, is to establish consequence scales, define risk functions and establish the risk evaluation criteria. We conduct these tasks in the same way as we conducted the corresponding tasks for the direct assets during Step 4.

More specifically, we conduct the tasks by addressing one CORAS asset diagram in turn, and thereby each of the parties of the risk analysis in turn. The assets, both direct and indirect, were identified and documented as a separate task during Step 3.

In the same way as for the direct assets, the consequence scales can be quantitative or qualitative. In order to harmonise the analysis of the indirect assets with the analysis of the direct assets, we should if possible operate with the same kind of scale for the direct and indirect assets.

Indirect assets are often less tangible than direct assets. Typical examples are assets such as goodwill, public image, trust and reputation. One possibility that we may consider for establishing the consequence scales is to assign values to these assets on a scale, say from 1 to 10, where 10 is the most positive. Harm to the indirect assets can then be described in terms of decrease of these values. This requires, however, that the target team has a fairly good insight into the current situation in this respect, and that unwanted incidents can be assigned consequences using these terms.

Experts on the domain of marketing, public relations, information advising, and so forth may also be consulted if the customer has such expertise in-house or can bring it in as consultants. These domain experts are trained and skilled in assessing incidents that affect reputation, trust and so forth, and may also be able to describe the consequences of such incidents in monetary terms. Alternatively, the consequences can be described in terms of loss of customer base, loss of market shares, decrease in yearly turnover, and so forth.

As for the direct assets, once we have defined a consequence scale for an indirect asset, we may find that some of the unwanted incidents cannot be adequately

11.2 Conducting the Tasks of Step 7

Table 11.7 Consequence scale for reputation

Consequence	Description
Catastrophic	Full negative exposure in virtually all media
Major	Wide negative exposure in most media
Moderate	Some negative exposure in most media
Minor	Minor notice of incident in some media
Insignificant	Incident virtually unnoticed in media

described in the terms that we use for this scale. For example, if we choose to operate with a consequence scale for reputation in terms of negative media exposure, these terms are not adequate for incidents that harm reputation, yet do not reach the media. However, the purpose of the description of the various consequences is only to have a reference point in order to properly understand the meaning and severity of consequences such as *insignificant* and *major*.

Example 11.3 The analysis leader presents the asset diagram of Fig. 7.6 and explains and motivates the objective of identifying risks with respect to the indirect assets. The indirect assets of the risk analysis are *Customer satisfaction*, *Reputation* and *Sub-supplier's trust*. The consequence scales for the direct assets uses intervals and ranges from *insignificant* to *catastrophic*. The analysis leader suggests that the same values are used for the indirect assets. The analysis team has prepared tables ready for being filled in with descriptions of the consequences.

For each of the indirect assets, the analysis leader invites the target team to discuss in what terms harm to these assets should be described. For each of the consequences such as *insignificant* and *moderate* they need to come up with a suitable description that characterises the severity of asset damage.

For *Customer satisfaction*, they consider describing consequences in terms of loss of customer base or decrease in yearly revenue. As to *Reputation*, the target team agrees to describe the consequence of unwanted incidents in terms of type of negative media exposure. The values then range from *Incident virtually unnoticed in media* as *insignificant* to *Full negative exposure in all media* as *catastrophic*. For *Sub-supplier's trust*, they choose to describe the consequences by intervals of loss of sub-suppliers measured in percentage. The resulting consequence scale for *Reputation* is shown in Table 11.7. Similar scales are also defined for *Customer satisfaction* and *Sub-supplier's trust*.

Having defined the consequence scale for each of the indirect assets of each party, we next need to define the risk function. The risk function yields for each combination of a consequence and a likelihood the resulting risk level. Again, we address each of the parties in turn and define a risk function for each of the consequence scales for the indirect assets.

For a given asset, it is the associated party of the analysis that determines the severity of the risks. The risk function can therefore only be defined by consulting

Table 11.8 Risk function for reputation

		Consequence				
		Insignificant	Minor	Moderate	Major	Catastrophic
Frequency	Rare					
	Unlikely					
	Possible					
	Likely					
	Certain					

the relevant party, or by consulting decision makers or other personnel that can speak on behalf of the party.

We conduct the definition of the risk function as we did for the direct assets during Step 4, and therefore refer to Chap. 8 for the details.

Example 11.4 The AutoParts target team, and particularly the decision maker, needs to come up with a risk function for each of the three consequence scales for the indirect assets. Since both the consequence scales and the likelihood scale use intervals, they choose to use risk matrices for representing the risk function, just as they did for the direct assets. For each consequence scale in turn, the analysis leader displays the risk matrix with blank entries and invites the target team to fill the entries with the colours green, yellow, orange and red, where the colours represent the risk levels *very low*, *low*, *high* and *very high*, respectively.

The analysis leader first brings the attention to the extremes since these cases are often the easiest to determine. Concerning the consequence *catastrophic*, they agree that for all likelihoods, except perhaps *rare*, the risk value is *very high* for the indirect asset *Reputation*. The likelihood *rare* refers to a frequency of less than once per ten years. Although this is very rare compared to time horizon of AutoParts, a catastrophic incident with respect to reputation is very severe. After some discussion, they decide to assign the matrix entry in question the risk value *high*.

Reasoning in a similar manner they fill in the matrix cell by cell, focusing on where the risk level shifts from one value to another. The resulting matrix for the reputation asset is shown in Table 11.8.

The exercise is repeated for the indirect assets of *Customer satisfaction* and *Sub-supplier's trust*, the result of which is documented in separate risk function matrices.

Also deciding the risk evaluation criteria, which is our next task, is conducted in the same way for the indirect assets as for the direct assets. The criteria is a specification of the risk level the parties of the analysis are willing to accept, and they must be decided by the parties themselves or someone that can speak on behalf of them. For further details, we refer to Chap. 8 where the task is described for the direct assets.

Example 11.5 For the direct assets of the analysis, the target team defined three different risk matrices that each maps combinations of a consequence and a likelihood to a risk value, where the risk values are *very low*, *low*, *high* and *very high*. These values were then interpreted uniformly for all of the assets in terms of sever-

11.2 Conducting the Tasks of Step 7

ity, which resulted in a combined representation of the risk evaluation criteria for all the assets. These risk evaluation criteria are documented in Table 8.16 in Chap. 8.

The analysis leader displays these criteria to the target team and asks whether they are suitable for the indirect assets also. If not, they may choose either to reconsider and adjust the risk functions for the indirect assets, or to operate with separate risk evaluation criteria. The target team agrees, however, that the criteria are suitable also for the indirect assets. Risk against the indirect assets of the values *very low* and *low* are therefore held as acceptable, whether risks of the values *high* and *very high* must be evaluated for possible treatment.

When we have finalised the consequence scales, the risk functions and the risk evaluation criteria for the indirect assets, we turn to the identification and estimation of the risks with respect to these assets.

In principle, we have already completed the identification of the risks with respect to the indirect assets through the identification of the risks with respect to the direct assets. We therefore use the finalised CORAS threat diagrams in combination with the CORAS asset diagrams to derive the risks against the indirect assets. More specifically, a risk against a direct asset is also a risk against all the indirect assets that are harmed via harm to the direct asset. During Step 3, we documented the indirect assets that are harmed through direct assets using the harms relation in the asset diagrams.

What we have yet to do is for each asset diagram to define the mapping from consequences of the direct assets to the consequences of the indirect assets for each relevant unwanted incident. Importantly, we must take into account that the harms relation is transitive. In a risk analysis for a hospital, for example, we may have the assets *Confidentiality of health records*, *Hospital's reputation* and *Patients' trust in hospital*. If harm to *Confidentiality of health records* implies harm to *Hospital's reputation*, and harm to *Hospital's reputation* in turn implies harm to *Patients' trust in hospital*, then harm to the former implies harm to the latter because of transitivity.

In some cases, depending on the unwanted incident, harm to one asset that leads to harm to a second asset is less severe for the second asset. In other cases, it is more severe. For a hospital case, for example, we may have that harm to *Confidentiality of health records* and harm to *Integrity of health records* both lead to harm to *Hospital's reputation*. However, a very severe harm to integrity may imply a less severe harm to reputation, whereas a less severe harm to confidentiality may imply a more severe harm to reputation. We must therefore, for each pair of an unwanted incident and an indirect asset, address one direct asset in turn and determine how a consequence with respect to the latter maps to a consequence with respect to the former.

We moreover need to take into account that an indirect asset may be harmed by an unwanted incident via several direct assets. Assume, for example, that harm to the direct assets *Confidentiality of health records* and *Integrity of health records* both lead to harm to the indirect asset *Hospital's reputation*. For an unwanted incident that impacts both these direct assets, we must first determine how much the indirect asset is harmed via harm to each of the direct assets. Subsequently, we must combine these two consequences in order to determine the full consequence of the unwanted incident in question regarding the indirect asset.

Fig. 11.4 Unwanted incident relevant for the indirect asset of reputation

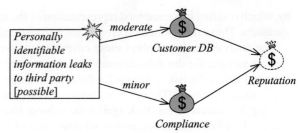

We use the finalised threat diagrams that document the risks with respect to the direct assets to identify and estimate the risks with respect to the indirect assets. Recall that the risk with respect to the indirect assets are only with respect to the target and scope of analysis. There may therefore generally be additional risks against the indirect assets, but since the causes of these additional risks are outside the scope of the analysis we do not document and analyse them here.

The threat diagrams that we made during the risk identification and estimation give the full risk picture with respect to the direct assets, and therefore implicitly also with respect to the indirect assets. In order to identify the relevant risks against an indirect asset, we first use the CORAS asset diagram in which this indirect asset is defined to identify the direct assets via which it may be harmed. Then we use the CORAS threat diagrams in which these direct assets occur to identify the unwanted incidents of relevance for the indirect asset; in other words, we identify the unwanted incidents with an impacts relation to the identified direct assets.

The likelihoods remain the same since these are the likelihoods of the unwanted incidents to occur, irrespective of which assets being harmed. The consequences are determined by the target team by estimating the extent to which the indirect assets are harmed via the direct assets for each unwanted incident in question. Once we have identified the risks with respect to the indirect assets and estimated the risk values, we should quality check the result. In case the result seems erroneous or does not match reality as perceived by the target team, we may have to reconsider the asset diagrams and the relations between the assets, or we may have to reconsider the consequence estimates.

Example 11.6 In order to identify and estimate the risks with respect to the indirect assets, the analysis team has already identified the relevant unwanted incidents for each indirect asset by first identifying the direct assets via which it may be harmed, and then the unwanted incidents that may harm these assets.

For each pair of an unwanted incident and indirect asset, the analysis secretary displays to the target team a separate diagram fragment that documents all possible ways in which the unwanted incident may harm the indirect asset in question via harm to direct assets.

In the case of the indirect asset *Reputation* and the unwanted incident *Personally identifiable information leaks to third party*, the target team is presented the fragment of the threat diagram shown in Fig. 11.4. This unwanted incident is documented in Fig. 10.6, and the full threat diagram is shown in Fig. 10.2.

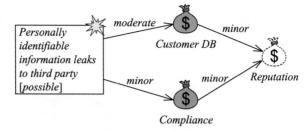

Fig. 11.5 Harm to indirect asset of reputation via direct assets

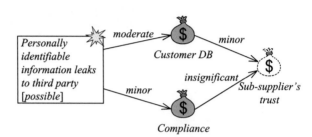

Fig. 11.6 Harm to indirect asset of sub-supplier's trust via direct assets

The analysis leader has the target team to discuss how much the indirect asset is harmed by the unwanted incident via harm to the direct assets. The consequence with respect to the confidentiality of the customer database is *moderate* as annotated on the relation from the unwanted incident to the direct asset *Customer DB*. The target team explains that the extent to which reputation is harmed via the confidentiality breach depends on the sensitivity of the data, and whether the data can and will be misused. Since confidentiality breaches often do not have further consequences they decide that *minor*, as defined in Table 11.7, most adequately represents the consequence, although the consequence in some cases is only *insignificant*. The analysis secretary annotates the diagram accordingly as shown in Fig. 11.5. The harm of the unwanted incident to *Reputation* via *Compliance* is estimated in the same way. Also this consequence is estimated to *minor*.

The relations between the assets in the asset diagram of Fig. 7.6 shows that all three indirect assets may be harmed by the unwanted incident *Personally identifiable information leaks to third party* via one or both of the direct assets *Customer DB* and *Compliance*. The analysis leader therefore in turn has the target team to address also the indirect assets *Customer satisfaction* and *Sub-supplier's trust* regarding this unwanted incident.

The two diagram fragments of Figs. 11.5 and 11.6 show the result of estimating the harm to, respectively, *Reputation* and *Sub-supplier's trust*. A similar diagram fragment for the unwanted incident in question is made for the indirect asset *Customer satisfaction*.

In order to determine the risk that a given pair of an unwanted incident and an indirect asset represents, we need to accumulate the consequences via all the relevant direct assets. This accumulated consequence then represents the harm of the unwanted incident in question on the indirect asset in question. Once we have ac-

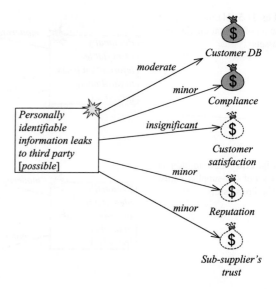

Fig. 11.7 Consequence of unwanted incident on indirect asset reputation

cumulated the consequence for each identified pair of an unwanted incident and an indirect asset, we document the result by placing the indirect assets with the estimated consequence into the finalised threat diagrams for the direct assets. We thereafter update the risk diagrams accordingly to document the risks with respect to the indirect assets.

Example 11.7 The analysis leader shows the diagram fragment of Fig. 11.5 and explains to the target team that they need to estimate the full consequence of the unwanted incident *Personally identifiable information leaks to third party* regarding the indirect asset *Reputation* by considering the combined harm via the two direct assets *Customer DB* and *Compliance.* The harm via the two direct assets together is of course higher than the individual consequences on the harms relations, but the target team agrees that the consequence *minor* is adequate also for the combination.

The analysis secretary documents the result in the relevant threat diagram by drawing a relation from the unwanted incident directly to the indirect asset *Reputation* and annotating the relation with the estimated consequence as shown by the diagram fragment in Fig. 11.7. The diagram also shows the result for the other two indirect assets regarding this unwanted incident. For the purpose of explicitly showing the consequences of the unwanted incidents on the indirect assets, the analysis secretary omits the harms relations from the direct to the indirect assets.

For each pair of identified unwanted incident and indirect asset, the same procedure is carried out. The result regarding the unwanted incident *Online store down*, for example, is shown in Fig. 11.8. All three indirect assets may be harmed via harm to the direct asset *Online store.* The consequence of the unwanted incident is therefore estimated for each of the indirect assets. The full documentation regarding this unwanted incident is given in the threat diagrams of Figs. 10.3 and 10.4.

Fig. 11.8 Consequence of unwanted incident on indirect assets

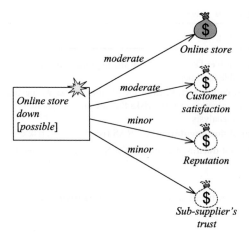

Fig. 11.9 Risk with respect to reputation

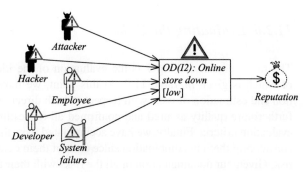

Each pair of an unwanted incident and an indirect asset that is harmed by the unwanted incident via a direct asset represents a risk with respect to the indirect asset. Having completed the estimation of the consequences of the unwanted incidents on the indirect assets, we are finally ready to estimate the risks. To this end, we use the threat diagrams completed with the indirect assets together with the risk functions defined for the indirect assets. This is conducted in the same way as the risk estimation with respect to the direct assets as described in Step 6.

Example 11.8 In order to derive the resulting risk levels regarding the indirect assets, the analysis leader refers to the risk functions for these assets. For the indirect asset *Reputation*, for example, the likelihood of the unwanted incident *Online store down* in Fig. 11.8 is *possible*, whereas the consequence is *minor*. By the risk function defined in the matrix of Table 11.8, this yields the risk value *low*.

The analysis leader finally gives an overview of all the risks with respect to the indirect assets with the risk values using CORAS risk diagrams. This is exemplified in the risk diagram of Fig. 11.9 for the indirect asset *Reputation*. Since the abbreviation *OD* has already been used to name the risk represented by the unwanted incident *Online store down* with respect to the direct asset *Online store*, the analysis leader uses the abbreviations *OD(I1)*, *OD(I2)* and *OD(I3)* for the risks with respect

Table 11.9 Overview of Step 7f

Evaluating the risks

- **Objective:** Decide which of the identified risks are acceptable and which must be evaluated for possible treatment
- **How conducted:** Each of the identified risks with estimated risk value is compared against the risk evaluation criteria; risks that are not acceptable are considered for treatment in the next step of the analysis
- **Input documentation:** CORAS risk diagrams; risk evaluation criteria
- **Output documentation:** CORAS risk diagrams annotated with risk evaluation result

to the indirect assets *Customer satisfaction*, *Reputation* and *Sub-supplier's trust*, respectively.

11.2.6 Evaluating the Risks

The final subtask of Step 7 is the evaluation of the identified risks, an overview of which is given in Table 11.9. At this point, we have completed the identification and estimation of risks with respect to both direct and indirect assets. We have furthermore quality assured and confirmed all our estimations, as well as the risk evaluation criteria. Finally, we have accumulated risks that can only be evaluated by considering them in combination since each of them contributes to the same overall risk. Given our documentation of all the risks with their risk level, together with the risk evaluation criteria, the risk evaluation is quite straightforward.

Because we have documented all the risks using CORAS risk diagrams, we conduct the risk evaluation by simply comparing the risk values against the risk evaluation criteria. This can be conducted as a walk-through of the risk diagrams where we annotate the risks with *acceptable* or *unacceptable* depending on the risk evaluation criteria. The risks that are deemed unacceptable need to be evaluated for possible treatment.

We document the result using CORAS risk diagrams for both the direct and indirect assets where we replace the risk levels with the annotation *acceptable* or *unacceptable*. This gives an overview of the result of the evaluation, and shows which risks must be considered for mitigation during the treatment identification of Step 8. If we have used the matrix format for defining the risk function, we may also conveniently represent the result of the risk evaluation by plotting the risks into the matrices. This shows not only which of the risks are acceptable and which are not, but also the level of each of the risks.

Example 11.9 The analysis leader explains that after the completion of the identification and estimation of risks with respect to the indirect assets, they are ready to conclude by evaluating the risks and determine which of them must be considered for treatment. Since all the results are in place, she conducts the risk evaluation as

Fig. 11.10 Evaluation of risks towards inventory DB

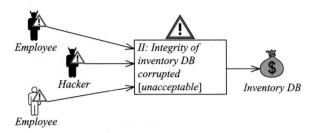

Fig. 11.11 Evaluation of risks towards customer DB and compliance

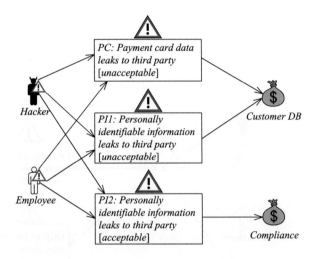

a presentation of a summary of the results. With reference to the risk evaluation criteria for both the direct assets and the indirect assets, she presents each of the finalised CORAS risk diagrams in turn. For each of the risks she indicates whether it is acceptable or not, and the analysis secretary annotates the diagram accordingly.

The CORAS risk diagram of Fig. 11.10 shows that the risk *II: Integrity of inventory DB corrupted* is unacceptable. The threat diagram of Fig. 9.7 shows the full documentation of this risk that during the risk evaluation were assigned the likelihood *possible* and the consequence *moderate*. The risks with respect to the indirect assets regarding the unwanted incident *Integrity of inventory DB corrupted* are presented similarly with a separate risk for each indirect asset.

The risk diagram of Fig. 11.11 shows two risks towards *Customer DB* and one risk against *Compliance*. The former two are both unacceptable as concluded from the risk identification and risk estimation documented in Fig. 10.2.

Figure 11.12 shows the evaluation of the accumulated risks for the online store to go down. The analysis leader explains that if the separate risks for the online store to go down were evaluated in isolation, they would all be held as acceptable according to the risk evaluation criteria. The accumulated risk level is, however, not acceptable, and these risks must therefore be considered for treatment.

Similarly for all of the other identified and estimated risks, both with respect to the direct and the indirect assets, the analysis leader displays the risk diagram

Fig. 11.12 Evaluation of risks towards online store

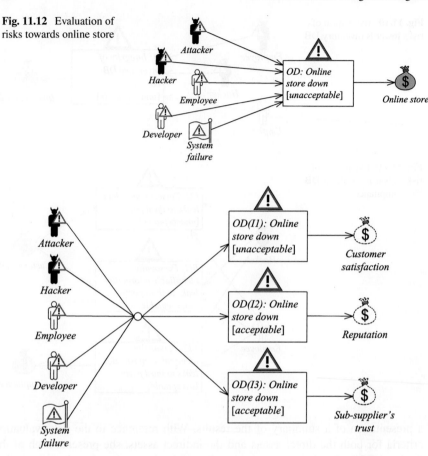

Fig. 11.13 Evaluation of risks towards indirect assets

and shows the result of the risk evaluation. The evaluation of risks with respect to the indirect assets is exemplified by the risk diagram of Fig. 11.13. The analysis secretary chooses to use a junction point on the initiates relations between the threats and the risks to ease the readability of this many-to-many relation.

Example 11.10 The CORAS risk diagrams are suitable for providing a brief overview of the results of the risk evaluation. However, they only indicate whether the identified risks are acceptable or not. The analysis leader therefore decides also to summarise the results by plotting the risks into the risk matrices. Instead of dividing the matrices into partitions according to risk levels, however, she presents bisected matrices. The matrices are now partitioned according to the risk evaluation criteria. The analysis leader uses the colour green to indicate the acceptable risks, and the colour red to indicate the risks that must be further evaluated.

The matrix depicted in Table 11.10 shows the evaluation of the risks towards the inventory database and the customer database that were documented in the risk

11.3 Summary of Step 7

Table 11.10 Evaluating risks towards databases

	Consequence				
Frequency	Insignificant	Minor	Moderate	Major	Catastrophic
Rare					*PC*
Unlikely					
Possible			*APU*		
Likely					
Certain					

diagrams of Figs. 11.10 and 11.11, respectively. An advantage of this representation is that it shows how far an unacceptable risk is from the acceptable level.

11.3 Summary of Step 7

Step 7 of the CORAS risk analysis method is summarised by the overview presented in Table 11.11.

Table 11.11 Summary of Step 7

Risk evaluation using risk diagrams

Analysis tasks:

- Likelihood and consequence estimates should be confirmed or adjusted
- If needed, final adjustments of the acceptable risk levels should be made
- An overview of the risks are given in risk diagrams
- Risks that contribute to the same overall risk are accumulated
- Risks with respect to the indirect assets are identified and estimated
- The identified risks are evaluated

People that should participate:

- Analysis leader (required)
- Analysis secretary (required)
- Representatives of the customer:
  - Decision makers (required)
  - Technical expertise regarding the target (optional)
  - Users (optional)

Modelling guideline:

- Risk diagrams with evaluation results:
  1. For the direct assets, use the risk diagrams from Step 6 and replace the risk value with *acceptable* or *unacceptable*, depending on the result of the risk evaluation
  2. For the indirect assets, extend the risk diagrams for the direct assets by adding the implied risks for the indirect assets
  3. Give each indirect risk a unique name or id but keep the risk description; annotate the risks with *acceptable or unacceptable*, depending on the result of the risk evaluation

Chapter 12
Risk Treatment Using Treatment Diagrams

Now that we have completed the evaluation of the risks and determined which of them are not acceptable for the parties of the analysis, we move to the final step of the analysis, which is risk treatment. Our objective with Step 8 is to identify strategies and action plans the implementation of which reduces risks to an acceptable level. Since the implementation of treatments may be costly, we furthermore conduct a cost-benefit analysis so as to identify cost-effective treatments.

Step 8 is typically organised as a workshop similar to the risk identification workshop of Step 5, and the recommended target team members are also the same. That is, we conduct the treatment identification as a structured brainstorming involving a target team consisting of personnel of various backgrounds and expertise with the aim of identifying as many treatments as possible. A three hours workshop should suffice for conducting the tasks of the risk treatment.

12.1 Overview of Step 8

Based on the risk diagrams from Step 7 annotated with the result of the risk evaluation, we pick out all the threat diagrams that document the unacceptable risks with respect to both direct and indirect assets. These threat diagrams show all the threats, threat scenarios and vulnerabilities that in combination give rise to the unacceptable risks, and thereby also show where we can direct measures for risk treatment.

Table 12.1 gives an overview of Step 8. The first subtask is to group various risks that may be treated by the same measures. This may, for example, be risks that are caused by the same threats or that arise because of the same vulnerabilities. We should conduct this subtask offline as part of the preparations for Step 8. This grouping of risks may facilitate the identification of the most effective and beneficial treatments, and also serve to structure the treatment identification brainstorming. As part of the preparations for the treatment identification, we furthermore, based on all the finalised threat diagrams and the risk evaluation, make preliminary treatment diagrams that serve as a starting point for the brainstorming. These preliminary

Table 12.1 Overview of Step 8

Risk treatment using treatment diagrams

- **Objective:** Identify cost effective treatments for the unacceptable risks
- **How conducted:** Conducted as a brainstorming session involving a target team consisting of personnel of various backgrounds; treatments are identified by a walk-through of the threat diagrams that document the unacceptable risks and their causes
- **Input documentation:** CORAS risk diagrams and CORAS threat diagrams documenting the unacceptable risks; preliminary treatment diagrams prepared by the analysis team
- **Output documentation:** CORAS treatment diagrams documenting the identified treatments for the risks with respect to direct and indirect assets
- **Subtasks:**
  a. Grouping of risks
  b. Treatment identification
  c. Treatment evaluation

treatment diagrams may include treatment suggestions that we can use to stimulate and bring forward the discussions if necessary.

The second subtask is the identification of possible treatments for the unacceptable risks. We conduct the treatment identification as a structured brainstorming organised as a walk-through of the preliminary treatment diagrams that we have prepared. For each of the diagrams, we have the target team to discuss how and where treatments can be implemented in order to mitigate the risks. As we proceed, we document the suggestions by inserting treatment scenarios into the diagrams. As a result, we gradually and on-the-fly make the treatment diagrams that document the results of the treatment identification task.

The third subtask is to evaluate the identified treatments with respect to cost and benefit. During the treatment identification, we focus on identifying as many strategies and plans for mitigating risks as possible without worrying much about the feasibility or cost of implementing them. Treatments may, however, be very costly, and if the cost of implementing them is higher than the benefit, they should obviously not be implemented. Alternatively, we may try to identify other and cheaper treatment strategies.

As a result of the treatment identification and evaluation, we come up with a recommendation for risk treatment.

12.2 Conducting the Tasks of Step 8

In this section, we go through the tasks of treatment identification and treatment evaluation in more detail, explaining how to conduct them and how to document the results.

12.2 Conducting the Tasks of Step 8

Table 12.2 Overview of Step 8a

Grouping of risks

- **Objective:** Identify several risks that may be treated by the same measures in order to facilitate identification of the most efficient treatments
- **How conducted:** Conducted by the analysis team as part of the preparations for Step 8; by going through the threat diagrams showing unacceptable risks, common elements of the various diagrams such as threats, vulnerabilities and threat scenarios are identified; such common elements may indicate that risks also may have common treatments; at the same time the analysis team prepares preliminary treatment diagrams documenting all the unacceptable risks and ready to be filled in with treatment scenarios
- **Input documentation:** CORAS threat diagrams showing only unacceptable risks and their causes
- **Output documentation:** An overview of common elements with respect to which risk may be grouped; preliminary treatment diagrams ready to be filled in with treatment scenarios

12.2.1 Grouping of Risks

In order to come up with a recommended treatment plan, we aim at identifying the treatments that give most value for money. A good strategy is therefore to search for treatments that each simultaneously mitigates several risks.

As part of the preparations for the risk identification of Step 5, we chose a categorisation of threat diagrams in order to structure the risk identification process. During a risk analysis, we usually focus on these categories separately. The various threat diagrams we make, however, are often overlapping in the sense that certain threats, vulnerabilities and threat scenarios reappear in several threat diagrams and for several categories of threat diagrams. Such overlaps indicate that various risks to a greater or lesser extent arise due to the same causes. This may also indicate that various risks can be treated by the same measures.

Before we begin the treatment identification, we therefore carefully go through the threat diagrams and identify common elements. Subsequently, we use the list of common elements to group the threat diagrams, and thereby also the risks. An overview of this task is given in Table 12.2.

It may of course be that the various common elements result in different groupings of the identified risks. A given threat, for example, may be common for one set of risks, whereas a given vulnerability yields a different set. For the list of common elements we identify, we therefore identify various possible ways of grouping the risks. During the treatment identification, we can use the list to structure the brainstorming and to present various possible viewpoint on the risk picture from a treatment perspective. Most importantly, however, we use this grouping of risks to identify the treatments that give the most benefit for the least cost.

Example 12.1 Before the meeting with the AutoParts target team, the analysis team prepares by setting up a plan for how to structure the brainstorming. They furthermore try to see if there are any obvious treatment strategies that should be considered and discussed by the target team.

In particular, they go through the documentation to identify the threats, vulnerabilities and so on that are more prevalent than others. An element that reoccurs in several diagrams may indicate a feature or aspect of the target that has much significance on the risk picture since it is relevant for several risks.

One of the obvious strategies is to consider the assets of the inventory database and the customer database together when identifying treatments, since the risk picture is basically the same for both of them.

Some of the risks that were identified are documented in the threat diagrams of Figs. 9.6 through 9.10. One of the elements that reoccur the most is the threat of a hacker, and the analysis team therefore makes a list of the risks and the threat diagrams where this threat is relevant. They also look for vulnerabilities that may be exploited by hackers to cause several risks. The vulnerability *Insufficient virus protection*, for example, may be a good target for risk treatment.

The analysis team similarly groups the risks with respect to other common elements and features, and prepares suggestions for treatment strategies to be proposed for discussion at the meeting.

The basis for the treatment identification is the full documentation of the risks that are not acceptable. The finalised threat diagrams from Step 7 for both direct and indirect assets implicitly provide this documentation by the likelihood and consequence estimates for all the unwanted incidents. During the treatment identification, however, we wish to present to the target team the full documentation of the risk picture where all the unacceptable risks are explicit.

We do this by replacing each unwanted incident in the threat diagrams with the risks that the unwanted incident represents. More precisely, each unwanted incident is replaced by one risk for each impacts relation it has towards an asset. We moreover remove all the acceptable risks to ensure that the risk identification focuses on the most urgent issues and on the threats, threat scenarios, vulnerabilities and so forth that show how unacceptable risks arise. We also remove all the diagram elements that are not related to the unacceptable risks. The risk diagrams from Step 7 document the unacceptable risk and we use these as input when identifying the risks to be removed from the diagrams.

The resulting diagrams are preliminary treatment diagrams each of which documents unacceptable risks and is ready for being filled in with treatment scenarios during the treatment identification. The diagrams are syntactically correct treatment diagrams, only without occurrences of treatment scenarios.

Example 12.2 As part of the preparations for Step 8, the analysis team goes through all the finalised risk diagrams and all the finalised threat diagrams from Step 7. For each of the acceptable risks, as documented by the risk diagrams annotated with the results of the risk evaluation, they clean up the corresponding threat diagram by removing elements that are related to the acceptable risks only. What remains is then the documentation of the risks for which treatments should be identified and evaluated, with respect to both direct and indirect assets.

The risk diagram of Fig. 11.10, for example, shows that the risk *II: Integrity of inventory DB corrupted* is unacceptable. The threat diagram of Fig. 9.7 documents this risk in detail, the consequence and likelihood of which were estimated to *moderate* and *possible*, respectively. Since the threat diagram does not document any acceptable risks, none of the diagram elements are removed. The diagram represents only one risk that the analysts insert in the place of the unwanted incident. The resulting diagram is shown in Fig. 12.1. The annotation indicating that the risk is unacceptable is omitted since the acceptable risks are removed.

The analysis team prepares corresponding preliminary treatment diagrams based on the unacceptable risks of all the other threat diagrams. This includes the risks with respect to both the direct and the indirect assets. They bring the resulting diagrams to the treatment identification workshop, ready to be used as input to the brainstorming session. At the same time, the diagrams serve as starting points for making the treatment diagrams that will document the results of the treatment identification.

12.2.2 Treatment Identification

During the treatment identification, an overview of which is given in Table 12.3, we aim at identifying as many treatments as possible. Our strategy for treatment identification is very similar to the strategy for risk identification we followed during Step 5; we conduct a structured brainstorming involving a target team consisting of personnel with various backgrounds, expertise, interests and so forth. This ensures that we get several different viewpoints to the problem, and that we are able to identify more treatments than by involving a more homogeneous group. The same personnel that participated at the risk identification workshop can with advantage participate also at the treatment identification workshop, since they have good insight into the problems. We conduct the brainstorming as a structured walk-through of all the preliminary treatment diagrams that show the unacceptable risks and how they may arise.

> **Remark** A risk *treatment* is an appropriate measure to reduce risk level.

The level of a risk is determined by its likelihood and the consequence for an asset. Therefore, when we seek to bring down the level of a risk from an unacceptable level to an acceptable level, we need to find means to reduce its likelihood and/or consequence. When we address a particular unacceptable risk, we need to carefully consider all the relevant threats, vulnerabilities and threat scenarios and try to identify means to remove these elements or to reduce their significance or impact.

The options for treating risks by reducing likelihood and reducing consequence are two of several so-called treatment categories. A treatment category is collection of treatments that each represents a kind of approach to the reduction or removal of risks.

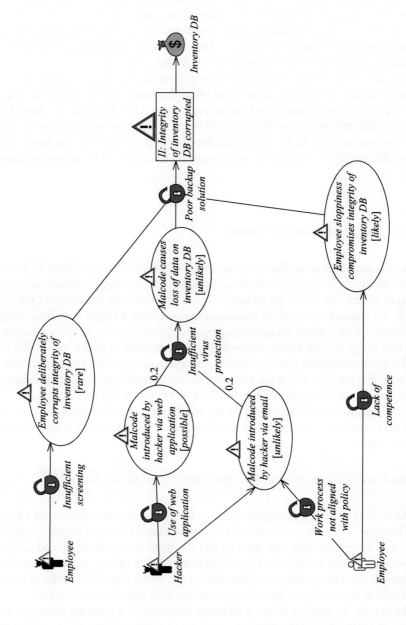

Fig. 12.1 Preliminary treatment diagram showing unacceptable risk towards inventory DB

12.2 Conducting the Tasks of Step 8

Table 12.3 Overview of Step 8b

Treatment identification

- **Objective:** Identification of possible treatments for the unacceptable risks
- **How conducted:** Conducted as a brainstorming involving a target team consisting of personnel of various backgrounds; preliminary treatment diagrams showing only the unacceptable risks are displayed, and possible treatments are identified through brainstorming; risks may be reduced or removed by reducing the likelihood and/or consequence of unwanted incidents, by transferring the risk to another party (for example through insurance or outsourcing), or by avoiding the activity that leads to the risk
- **Input documentation:** Preliminary CORAS treatment diagrams showing only unacceptable risks and their causes ready to be filled in with treatment scenarios
- **Output documentation:** CORAS treatment diagrams showing the identified treatments for the unacceptable risks

Remark A *treatment category* is a general approach to treating risks. The categories are avoid, reduce consequence, reduce likelihood, transfer and retain.

In addition to the treatment categories of consequence reduction and likelihood reduction, we can seek for risk treatments in the categories we refer to as "avoid", "transfer" and "retain". The avoid category refers to risk treatment by simply refraining from carrying out the activities that may lead to risks. The risk of installing Trojan horses, for example, can be reduced by avoiding downloading software from the Internet. The transfer category refers to risk treatment by transferring the risk, or some of the risk, to a third-party. In a partnership between two enterprises, for example, certain risks can be shared by agreements in legally binding contracts. Buying insurance for assets is another typical example of risk transfer. The retain category refers to the decision to accept risks. This may be relevant when sufficiently reducing the risk, or avoiding the activities that may lead to them, is not an option. Accepting risks, for example by offering Internet services or entering a business, may be rational if the potential gain or opportunities, such as the possibility of earning profits, are sufficiently high as compared to the risk level.

As an introduction to the brainstorming session, we present to the target team the various treatment categories and explain how we can go by to mitigate risks. It is important that we stress that all the elements of the relevant preliminary treatment diagrams must be considered. When we seek treatments to risks, we are often inclined to focus on vulnerabilities, but there may also be good and adequate treatments that can be directed towards threats or threat scenarios, and also directly towards the risks. This broad approach to the risk picture may ensure that we identify treatments that otherwise would not be thought of. Most of the treatments that the target team comes up with are nevertheless likely to be related to the vulnerabilities since vulnerabilities are properties or aspects that opens for threats to harm assets.

In addition to preparing the target team by this introduction, we should ourselves prepare for the workshop by coming up with treatment suggestions. The suggestions can be based on our own experience from previous analyses, and we can also consult domain experts, best practice descriptions, and so forth. We may present these suggestions to trigger discussions, or we may propose them as recommendations if we believe that they should be considered for implementation.

Although risk treatment is a separate step of the analysis, we often experience that the target team comes up with and mentions various treatment options throughout all the steps of the analysis. This is because much of the analysis is based on free and open discussion, and discussions about risks often lead to potential treatments being mentioned and proposed. To the extent that treatments are mentioned throughout the analysis, we should write them down on the side and bring them forward during this step for further and more detailed discussion.

Example 12.3 The analysis leader from CORAS Risk Consulting presents in turn each of the preliminary treatment diagrams showing the unacceptable risks. For each of the diagrams she first calls attention to the risks, before she leads the target team through each of the elements of the diagram, roughly one path at the time.

For the diagram of Fig. 12.1, the target team is foremost preoccupied with the hacker and how to reduce the possibility of this threat to cause risks. The very use of the web application is an important vulnerability in this respect, but its removal is obviously no option for AutoParts. The target team agrees that any introduction of malcode is intolerable and proposes as a treatment that the virus protection should always be state of the art and up to date. The analysis secretary inserts the treatment scenario and relates it to the appropriate diagram element, namely the vulnerability *Insufficient virus protection*, as depicted in Fig. 12.2.

An efficient virus protection will also mitigate the risks represented by malcode that is introduced because of employees being careless about or unaware of malicious contents in emails. The target team is aware that carelessness or sloppiness among employees, as well as insufficient security awareness, can be the source of significant incidents. Irrespective of the virus protection, some of the target team members therefore argue that AutoParts should have good routines for increasing the security awareness among the employees. This is therefore suggested as a separate treatment, as depicted in Fig. 12.2. The treatment is relevant both with respect to adherence to AutoParts' security policy, and with respect to avoiding accidental corruption of data. The treatment is therefore related to the two vulnerabilities *Work process not aligned with policy* and *Lack of competence*.

The data security officer of AutoParts explains that recovery from loss of integrity of the databases often can be ensured by performing rollbacks. Additionally, they do have backup solutions to recover from such incidents. He nevertheless agrees that maintaining reliable backup solutions should always be a priority, and that this should be considered as a recommended treatment. As shown in Fig. 12.2, this treatment is inserted into the diagram accordingly.

Although one can never know for sure, the target team members find it hard to believe that AutoParts has malicious insiders. This is reflected by the likelihood *rare* of the threat scenario *Employee deliberately corrupts integrity of inventory DB*. It is therefore decided to not take any measures in this respect.

12.2 Conducting the Tasks of Step 8

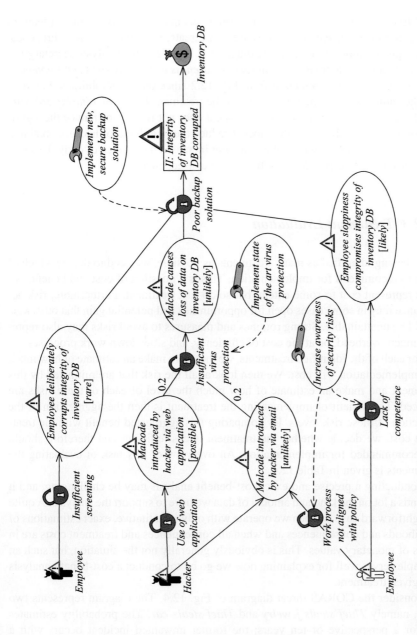

Fig. 12.2 Treatments of risks with respect to integrity of inventory DB

The diagram of Fig. 12.3 depicts some of the identified treatments of risks with respect to the online store. As shown by the diagram, these treatments additionally represent treatments of risks with respect to the indirect asset of customer satisfaction.

Different risks with respect to different assets may often have common features to a grater or lesser extent in the sense that threats, vulnerabilities and threat scenarios are common. As regards the diagram of Fig. 12.3, the analysis secretary immediately inserts the treatments *Increase awareness of security risks* and *Implement state of the art virus protection* from Fig. 12.2 since the vulnerabilities they treat are the same. Worth noticing is also that the treatment *Install monitoring software* is related to a threat scenario. This treatment does not remove or reduce the significance of vulnerabilities, neither does it reduce the likelihood of hackers obtaining access to databases. However, the treatment increases the chance of early detection of intrusion and thereby the reduction of the consequence.

12.2.3 Treatment Evaluation

After having identified as many treatments as possible, we need to decide which of them to recommend for implementation. This is a question of cost and benefit. All risks represent cost to some extent, but so do risk treatment. Furthermore, risk acceptance may in some cases open for opportunities and potential gain that otherwise would be unattainable. Strong routines and measures to avoid risks may also represent much overhead, be at the cost of efficiency, and slow down work processes.

For each of the identified treatments, we therefore make an estimate of how much the implementation will cost. We then look at all the risk that are mitigated by this treatment and make an estimate of how much the level of each of these risks are reduced. The benefit of implementing the treatment is then the aggregation of the reduction of these risk levels. By comparing this aggregated benefit with the treatment cost, we decide whether the treatment is cost-efficient and therefore should be recommended for implementation. An overview of this task of evaluating the treatments is given in Table 12.4.

Conducting a precise and exact cost-benefit analysis may be challenging, and it depends a lot on the kind and amount of data we have to support the analysis. A quite straightforward case is when we operate with only quantitative, exact estimations of likelihoods and consequences and when all consequences and treatment costs are in terms of monetary values. This is obviously generally not the situation, but such an example serves well for explaining how we go by to conduct a cost-benefit analysis of a given treatment.

Consider the CORAS threat diagram of Fig. 12.4. The diagram represents two risks, namely *Thief steals jewelry* and *Thief steals car*. The probability estimates are for a perspective of ten years; the former unwanted incident occurs with a probability of 0.2 over a period of ten years, and the latter with a probability of 0.1. The consequences are purely monetary, for example given in Euro. The level

12.2 Conducting the Tasks of Step 8

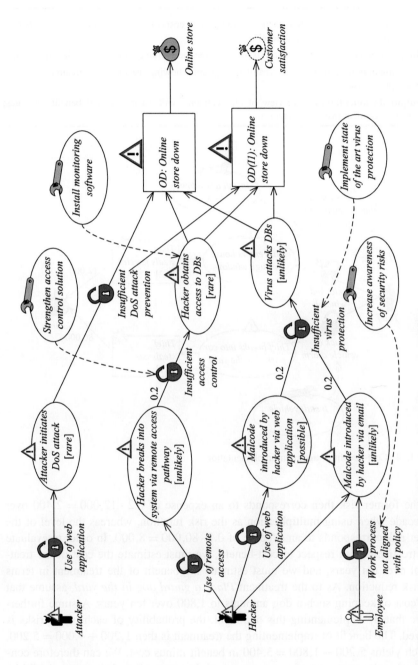

Fig. 12.3 Treatments of risks with respect to the online store and customer satisfaction

Table 12.4 Overview of Step 8c

Treatment evaluation

- **Objective:** Identify the treatments the implementation costs of which are lower than their benefit with respect to reducing the risk level
- **How conducted:** Cost-benefit analysis to determine the usefulness of the identified treatments
- **Input documentation:** CORAS treatment diagrams showing the identified treatments for the unacceptable risks
- **Output documentation:** Overview of all treatments with their cost and benefit estimates; CORAS treatment overview diagrams

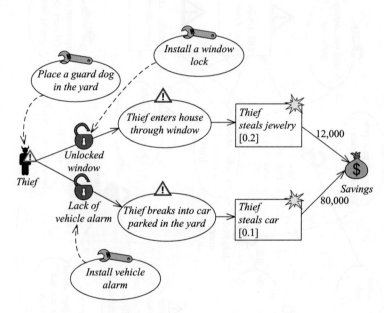

Fig. 12.4 Cost-benefit of treatment implementation

of the former risk then corresponds to an expense of $0.2 \cdot 12,000 = 2,400$ over a decade when using multiplication as the risk function, whereas the level of the latter risk corresponds to an expense of $0.1 \cdot 80,000 = 8,000$. In order to evaluate the treatment with respect to cost-benefit, we must estimate the cost of the treatment over ten years, and we must estimate the benefit of the treatment in terms of risk reduction. As to the treatment *Place a guard dog in the yard*, assume that the cost of keeping such a dog amounts to 1,800 over ten years. Assume furthermore that by implementing this treatment, the probability of each of the risks is halved. The benefit of implementing the treatment is then $1,200 + 4,000 = 5,200$, which yields $5,200 - 1,800 = 3,400$ in benefit minus cost. We can therefore conclude that the treatment of keeping a guard dog pays off and therefore should be implemented.

12.2 Conducting the Tasks of Step 8

Notice, importantly that if several treatments mitigate the same risk, this strategy for estimating the cost-benefit of one of the treatments may not yield the correct result. With respect to the diagram in Fig. 12.4, for example, if all three treatments are implemented the effect of each of them may be smaller than when considered in isolation. For example, the effect of keeping a guard dog may have little significance if a window lock and a vehicle alarm are also installed.

In many risk analyses, we operate with assets for which the consequences are not immediately comparable. If we furthermore operate with qualitative values for likelihoods and consequences, it may be challenging to precisely estimate the cost of risks. The identified treatments may also be difficult to assess with respect to precise costs, irrespective of whether implementing them is a non-recurring expense or a running expense. The objective with the cost-benefit analysis, however, is to select appropriate treatments and to come up with a recommendation. Such a recommendation can be found even when we do not know the exact costs and benefits; we only need to substantiate that the benefits outweighs the costs. Furthermore, given the result of the cost-benefit analysis and our recommendations, the customer or the parties of the analysis must next consider these recommendations and carefully make their own decisions. It may often also be that a customer of an analysis does not share all information with us, for example, with respect to business strategies and other known risks. Our recommendations should therefore be further evaluated by the customer.

When we operate with values for consequences, likelihoods and risk that denote intervals we must use such intervals as the basis for the cost-benefit analysis. A risk value can be interpreted as a specification of a certain cost over a certain period of time. When comparing risks to treatments, we can describe treatments in similar terms. Therefore, if we, for example, operate with a risk level scale of the values *very low*, *low*, *high* and *very high* we can use the same scale to describe the cost of implementing treatments. The cost-benefit analysis then amounts to comparing treatment costs over such a scale with benefits in terms of reduction of risk levels.

For the diagram in Fig. 12.4, for example, we may operate with a risk scale defined by the intervals *very low* $= [0, 500)$, *low* $= [500, 2000)$, *high* $= [2000, 5000)$ and *very high* $= [5000, \infty)$. The risk *Thief steals jewelry* has then risk level *high* and the risk *Thief steals car* has risk level *very high*. Operating with the same values for treatments, the treatment *Place a guard dog in the yard* has the cost *low*. By implementing the treatment, the former risk value is reduced from *high* to *low* and the latter risk value is reduced from *very high* to *high*. The benefit of implementing the treatment then outweighs its cost.

For the purpose of conducting and documenting a cost-benefit analysis where the risk scale is based on intervals, we may use the table format exemplified in Table 12.5. Each treatment is placed in the first column followed by the estimated treatment cost in the second column. For each treatment, the risks it mitigates are placed in the third column followed by the estimated risk reductions in the fourth column. For some of the treatments, we can immediately tell from this overview whether or not it is worth implementing. For the treatments that are less clear, we may have to look more carefully at the whole picture to draw a conclusion, or we may leave it to the customer to make the decision.

Table 12.5 Treatment evaluation

Treatment	Cost	Risk	Risk reduction
Strengthen access control solution	High	OD	High to High
		OD(I1)	High to High
Increase awareness of security risks	Low	OD	High to High
		OD(I1)	High to High
		I1	High to Low
Implement state of the art virus protection	Low	OD	High to Low
		OD(I1)	High to Low
		I1	High to High

Notice that in such a table we represent the cost-benefit of each treatment in isolation. We must therefore use the table as a basis for evaluating the effect of combining treatments that mitigate the same risks. As explained above, combining treatments may mean that the treatments to some extent neutralise the effect of each other. We also use the table to try to determine whether the implementation of one treatment is sufficient to mitigate given risks, or whether we need to combine several treatments to get the necessary effect.

Example 12.4 In order to have the target team from AutoParts to evaluate the treatments in terms of cost-benefit, the analysis leader presents each treatment in turn along with the risks the treatment mitigates. The task for the target team is then to come up with estimates for costs and benefits.

The risk scale they use are of the values *very low*, *low*, *high* and *very high*. Throughout the risk analysis, the target team has reached a good understanding of what these values mean in terms of significance and severity. The analysis leader suggests that they describe the cost of implementing treatments using the same terms. This means, for example, that a treatment of cost *low* that eliminates a risk of value *very high* should be implemented. Similarly, a treatment of cost *very high* that eliminates a risk of value *high* should not be implemented unless the treatment also mitigates other risks to a sufficient degree.

Table 12.5 shows the result of the evaluation of some of the treatments. As to the treatment *Strengthen access control solution*, the system developer of the target team claims that it is easily implemented and fairly cheap considering the expected lifetime of the online store solution they are about to launch. The data security officer and the system user are, however, not convinced. They explain that in order to properly evaluate this treatment they need to know how the access control will be strengthened. Will there be certain work processes that no longer can be performed by certain employees, for example? And will other work processes be slowed down because of less availability? There may be several such long term costs and overheads that need to be considered. The target team finally agrees that for the time being they should assign the cost *high* to this treatment.

12.2 Conducting the Tasks of Step 8

Two of the risks that may be mitigated by this treatment are *OD: Online store down* and *OD(I1): Online store down* as shown in the treatment diagram of Fig. 12.3. Both these risks are currently *high*. The latter risk is with respect to the indirect asset of customer satisfaction. The target team members believe that the treatment will have an effect on the level of these risks, but they are not sure whether the effect is sufficient to shift the risk levels from *high* to *low*. After all, the relevant threat scenario *Hacker obtains access to DBs* has likelihood *rare*. Since they already have decided that this treatment must later be more carefully considered with respect to cost, they assign no risk reduction for now. They must furthermore consider this treatment with respect to the full risk picture and take into account also the risks with respect to the customer database, the inventory database and the indirect assets that are affected.

The target team believes that the treatment *Increase security awareness of security risks* may not have much effect on incidents that cause the online store to go down. They believe, however, that there is a gain of this treatment with respect to the risk *II: Integrity of inventory DB corrupted* since the treatment is directed towards two vulnerabilities, and since carelessness and lack of competence among employees potentially may cause significant incidents. They furthermore agree that it should be possible to carry out this treatment at a quite low cost.

The discussion proceeds along similar lines with respect to the treatment *Implement state of the art virus protection*. They believe that the treatment may not have much effect on the risk *II: Integrity of inventory DB corrupted*, but that it should be considered with respect to the risk *OD: Online store down*.

The treatment evaluation table indicates that the treatments *Increase awareness of security risks* and *Implement state of the art virus protection* should be recommended for implementation. Whether or not to implement the treatment *Strengthen access control solution* must be decided by AutoParts at a later point in time, since it must be clarified more precisely how it should be implemented, what the costs are and what the effects are on the general risk picture.

When we have finalised the treatment identification and treatment evaluation, we may give an overview of the results using treatment overview diagrams. These diagrams are basically risk diagrams annotated with treatment scenarios.

The treatment overview diagrams do not provide any new information, but rather serve as a form of executive summary of Step 8. When we make these diagrams, we simply use the risk diagrams as input, removing the acceptable risks, as well as the threats and assets that are related to acceptable risks only. We then add the identified treatment scenarios and relate them to the risks they mitigate. Depending on how we want to document the results of the treatment identification and evaluation, we may choose either to represent all the identified treatments in the treatment overview diagrams, or to represent only the treatments that are recommended for implementation.

One and the same identified treatment is often relevant for several risks, and therefore appears in several treatment diagrams. Since treatment overview diagrams are a kind of condensed treatment diagrams, we may sometimes be able combine

Fig. 12.5 Treatment overview

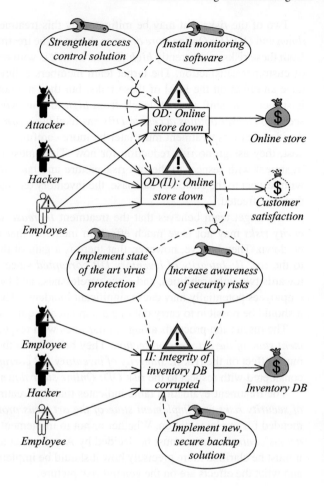

several diagrams such that each treatment appears only once. This is convenient since it immediately shows all the risks that are relevant for a given treatment

Example 12.5 Using the treatment diagrams as input, the analysis team makes treatment overview diagrams to document in an intuitive way the risks that may be mitigated by each treatment. They choose to do this offline after the last workshop, since it does not require any new information to be gathered. The diagrams will then be used as part of the risk analysis report that they later will deliver to AutoParts.

The treatment overview diagram of Fig. 12.5 combines the treatment diagrams of Figs. 12.2 and 12.3. Conveniently, the overview diagram then shows how the two treatments *Implement state of the art virus protection* and *Increase awareness of security risks* are relevant for risks in both of the treatment diagrams.

The junction points on the treats relations are simply shorthand notation in place of many-to-many relations between treatments and risks. The junction points may be used to ease readability.

12.3 Summary of Step 8

Step 8 of the CORAS risk analysis method is summarised by the overview presented in Table 12.6.

Table 12.6 Summary of Step 8

Risk treatment using treatment diagrams	

Analysis tasks:
• Group risks that may be treated by the same measures
• Identify treatments to unacceptable risks
• Estimate the cost-benefit of each treatment and decide which ones to implement; show treatments in treatment overview diagrams
	People that should participate:
	• Analysis leader (required)
	• Analysis secretary (required)
	• Representatives of the customer:
	  – Decision makers (required)
	  – Technical expertise (required)
	  – Users (required)

Modelling guideline:

- Treatment diagrams:
  1. Use the threat diagrams as a basis and replace each unwanted incident with the risks it represents; this yields one risk for each relation between an unwanted incident and an asset; acceptable risks may be removed along with the unwanted incidents, vulnerabilities, assets, threats and threat scenarios that do not constitute unacceptable risks
  2. Annotate the diagram with treatments, pointing to where they will be applied
  3. If several treatments point towards the same risks (a many-to-many relation), we recommend using "junction points" to avoid multiple, crossing lines

- Treatment overview diagrams:
  1. Use the risk diagrams as a basis; remove the acceptable risks (including threats and assets that are not associated with an unacceptable risk)
  2. Add treatments as specified in the treatment diagrams; treatments directed towards vulnerabilities or threat scenarios should point towards the risks they indirectly treat
  3. If several treatments point towards the same risks (a many-to-many relation), we recommend using "junction points" to avoid multiple, crossing lines

Part III
Selected Issues

Chapter 13
Analysing Likelihood Using CORAS Diagrams

Assigning likelihoods to the unwanted incidents of CORAS diagrams, either qualitatively or quantitatively, is a necessary prerequisite for risk estimation. To support likelihood estimation and documentation, we may furthermore assign likelihoods to the threat scenarios, as well as to the initiates and leads-to relations. Techniques for likelihood analysis are useful not only in order to deduce the likelihood of a given threat scenario or unwanted incident on the basis of previous estimations; likelihood analysis is also a useful means to identify uncertainties and unclarities in how the risk picture is understood by identifying inconsistencies. In this chapter, we explain how we may analyse and reason about CORAS diagrams with respect to likelihoods. In particular, we explain:

1. How to calculate the likelihood of an unwanted incident or a threat scenario from likelihood assignments elsewhere in the diagram.
2. How to check that the likelihood assignments are meaningful, in other words that they make sense given the semantics of CORAS diagrams.
3. How to restructure CORAS diagrams to simplify likelihood reasoning and facilitate an improved understanding of likelihood.

In order to determine the level of risk associated with an unwanted incident, we need to know its likelihood. Ideally, the likelihood can be obtained from historical data for the target of analysis. Unfortunately, in practice this is often not possible, for a number of reasons: The target may, for example, be a brand new system yet to be deployed; the target or its environment may have changed, rendering previous observations less relevant or incorrect; the unwanted incident may be extremely rare while having a high consequence, such as a nuclear reactor meltdown; or the unwanted incident may not be easily observable, such as an unfaithful employee revealing secret business information to a competitor.

In such cases, we need to estimate and assign likelihoods on the basis of incomplete or lacking evidence. Often, the best we can do is to use expert judgements.

This chapter was coauthored by Atle Refsdal, SINTEF ICT. Email: Atle.Refsdal@sintef.no

Clearly, there will be uncertainty associated with such estimates, so we should seek ways of identifying weaknesses of the estimates, and ways of validating the estimates. One way of doing this is to check whether they are consistent with respect to the diagrams we have drawn.

The rest of this chapter is structured as follows. Section 13.1 explains how to calculate likelihoods of unwanted incidents and threat scenarios based on the likelihood of related diagram elements, and based on likelihoods that are assigned to the relations. In Sect. 13.2, we then explain how to check the consistency of assigned likelihoods. Thereafter, in Sect. 13.3, we address the issue of analysing scenarios with logical connectives. In Sect. 13.4, we give pragmatic guidance on how to structure CORAS diagrams in order to fully exploit the potential for likelihood analysis. We conclude in Sect. 13.5 by summarising.

13.1 Using CORAS Diagrams to Calculate Likelihood

In this section, we explain how to use CORAS diagrams to calculate and reason about likelihoods. We start in Sect. 13.1.1 by giving an overview of how likelihoods can be specified in the CORAS language. We furthermore introduce some shorthand notation to simplify the reasoning about the diagrams, and we introduce an example that will run throughout this chapter. Likelihoods can be expressed in terms of probability or frequency. In Sect. 13.1.2, we present rules for calculating probabilities for CORAS diagrams, and in Sect. 13.1.3 we present corresponding rules for frequencies. Since one and the same diagram can be annotated with both probabilities and frequencies, we explain in Sect. 13.1.4 how to handle these cases, and in particular how to convert frequencies into probabilities. Finally, in Sect. 13.1.5 we show how to generalise the calculation rules to cope with likelihood intervals. We furthermore discuss the use of likelihood distributions.

13.1.1 Specifying Likelihood Using CORAS Diagrams

There are two kinds of CORAS language elements and two kinds of relations between elements to which likelihoods can be assigned. The language elements are unwanted incidents and threat scenarios, and the relations are the initiates relation and the leads-to relation. The ultimate goal is to assign likelihoods to the identified unwanted incidents, since the unwanted incidents represent the risks. It is, however, in many situations helpful to estimate the likelihoods of the threat scenarios and the relations also in order to facilitate the estimation of the likelihood of unwanted incidents, to document the most important origins of risks, and to identify possible unclarities in the understanding of the risk picture.

The CORAS diagram of Fig. 13.1 shows a fragment of the risk modelling that is conducted in the risk analysis of some information system. This particular diagram

13.1 Using CORAS Diagrams to Calculate Likelihood

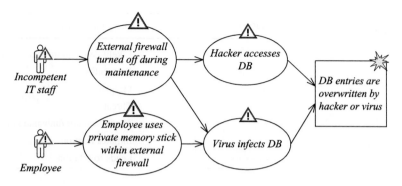

Fig. 13.1 A diagram used as a basis for likelihood estimation

addresses risks with respect to a database *DB*, and documents the unwanted incident of database entries being overwritten by a hacker or a virus.

The example diagram of Fig. 13.1 is not annotated with likelihoods. Throughout this chapter, we introduce various rules and strategies for conducting the likelihood estimations and for validating the result. In order to simplify the presentation, we will use the following abbreviations for the names of the threats, the descriptions of the threat scenarios and the description of the unwanted incident of the example:

$ii = $ *Incompetent IT staff*

$em = $ *Employee*

$ef = $ *External firewall turned off during maintenance*

$eu = $ *Employee uses private memory stick within external firewall*

$ha = $ *Hacker accesses DB*

$vi = $ *Virus infects DB*

$de = $ *DB entries are overwritten by hacker or virus*

For arbitrary diagram constructs and diagram annotations of relevance for this chapter, we use the syntactic variables shown in Table 13.1. Notice that although → ranges over both initiates relations and leads-to relations, the kinds of relations are distinguished by the source of the relation; threats are always the source of an initiates relation and never the source of a leads-to relation. Notice also that although both the initiates relation and the leads-to relation may point at a risk, we will in the context of this chapter let e range over threat scenarios and unwanted incidents only. This is because these relations are never annotated with likelihoods when they have risks as target. Notice also that as already pointed out in Sect. 10.2 of Chap. 10, from a practical point of view we do not consider conditional likelihoods in terms of frequencies. We therefore always represent conditional likelihoods by the syntactic variable l.

Table 13.2 shows the shorthand notation that we use in this chapter for the likelihood specifications in CORAS diagrams. In particular, we will for presentation and

Table 13.1 Naming conventions

Variable	Diagram construct
l	Likelihood, conditional likelihood
p	Probability
f	Frequency
t	Threat
e	Threat scenario/unwanted incident
$t \to e$	Initiates relation
$e_1 \to e_2$	Leads-to relation

Table 13.2 Denoting likelihood

Likelihood specification	Interpretation
$e(l)$	e occurs with likelihood l
$e(p)$	e occurs with probability p
$e(f)$	e occurs with frequency f
$t \xrightarrow{l} e$	t initiates e with likelihood l
$t \xrightarrow{p} e$	t initiates e with probability p
$t \xrightarrow{f} e$	t initiates e with frequency f
$e_1 \xrightarrow{l} e_2$	e_1 leads to e_2 with conditional likelihood l

readability purposes use this shorthand notation in the definitions of the rules for calculating likelihood.

With this overview of the means for specifying likelihood using the CORAS language, we turn to the rules for likelihood calculation.

13.1.2 Rules for Calculating Probability in CORAS Diagrams

We now present rules for calculating and reasoning about probabilities as specified in CORAS diagrams, and we illustrate the application of the rules on the running example. The rules are of the following form:

$$\frac{P_1 \quad P_2 \quad \ldots \quad P_n}{C}$$

We refer to $P_1, \ldots, P_n$ as the premises and to C as the conclusion. The interpretation is that if the premises are valid, so is the conclusion.

The initiate rule captures the semantics of the initiates relation. The probability of the occurrences of scenario/incident e due to threat t is equal to the probability with which t initiates e. This is expressed as follows, where $t \sqcap e$ can be understood

13.1 Using CORAS Diagrams to Calculate Likelihood

as the instances of scenario/incident e that are initiated by threat t, in other words $t \sqcap e$ is a subset of e.

Rule 13.1 (Initiates)   For threat t and scenario/incident e related by the initiates relation, we have:

$$\frac{t \xrightarrow{p} e}{(t \sqcap e)(p)}$$

Next, the leads-to rule captures the conditional likelihood semantics embedded in the leads-to relation. The probability of the occurrences of e_2 that are due to e_1 is equal to the probability p of e_1 multiplied with the conditional likelihood l that e_1 will lead to e_2 given that e_1 occurs. This is captured as follows, where $e_1 \sqcap e_2$ can be understood as the subset of the scenarios/incidents e_2 that are preceded by e_1. Note that this means that $\sqcap$ is not commutative.

Rule 13.2 (Leads-to)   For the scenarios/incidents e_1 and e_2 related by the leads-to relation, we have:

$$\frac{e_1(p) \quad e_1 \xrightarrow{l} e_2}{(e_1 \sqcap e_2)(p \cdot l)}$$

The initiate rule and the leads-to rule concern only calculations on a single branch of a threat diagram. The next rules show how to do probability calculations on parallel branches.

If two events are mutually exclusive, the probability of their union is equal to the sum of their respective probabilities. When throwing a die, for example, the outcomes are mutually exclusive. Since the probability of each outcome is $\frac{1}{6}$, the probability of two different outcomes in the first throw, for example five or six, is $\frac{2}{6}$. This is captured by the following rule, where $e_1 \sqcup e_2$ denotes all instances of e_1 and e_2. In this rule, this means that either e_1 or e_2 occurs, but not both since they are mutually exclusive.

Rule 13.3 (Mutually exclusive scenarios/incidents)   If the scenarios/incidents e_1 and e_2 are mutually exclusive, we have:

$$\frac{e_1(p_1) \quad e_2(p_2)}{(e_1 \sqcup e_2)(p_1 + p_2)}$$

If two events are statistically independent, the probability of the one has no relation to the probability of the other. When throwing two dice, for example, the probability of the separate outcomes are independent of each other. Given statistic independence, the probability of the union of two events is equal to the sum of their individual probabilities minus the product of these. The probability of the intersection must be subtracted such that common outcomes are counted only once. For example, when calculating the probability of getting (at least one) six as the outcome when throwing two dice, we need to subtract the probability of getting six on both. The following rule captures statistical independence.

Rule 13.4 (Independent scenarios/incidents) If the scenarios/incidents e_1 and e_2 are statistically independent, we have:

$$\frac{e_1(p_1) \quad e_2(p_2)}{(e_1 \sqcup e_2)(p_1 + p_2 - p_1 \cdot p_2)}$$

Example 13.1 We now demonstrate the application of these probability rules on our example. Figure 13.2 shows the result of applying the rules to calculate the probabilities for the four threat scenarios, as explained below. We use the probability annotations on the relations as input data to the calculations, assuming that these values have already been estimated.

To simplify the presentation, we overload the diagram abbreviations listed on p. 209 and let the description of the scenarios and the unwanted incident denote their instances. Hence, we let ef denote all the possible occurrences of the threat scenario *External firewall turned off during maintenance*, we let $ii \sqcap ef$ denote the subset of ef that is initiated by *Incompetent IT staff*, and so forth.

Furthermore, in order to keep the example simple we make the following assumptions:

- The diagram is complete in the sense that no other threats, threat scenarios or unwanted incidents than the ones explicitly shown lead to any of the threat scenarios or the unwanted incident in the diagram. In particular, this means that
  - $ef = ii \sqcap ef$
  - $eu = em \sqcap eu$

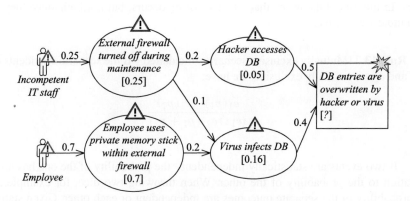

Fig. 13.2 A diagram used for calculating the probability of threat scenarios

13.1 Using CORAS Diagrams to Calculate Likelihood

- $ha = ef \sqcap ha$
- $vi = (ef \sqcap vi) \sqcup (eu \sqcap vi)$
- $de = (ha \sqcap de) \sqcup (vi \sqcap de)$

- The threat scenarios ef and eu, as well as the conditional likelihoods on the leads-to relations going from these scenarios to vi, are statistically independent. This means that also $ef \sqcap vi$ and $eu \sqcap vi$ are statistically independent.

Starting from the left, we use the probability annotation 0.25 on the initiates relation from ii to ef and apply Rule 13.1 to calculate the probability of the threat scenario ef. By assumption, we have $ef = ii \sqcap ef$, which means that the probability of ef is equal to the probability of $ii \sqcap ef$. Therefore, the probability 0.25 has been assigned to ef. A similar consideration leads to the assignment of the probability 0.7 for eu, which is initiated by em.

In order to calculate the probability for ha, we apply Rule 13.2. From the probability 0.25 assigned to ef and the conditional likelihood 0.2 previously estimated for the leads-to relation from ef to ha, we obtain the probability 0.05 for $ef \sqcap ha$. Since $ha = ef \sqcap ha$, the probability of ha is then also 0.05.

In order to calculate the probability for vi, we first calculate the probabilities for $ef \sqcap vi$ and $eu \sqcap vi$. Applying Rule 13.2 in the same way as when calculating the probability for $ef \sqcap ha$ above yields the probability 0.025 for $ef \sqcap vi$ and 0.14 for $eu \sqcap vi$. As $ef \sqcap vi$ and $eu \sqcap vi$ are statistically independent, we may apply Rule 13.4 in order to calculate the probability for $(ef \sqcap vi) \sqcup (eu \sqcap vi)$. This gives the probability $0.025 + 0.14 - 0.025 \cdot 0.14 = 0.1615$. Since $vi = (ef \sqcap vi) \sqcup (eu \sqcap vi)$, this means that the probability of vi is also $0.1615 \approx 0.16$ as annotated in the diagram of Fig. 13.2.

Next, we would like to calculate the probability of the unwanted incident de from the probabilities calculated for ha and vi, as well as the conditional likelihoods previously assigned to the leads-to relations from these threat scenarios. Unfortunately, this is not possible from the above rules. The reason is that ha and vi are neither statistically independent nor mutually exclusive. We have therefore inserted a question mark in place of the probability for the unwanted incident. In Sect. 13.4, we show how to restructure a diagram in order to cope with this kind of situation.

However, even if we are not able to calculate the exact probability for de, we may calculate the maximum and minimum value. The assumption that $de = (ha \sqcap de) \sqcup (vi \sqcap de)$ means that there are no other scenarios than ha and vi that may lead to de. It follows that the probability of de cannot be higher than the sum of the probabilities of $(ha \sqcap de)$ and $(vi \sqcap de)$. Applying Rule 13.2 for each of these, we obtain the probability $0.05 \cdot 0.50 = 0.025$ for $(ha \sqcap de)$ and $0.1615 \cdot 0.40 = 0.0646$ for $(vi \sqcap de)$. This means that the maximum probability of de is $0.025 + 0.0646 = 0.0896$. Furthermore, the minimum probability of de can obviously not be lower than the maximum of the probabilities of $(ha \sqcap de)$ and $(vi \sqcap de)$. This means that the minimum probability of de is 0.0646. Approximating the maximum and minimum values, the probability of the unwanted incident de is therefore in the interval [0.065, 0.090].

When conducting risk analyses in practice, the diagrams we make are often incomplete. Usually we focus on the most important threats and threat scenarios that lead to risks, as it is too time and resource consuming, sometimes even infeasible, to think of, address and document every detail of the risk picture such that it becomes complete. For practical reasons, we should not spend time on insignificant sources of risks. We may view a particular source of risk as insignificant if its individual contribution has no effect on the risk evaluation or our decisions about risk treatments. Nevertheless, all the details that we choose not to document may in combination have a degree of impact that matters and that we must take into account when analysing probabilities, and also likelihoods in general. In particular, when diagrams are incomplete, we can deduce only the lower bounds of the probabilities.

Example 13.2 Let us consider again the threat diagram of Fig. 13.2. If the diagram is incomplete, we need to be more careful when estimating the probabilities of the various threat scenarios; for each of them we must consider to what extent the diagram documents how it may arise.

Starting with the threat scenario ef, we know that it may be initiated by the threat ii, and that this happens with a probability of 0.25. If we know that it is only the IT staff that can turn off the firewall, we can straightforwardly use Rule 13.1 and assign the probability 0.25 to ef. On the other hand, if the diagram does not completely document all the possible causes of this threat scenario to occur, we can only deduce that its probability is at least 0.25.

Considering that there are many ways of attacking a system, also when the external firewall is turned on, it seems obvious that the diagram does not completely document the causes of a hacker accessing the database. This is so even in the case that only IT staff can turn off the firewall. Since the minimum probability of ef is 0.25, we get that the probability of $ef \sqcap ha$ is at least 0.05 by using Rule 13.2 and the conditional likelihood 0.2 on the leads-to relation. Hence, it follows that the probability of ha is at least 0.05 since $ef \sqcap ha$ is contained in ha.

We can use expert judgement, historical data or the like to estimate the probability of ha, and use the information that the probability of $ef \sqcap ha$ is 0.05 as part of the evidence. Obviously, if the estimation is significantly higher than 0.05, we must consider to elaborate on and extend the diagram in order to specifically address and document the identified risks.

In a similar way, we go through the other threat scenarios of the diagram and judge to what extent the documentation is complete; we apply the rules as shown in the previous example, but for incomplete diagrams the estimations are only lower bounds.

In many situations, it is useful to compose two threat scenarios into a single threat scenario, or likewise for unwanted incidents, for example, to obtain a more compact diagram. In these cases, we want to deduce the conditional likelihoods of the leads-to relations terminating or originating in the composite scenario or incident from the conditional likelihoods of the relations connected to the original scenarios or incidents. As before, we distinguish between the mutually exclusive and the statistically independent cases.

The first two rules address the case where two relations from the same scenario/incident terminate in each of the two scenarios/incidents that are composed.

13.1 Using CORAS Diagrams to Calculate Likelihood

Notice, importantly, that whereas the above rules are based on assumptions of mutual exclusiveness and statistical independence of threat scenarios and incidents, the following two rules are based on such assumptions for leads-to relations.

Two leads-to relations $e \rightarrow e_1$ and $e \rightarrow e_2$ originating in the same scenario or incident e are mutually exclusive if e never leads to e_1 and e_2 simultaneously. The following rule is for calculating the conditional likelihood of the composition of mutually exclusive leads-to relations.

> **Rule 13.5** (Composing mutually exclusive relations)   If the leads-to relations $e \rightarrow e_1$ and $e \rightarrow e_2$ are mutually exclusive, we have:
>
> $$\frac{e \xrightarrow{l_1} e_1 \quad e \xrightarrow{l_2} e_2}{e \xrightarrow{l_1+l_2} (e_1 \sqcup e_2)}$$

Two leads-to relations $e \rightarrow e_1$ and $e \rightarrow e_2$ are statistically independent if the probability of e leading to e_1 has no relation to the probability of e leading to e_2, and vice versa. The following rule is for calculating the conditional likelihood of the composition of statistically independent leads-to relations.

> **Rule 13.6** (Composing statistically independent relations)   If the leads-to relations $e \rightarrow e_1$ and $e \rightarrow e_2$ are statistically independent, we have:
>
> $$\frac{e \xrightarrow{l_1} e_1 \quad e \xrightarrow{l_2} e_2}{e \xrightarrow{l_1+l_2-l_1 \cdot l_2} (e_1 \sqcup e_2)}$$

Example 13.3 As an example of the composition of threat scenarios, consider the threat diagram of Fig. 13.3 that shows possible ways of data being disclosed to

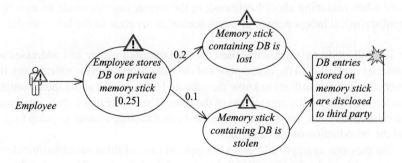

Fig. 13.3 A diagram used for calculating the probability of composed leads-to relations from one threat scenario

third parties due to storage of database on irregular media. We use the following abbreviations to simplify the presentation:

$em = $ *Employee*

$es = $ *Employee stores DB on private memory stick*

$ml = $ *Memory stick containing DB is lost*

$ms = $ *Memory stick containing DB is stolen*

$dd = $ *DB entries stored on memory stick are disclosed to third party*

Assuming that the leads-to relations to *ml* and *ms* are statistically independent, we can use Rule 13.6 to calculate the conditional likelihood of the leads-to relation from *es* to the composition of *ml* and *ms*, in other words from *es* to $ml \sqcup ms$. We can read this result of the composition as the treat scenario *Memory stick containing DB is lost or memory stick containing DB is stolen*, in other words as a disjunction.

The conditional likelihood of the leads-to relation from *es* to *ml* is 0.2, and the conditional likelihood from *es* to *ms* 0.1. Using Rule 13.6, we get that the conditional likelihood of the composed leads-to relation is $0.2 + 0.1 - 0.2 \cdot 0.1 = 0.28$.

Having calculated the conditional likelihood of the composed leads-to relation, we can use Rule 13.2 and the probability 0.25 of threat scenario *es* to calculate the probability $0.25 \cdot 0.28 = 0.07$ of the composed threat scenario $ml \sqcup ms$, assuming that the diagram is complete.

Notice, importantly, that we made no assumption about statistical relationships between the threat scenarios *ml* and *ms* in this example, but only about the statistical independence of the two leads-to relations in the diagram of 13.3. By instead assuming statistical independence of $es \sqcap ml$ and $es \sqcap ms$, we could use Rule 13.2 to calculate the probability of these threat scenarios separately, and then use Rule 13.4 to calculate the probability of the composed threat scenario $(es \sqcap ml) \sqcup (es \sqcap ms)$. The latter equals the threat scenario $ml \sqcup ms$ if we assume that the threat diagram is complete. The reader can verify that the resulting probability of the composed threat scenario in that case is 0.07375, which is slightly higher than the probability 0.07 that we deduced above.

This demonstrates the importance of being careful about the assumptions we can make when reasoning about likelihood; in the former case we made no assumption about statistical independence of threat scenarios, whereas in the latter we did.

The following passage is intended for the interested reader and addresses some subtleties in relation to the assumptions of the rules for probability calculation. In the practical setting, it suffices to know the rules and to be careful about the assumptions that are required for the application of the rules. The reader that is only interested in the calculation rules and how they are used can therefore choose to jump to p. 218 and the introduction of the next rules.

The previous example addresses the general case of three scenarios/incidents e, e_1 and e_2 where the former may lead to any of the latter two. The probability $e(p)$ is furthermore known, as well as the conditional likelihoods $e \xrightarrow{l_1} e_1$ and $e \xrightarrow{l_2} e_2$.

13.1 Using CORAS Diagrams to Calculate Likelihood

How to approach this general case in order to calculate the probability of the composed scenario/incident $e_1 \sqcup e_2$ depends on the assumptions we can make in each particular case. As demonstrated by the previous example, the assumptions we make determine which rules are applicable and thereby the result.

In the example, we first made the assumption that the leads-to relations $e \xrightarrow{l_1} e_1$ and $e \xrightarrow{l_2} e_2$ are statistically independent in order to calculate the probability of $e_1 \sqcup e_2$. By first applying Rule 13.6 and then Rule 13.2, we deduced the result $p \cdot (l_1 + l_2 - l_1 \cdot l_2)$ which equals $p \cdot l_1 + p \cdot l_2 - p \cdot l_1 \cdot l_2$. Subsequently, we made the assumption that e_1 and e_2 are statistically independent after which we first applied Rule 13.2 to each of the leads-to relations and then Rule 13.4 to deduce the result $p \cdot l_1 + p \cdot l_2 - p \cdot l_1 \cdot p \cdot l_2$. This explicitly shows that under the latter assumption of statistically independent scenarios/incidents, the probability p of e is multiplied twice into the part that is subtracted, whereas under the former assumption of statistically independent leads-to relations, the probability p is multiplied only once into the subtracted part. In order to understand this subtle difference, we need to take a more careful look at the respective assumptions.

To simplify the presentation, we introduce some new notation. For scenarios/incidents e_1 and e_2 the expression $e_1 \sqcap e_2$ denotes the occurrence of both e_1 and e_2. P is a function from a scenario/incident e to the probability of e, such that for a specification $e(p)$ we have $P(e) = p$.

Assuming that the leads-to relations are statistically independent, we can represent the conditional likelihood of e leading to both e_1 and e_2 occurring as follows:

$$\frac{P((e \sqcap e_1) \sqcap (e \sqcap e_2))}{P(e)} = l_1 \cdot l_2 \tag{13.1}$$

Since we know that $P(e) = p$, we further get the following:

$$P((e \sqcap e_1) \sqcap (e \sqcap e_2)) = l_1 \cdot l_2 \cdot P(e) = l_1 \cdot l_2 \cdot p \tag{13.2}$$

Our goal is to calculate the probability of $(e \sqcap e_1) \sqcup (e \sqcap e_2)$ (which equals $e_1 \sqcup e_2$ when the diagram is complete). The probability of this composition is the sum of the respective probabilities minus the probability of the occurrence of both $e \sqcap e_1$ and $e \sqcap e_2$. This can be expressed as follows:

$$P((e \sqcap e_1) \sqcup (e \sqcap e_2)) = P(e \sqcap e_1) + P(e \sqcap e_2) - P((e \sqcap e_1) \sqcap (e \sqcap e_2)) \tag{13.3}$$

By Rule 13.2, we have that $P(e \sqcap e_1) = p \cdot l_1$ and $P(e \sqcap e_2) = p \cdot l_2$. Together with (13.2), the previous equation yields the result for the assumption of statistically independent leads-to relations:

$$P((e \sqcap e_1) \sqcup (e \sqcap e_2)) = p \cdot l_1 + p \cdot l_2 - p \cdot l_1 \cdot l_2 \tag{13.4}$$

We now turn to the case of statistically independent scenarios/incidents. Usually we cannot in the general case straightforwardly assume that the scenarios/incidents e_1 and e_2 are statistically independent when the same scenario/incident e can lead to both of them. (Such a statistical independence between e_1 and e_2 can be the case if e is actually the composition of two independent scenarios where the one can lead to e_1 and the other can lead to e_2.) For now, we assume that $e \sqcap e_1$ and $e \sqcap e_2$ are statistically independent as captured by the following:

$$P\big((e \sqcap e_1) \sqcap (e \sqcap e_2)\big) = P(e \sqcap e_1) \cdot P(e \sqcap e_2) \qquad (13.5)$$

By Rule 13.2, this yields the following:

$$P\big((e \sqcap e_1) \sqcap (e \sqcap e_2)\big) = p \cdot l_1 \cdot p \cdot l_2 \qquad (13.6)$$

Together with (13.3), the previous equation yields the result for the assumption of statistically independent scenarios/incidents:

$$P\big((e \sqcap e_1) \sqcup (e \sqcap e_2)\big) = p \cdot l_1 + p \cdot l_2 - p \cdot l_1 \cdot p \cdot l_2 \qquad (13.7)$$

The same case of a scenario/incident e that may lead to both e_1 and e_2 is more straightforward given mutual exclusiveness. When the leads-to relations $e \xrightarrow{l_1} e_1$ and $e \xrightarrow{l_2} e_2$ are mutually exclusive we can first apply Rule 13.5 and then Rule 13.2 to calculate the probability of $e_1 \sqcup e_2$. When instead the scenarios/incidents e_1 and e_2 are mutually exclusive, we can first use Rule 13.2 to each of the leads-to relations and then Rule 13.3 to deduce the result. The reader can verify that in both cases the result equals $p \cdot l_1 + p \cdot l_2$.

This concludes our passage about the subtleties in relation to the assumptions of statistical independence, and we continue by the introduction of two further rules.

The case where two relations to the same scenario/incident originate in each of the two scenarios/incidents that are composed is covered by the next two rules.

Rule 13.7 (Composing relations from mutually exclusive scenarios/incidents) If the scenarios/incidents e_1 and e_2 are mutually exclusive, we have:

$$\dfrac{e_1(p_1) \qquad e_2(p_2) \qquad e_1 \xrightarrow{l_1} e \qquad e_2 \xrightarrow{l_2} e}{(e_1 \sqcup e_2) \xrightarrow{\frac{p_1 \cdot l_1 + p_2 \cdot l_2}{p_1 + p_2}} e}$$

Rule 13.8 (Composing relations from statistically independent scenarios/incidents) If the scenarios/incidents e_1 and e_2 are statistically independent, we have:

$$\dfrac{e_1(p_1) \qquad e_2(p_2) \qquad e_1 \xrightarrow{l_1} e \qquad e_2 \xrightarrow{l_2} e}{(e_1 \sqcup e_2) \xrightarrow{\frac{p_1 \cdot l_1 + p_2 \cdot l_2 - p_1 \cdot l_1 \cdot p_2 \cdot l_2}{p_1 + p_2 - p_1 \cdot p_2}} e}$$

13.1 Using CORAS Diagrams to Calculate Likelihood

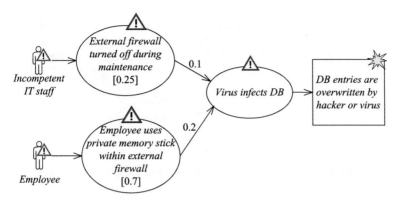

Fig. 13.4 A diagram used for calculating the probability of composed leads-to relations from two threat scenarios

Notice that the latter two rules are redundant in the sense that Rule 13.7 can be deduced from Rule 13.2 and Rule 13.3, whereas Rule 13.8 can be deduced from Rule 13.2 and Rule 13.4. To see this, observe that by the premises of Rule 13.7 we can use Rule 13.2 on each of the leads-to relation to calculate the probabilities $p_1 \cdot l_1$ and $p_2 \cdot l_2$ of $e_1 \sqcap e$ and $e_2 \sqcap e$, respectively. By Rule 13.3, we get the probability $p_1 \cdot l_1 + p_2 \cdot l_2$ for $(e_1 \sqcap e) \sqcup (e_2 \sqcap e)$. Next, we can use Rule 13.3 to calculate the probability $p_1 + p_2$ of $e_1 \sqcup e_2$. Knowing the probability of $e_1 \sqcup e_2$ and $(e_1 \sqcap e) \sqcup (e_2 \sqcap e)$, we can then use Rule 13.2 to deduce the conditional likelihood of the former leading to the latter. The argument for Rule 13.8 is symmetric. Although these rules in a sense are redundant, it is nevertheless useful to state them explicitly as separate rules in order to facilitate the reasoning about the composition of conditional likelihoods.

Example 13.4 The threat diagram of Fig. 13.4 shows two leads-to relations ending in the threat scenario vi. Assuming that the threat scenarios ef and eu are statistically independent, we can use Rule 13.8 to calculate the conditional likelihood of the leads-to relation that results from composing the threat scenarios ef and eu into $ef \sqcup eu$.

The probabilities and conditional likelihoods of the example diagram can be straightforwardly plotted into the schema of Rule 13.8 to deduce the conditional likelihood 0.208 of the composed leads-to relation from the composed threat scenario $ef \sqcup eu$ to the threat scenario vi.

Notice that Rule 13.7 and Rule 13.8 make assumptions about mutual exclusiveness and statistical independence, respectively, of threat scenarios and unwanted incidents. We can therefore proceed by using Rule 13.4 to calculate the probability of the composed threat scenario $ef \sqcup eu$, and finally Rule 13.2 to calculate the probability of 0.16 $(ef \sqcup eu) \sqcap vi$, which equals vi in case the diagram is complete. Recall that we calculated the exact same probability for vi in Example 13.1, however by an alternative approach.

In Sect. 13.4, we give further demonstrations of the use of the rules for composition of relations.

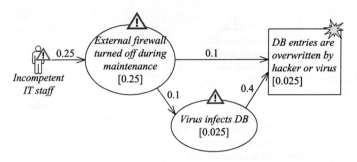

Fig. 13.5 A diagram used for calculating probability in case one path is included in another

The flexibility of the CORAS language allows us to model and study several ways in which a threat can initiate a threat scenario, a threat scenario can lead to an unwanted incident, and so forth. When there are several paths from, for example, one threat scenario e_1 to another threat scenario e_2 it may be the case that one path is a special case of another. The simplest case is when we initially have specified the leads-to relation $e_1 \rightarrow e_2$ and later extend the diagram by adding a threat scenario e' and the leads-to relations $e_1 \rightarrow e'$ and $e' \rightarrow e_2$. If the diagram is complete, that is, e_1 is the only threat scenario that can lead to both e_2 and e', then the path from e_1 to e_2 via e' is a special case of the direct leads-to relation $e_1 \rightarrow e_2$. In other words, the scenarios that may lead to e_2 via e' is contained in the scenarios that may lead to e_2 directly from e_1. This is best illustrated by an example.

Example 13.5 Consider the threat diagram of Fig. 13.5, ignoring the probability annotations for now. Assume that this threat diagram was initially specified without the threat scenario vi and the leads-to relations to and from vi. The diagram without vi documents and explains how the unwanted incident de may occur because of the threat ii and the threat scenario ef initiated by ii.

The leads-to relation from ef to de only explains that the threat scenario ef may lead to de, but it does not explain how. By adding the threat scenario vi in a separate path from ef to de the diagram explains one way in which de may arise as a result of the occurrence of the threat scenario ef. The path via vi is therefore a special case of the direct path.

If we assume that the diagram is complete, we have that $de = ef \sqcap de$, that is, the threat scenario ef accounts for all occurrences of the unwanted incident de, including the occurrences that are preceded by vi. Since $vi = ef \sqcap vi$, we can denote the occurrences of de that are caused by vi by $(ef \sqcap vi) \sqcap de$. Since $ef \sqcap de$ denotes all occurrences of de we have that $(ef \sqcap vi) \sqcap de$ denotes the subset of $ef \sqcap de$ that includes the scenario vi.

When we are reasoning about probability in such cases, we use the direct relation from the initial scenario/incident to the final scenario/incident since this path includes all ways in which the final scenario/incident may arise. At first sight, any additional path via other scenarios or incidents may therefore seem redundant when

13.1 Using CORAS Diagrams to Calculate Likelihood

we are calculating probabilities. This is, however, not so for two reasons. First, we can calculate the contribution of the additional paths to document their significance with respect to the risk level. Second, if we have insufficient evidence for estimating the conditional likelihood of the direct relation, we can use the probability calculations on the additional paths as a basis for reaching an estimate.

Example 13.6 Let us now see how to calculate the probabilities of the threat scenarios and the unwanted incident of Fig. 13.5 assuming that the diagram is complete. We furthermore assume that the probability of the initiates relation as well as the conditional likelihoods of the leads-to relations already have been estimated.

We use Rule 13.1 to calculate the probability of $ii \sqcap ef$, which yields 0.25. By assumption of completeness $ef = ii \sqcap ef$, and we can therefore assign the probability 0.25 to ef as shown in the diagram. By completeness, we furthermore have that $vi = ef \sqcap vi$ and therefore use Rule 13.2 to calculate the probability $0.1 \cdot 0.25 = 0.025$ for the threat scenario vi.

Finally, we calculate the probability of the unwanted incident de. There are two leads-to relations that have de as target, namely the leads-to relations that have ef and vi as source. We cannot use Rule 13.3 since ef and vi are obviously not mutually exclusive. We can furthermore not use Rule 13.4 since vi completely depends on ef. Rather, since ef accounts for all occurrences of de we can apply Rule 13.2 and use the probability 0.25 of ef and the conditional likelihood 0.1 of the leads-to relation from ef to de and calculate the probability $0.25 \cdot 0.1 = 0.025$ of the unwanted incident de.

Since the direct path from ef to de includes the scenario vi we cannot aggregate the contribution of vi to the probability of de into the result. The information about the probability of vi and the conditional likelihood of the leads-to relation from vi to de is, however, useful. By Rule 13.2, we can calculate the probability 0.01 for $vi \sqcap de$. This probability explicitly shows the contribution of virus infection to the overall probability for the database entries to be overwritten. This contribution is 0.01 to 0.025, that is, 40%.

If CORAS diagrams such as the one shown in Fig. 13.5, that is, where one path is included in another, are incomplete, we cannot calculate exact probabilities. In cases of incompleteness, there may be threats and threat scenarios that are not specified such that the actual probabilities are higher than what the information in the diagram accounts for. As for incomplete diagrams in general, we can, however, calculate minimum probability values.

Example 13.7 If we assume that the diagram of Fig. 13.5 is incomplete, we need to determine for each of the threat scenarios and for the unwanted incident whether there are further threats and threat scenarios that are not documented and that can cause them to occur.

As in Example 13.2, we can assign the probability 0.25 to ef if we know that only the IT staff can turn off the firewall. We can then use Rule 13.2 and calculate the probability 0.025 for $ef \sqcap de$, which is then the probability of the subset of de that is caused by ef. We then know that the probability of de is at least 0.025.

Since the diagram is incomplete, we also know that the probability of vi is at least 0.025. This is because the probability of $ef \sqcap vi$ is 0.025. If the actual probability of vi is higher than 0.025, there are then undocumented threat scenarios or unwanted incidents that may lead to vi. If the actual probability of the unwanted incident de is higher than the contribution of 0.025 from ef, this may partly be due to these.

Having completed the presentation of the rules for probability calculation, we present in the following some guidelines for how to use the rules. These guidelines should be used in conjunction with the rules and do not address the cases in which the rules are self-contained and can be straightforwardly used.

Notice that Rule 13.3 through Rule 13.8 are binary in the sense that they are applicable to pairs of scenarios/incidents or pairs of relations. For the application of one rule on multiple premises, the rule must simply be applied several times in a stepwise manner. For example, applying Rule 13.3 to the three scenarios/incidents $e_1(p_1)$, $e_2(p_2)$ and $e_2(p_2)$ yields the result $((e_1 \sqcup e_2) \sqcup e_2)((p_1 + p_2) + p_3)$.

The guidelines for calculating probability in any CORAS diagram, complete or not, are given in Table 13.3. Notice that for the cases in which we cannot calculate exact values, the maximum probability should of course never exceed 1.

Table 13.4 gives additional guidelines for cases of complete diagrams. Notice, however, that although a given diagram is incomplete, there may nevertheless be fragments of the diagram that are complete and for which these guidelines apply.

13.1.3 Rules for Calculating Frequency in CORAS Diagrams

We now present rules for calculating and reasoning about frequencies. Recall that we have restricted ourselves to always use probabilities to specify the conditional likelihood of leads-to relations. Operating with frequencies elsewhere therefore requires us to multiply frequencies with conditional probabilities. This is, however, quite straightforward. For example, multiplying the frequency *twice per year* with the probability 0.1 yields 0.2 *times per year*, which corresponds to *once per five years*. In more numeric terms, we can write this as $0.1 \cdot 2 : 1y = 0.2 : 1y = 1 : 5y$.

The initiate rule and the leads-to rule of Sect. 13.1.2, do not depend on whether the likelihoods are given as frequencies or probabilities. We can therefore simply substitute the occurrences of p with f in Rule 13.1 and Rule 13.2 as follows:

Rule 13.9 (Initiate)   For threat t and scenario/incident e related by the initiates relation, we have:

$$\frac{t \xrightarrow{f} e}{(t \sqcap e)(f)}$$

13.1 Using CORAS Diagrams to Calculate Likelihood

Table 13.3 Guidelines for probability calculation in all diagrams

How to calculate probabilities using any CORAS diagram

Input: $t \xrightarrow{p_1} e$
Output: $e(p_2)$
Calculation: Use Rule 13.1 to calculate $p_2 \geq p_1$

Input: $e_1(p_1)$ and $e_1 \xrightarrow{l} e_2$
Output: $e_2(p_2)$
Calculation: Use Rule 13.2 to calculate $p_2 \geq p_1 \cdot l$

Input: $e_1(p_1)$ and $e_2(p_2)$
Output: $(e_1 \sqcup e_2)(p)$
Calculation:
– If e_1 and e_2 are mutually exclusive: Use Rule 13.3 to calculate $p = p_1 + p_2$
– If e_1 and e_2 are statistically independent: Use Rule 13.4 to calculate $p = p_1 + p_2 - p_1 \cdot p_2$
– Else: $p \in [\max(p_1, p_2), \min(p_1 + p_2, 1)]$
Note: If we only know that the input values are at least p_1 and p_2, then the calculated result is at least p; for the else clause we then only know the lower bound $\max(p_1, p_2)$ of the interval

Input: $e_1(p_1)$, $e_2(p_2)$, $e_1 \xrightarrow{l_1} e$ and $e_2 \xrightarrow{l_2} e$
Output: $e(p)$
Calculation: Calculate $(e_1 \sqcap e)(p_1 \cdot l_1)$ and $(e_2 \sqcap e)(p_2 \cdot l_2)$ using Rule 13.2
– If $e_1 \sqcap e$ and $e_2 \sqcap e$ are mutually exclusive: Use Rule 13.3 to calculate $p \geq p_1 \cdot l_1 + p_2 \cdot l_2$
– If $e_1 \sqcap e$ and $e_2 \sqcap e$ are statistically independent: Use Rule 13.4 to calculate $p \geq p_1 \cdot l_1 + p_2 \cdot l_2 - p_1 \cdot l_1 \cdot p_2 \cdot l_2$
– Else: $p \in [\max(p_1 \cdot l_1, p_2 \cdot l_2), 1]$

Input: $e \xrightarrow{l_1} e_1$ and $e \xrightarrow{l_2} e_2$
Output: $e \xrightarrow{l} e_1 \sqcup e_2$
Calculation:
– If l_1 and l_2 are mutually exclusive: Use Rule 13.5 to calculate l
– If l_1 and l_2 are statistically independent: Use Rule 13.6 to calculate l
– Else: $l \in [\max(l_1, l_2), \min(l_1 + l_2, 1)]$

Table 13.4 Additional guidelines for probability calculation in complete diagrams

How to calculate probabilities using complete CORAS diagrams

Input: $e_1(p_1)$, $e_2(p_2)$, $e_1 \xrightarrow{l_1} e$ and $e_2 \xrightarrow{l_2} e$
Output: $e(p)$
Assumption: $e = (e_1 \sqcap e) \sqcup (e_2 \sqcap e)$
Calculation: Calculate $(e_1 \sqcap e)(p_1 \cdot l_1)$ and $(e_2 \sqcap e)(p_2 \cdot l_2)$ using Rule 13.2
– If $e_1 \sqcap e$ and $e_2 \sqcap e$ are mutually exclusive: Use Rule 13.3 to further calculate p
– If $e_1 \sqcap e$ and $e_2 \sqcap e$ are statistically independent: Use Rule 13.4 to further calculate p
– Else: $p \in [\max(p_1 \cdot l_1, p_2 \cdot l_2), \min(p_1 \cdot l_1 + p_2 \cdot l_2, 1)]$

Rule 13.10 (Leads-to) For the scenarios/incidents e_1 and e_2 related by the leads-to relation, we have:

$$\frac{e_1(f) \quad e_1 \xrightarrow{l} e_2}{(e_1 \sqcap e_2)(f \cdot l)}$$

Rule 13.3 through Rule 13.8 for probability calculation in Sect. 13.1.2 all make assumptions about statistical independence or mutual exclusiveness. The distinction between these two cases is equally important when we are reasoning about frequencies.

Two scenarios/incidents e_1 and e_2 are mutually exclusive if in any run or execution (of the target or system in question) in which e_1 occurs, e_2 cannot occur, and the other way around. We therefore get the following rule for frequencies in the case of mutual exclusion.

Rule 13.11 (Mutually exclusive scenarios/incidents) If the scenarios/incidents e_1 and e_2 are mutually exclusive, we have:

$$\frac{e_1(f) \quad e_2(f)}{(e_1 \sqcup e_2)(f)}$$

Two scenarios/incidents e_1 and e_2 are separate if they do not overlap in content. That e_1 and e_2 do not overlap in content means that no possible instance of the one can be an instance of the other. This also means that one scenario/incident cannot be a special case of the other. For example, since the scenario *Virus infects DB* is a special case of *Malware infects DB*, these two scenarios are not separate. Given separate events we get the following rule for statistical independence.

Rule 13.12 (Independent scenarios/incidents) If the scenarios/incidents e_1 and e_2 are separate and statistically independent, we have:

$$\frac{e_1(f_1) \quad e_2(f_2)}{(e_1 \sqcup e_2)(f_1 + f_2)}$$

Example 13.8 We now demonstrate the application of the frequency rules by revisiting Example 13.1. Since both the assumptions and the course of the example remain the same we do a more brief walk-through this time.

We use the diagram of Fig. 13.6 as basis for the frequency calculations and make the same diagram assumptions as in Example 13.1. We furthermore assume that the frequencies on the initiates relations as well as the conditional likelihoods have already been estimated. What remains is to explain how the frequencies of the threat scenarios and the unwanted incident are derived.

13.1 Using CORAS Diagrams to Calculate Likelihood

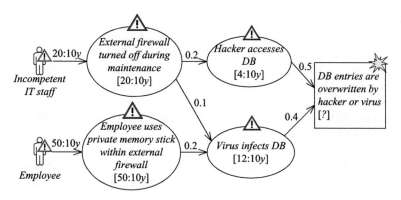

Fig. 13.6 A diagram used for calculating the frequency of threat scenarios

Since the diagram is complete we can use Rule 13.9 to deduce the frequencies $20:10y$ and $50:10y$ of the threat scenarios ef and eu, respectively, based on the frequencies on the initiates relations. In order to derive the frequency of ha we use Rule 13.10 and calculate $0.2 \cdot 20:10y = 4:10y$.

As to the frequency of vi we first calculate the frequencies of $ef \sqcap vi$ and $eu \sqcap vi$ by using Rule 13.10. This gives $0.1 \cdot 20:10y = 2:10y$ and $0.2 \cdot 50:10y = 10:10y$, respectively. By assumption of completeness of the diagram we have that $vi = (ef \sqcap vi) \sqcup (eu \sqcap vi)$. Given that $ef \sqcap vi$ and $eu \sqcap vi$ are separate we therefore use Rule 13.12 to calculate the frequency of threat scenario vi, which is $2:10y + 10:10y = 12:10y$. Note that this requires that $ef \sqcap vi$ and $eu \sqcap vi$ are counted as two separate events even if they occur at the same point in time.

As in Example 13.1, we cannot based on the diagram and the assumptions we have made calculate the exact frequency of the unwanted incident de. This is because the threat scenarios ha and vi that lead to de are not statistically independent or mutually exclusive. We can, however, deduce an approximation by calculating the maximum and minimum value.

By assumption of completeness, it is only ha and vi that can lead to de, so the frequency of de cannot be higher than the sum of the frequencies of $ha \sqcap de$ and $vi \sqcap de$. Applying Rule 13.10 for each of these yields the frequency $4:10y \cdot 0.5 = 2:10y$ for $ha \sqcap de$ and the frequency $12:10y \cdot 0.4 = 4.8:10y$ for $vi \sqcap de$. The maximum frequency of de is therefore $6.8:10y$, which corresponds to $17:25y$. Since the frequency of de cannot be lower than the highest of the frequencies of $ha \sqcap de$ and $vi \sqcap de$, the minimum frequency of de is $4.8:10y$, which corresponds to $12:25y$. The frequency of de is therefore in the interval $[12, 17]:25y$.

Example 13.9 As a further example of frequency calculation, consider again the diagram of Fig. 13.5 from Example 13.5. This CORAS diagram illustrates the special cases in which one path is included in another in the sense that the former path is a special case of the latter.

When we are calculating the likelihoods in such diagrams we proceed in the same way independent of whether the likelihoods are specified as probabilities or frequencies. Therefore, assuming that the frequency of the initiates relation from ii to ef is 20 : 10y then also the frequency of ef is 20 : 10y by Rule 13.9. By Rule 13.10, the frequency of vi is 2 : 10y. Finally, by Rule 13.10 the frequency of the unwanted incident de is 2 : 10y. Since the occurrences of de that are preceded by vi are included in the occurrences that are preceded by ef, we do not aggregate the contribution from vi into the frequency of de that we have calculated. Nevertheless, we can use Rule 13.10 to calculate the frequency 0.8 : 10y as the contribution of vi to the frequency of de.

Rule 13.5 and Rule 13.6 refer only to conditional likelihoods and therefore remain the same irrespective of whether the likelihoods of threat scenarios and unwanted incidents are given in terms of probability or frequency.

Rule 13.7 and Rule 13.8 have no corresponding rules for the cases in which the likelihoods are given in terms of frequencies. This is because the conditional likelihoods are specified in terms of probability. However, as we will explain next, there are methods for converting frequencies into probabilities and thereby making these rules applicable in the general case.

13.1.4 Likelihood as Probability or Frequency

Likelihood can be expressed in terms of either a probability, in other words a value between 0 and 1, or a frequency, such as *twice per year*. When a probability x is assigned to a threat scenario, for example, it can be taken to mean that the probability is x of this threat scenario to occur during some implicit observation period. Hence, a probability can be understood as a frequency where the unit of observation has not been explicitly stated. For example, if the implicit observation period is one day, then a probability of 0.05 corresponds to a frequency of about 18 times per year.

Conversely, by making the observation period implicit rather than explicit, a frequency can in certain cases be understood as a probability. This applies in cases where the observation period is sufficiently small, so that the probability of the scenario/incident to occur more than once during the observation period is close to 0, and the frequency is very low (typically less than 0.1) [2]. The choice of an observation period small enough to make all frequencies, and thereby the derived probabilities, less than 0.1 is simply a rule of thumb.

Converting frequencies to probabilities relies on the probability of events to occur more than once to be negligible, since probabilities that are higher than 1 make no sense.

By choosing a sufficiently small observation period, we can express a frequency as a probability. For example, by dividing with $365 \cdot 24$ we see that a frequency of 10 times per year corresponds to 0.00114 per hour, which can be interpreted as meaning that the probability is 0.00114 that the event in question will occur once during one hour.

13.1 Using CORAS Diagrams to Calculate Likelihood

Table 13.5 Guidelines for converting frequencies to probabilities

How to convert frequencies to probabilities in CORAS diagrams
Each frequency in the diagrams of a risk analysis is given by a number x of occurrences over a time unit T, and is denoted $x : T$. The frequencies can be converted to probabilities by the following three steps: 1. Choose a constant c such that $\frac{y}{c} < 0.1$ for the highest frequency $y : T$ in the diagrams   2. Convert each frequency $x : T$ in the diagrams to the equivalent frequency $\frac{x}{c} : \frac{T}{c}$ 3. Convert each resulting frequency $\frac{x}{c} : \frac{T}{c}$ to the probability $\frac{x}{c}$

Expressing frequencies as probabilities is useful not only because we have more support for reasoning about probabilities than for reasoning about frequencies; CORAS diagrams allow likelihoods to be specified both as frequencies and probabilities, even in one and the same diagram, and it can therefore be useful to be able to convert the values into a set of more uniform representations for which the calculation rules can be applied.

Example 13.10 Consider again the threat diagram of Fig. 13.6. In Example 13.8, we calculated the frequencies of the threat scenarios based on the previous estimations of the frequencies of the initiates relations and the conditional likelihood of the leads-to relations. By converting the frequencies of the initiates relations into probabilities, we can do the likelihood analysis on the basis of probabilities only.

The frequencies assigned to the leads-to relations are $20 : 10y$ and $50 : 10y$. These frequencies cannot immediately be interpreted as probabilities, since a probability of 20 or 50 does not make sense. We must therefore convert the time interval of 10 years to a time interval that is sufficiently small. This can be done by, for example, operating with a time interval of 5 days instead.

10 years equals 3650 days (ignoring leap years), so we need to divide the frequencies by 730 to obtain the number of occurrences per 5 days. The frequency of $20 : 10y$ then corresponds to the frequency of $0.0274 : 5d$ and is therefore converted to the probability 0.0274. The frequency of $50 : 10y$ is in the same way converted to the probability 0.0685.

With this conversion of frequencies to probabilities, we can do the likelihood analysis of the threat diagram by using Rule 13.1 through Rule 13.8.

In Table 13.5, we give guidelines for how to convert frequencies into corresponding probabilities.

13.1.5 Generalisation to Intervals and Distributions

So far in this chapter, we have explained how to reason about likelihood using CORAS diagrams, but the likelihoods have until now been exact. When conducting a risk analysis in practice, however, we are often forced to use intervals. The use of likelihood intervals is exemplified in Table 8.10 on p. 118 where for example *possible* is defined by the interval [6, 19] : 10y. In the following, we explain how the calculations performed on exact likelihoods as presented so far can be generalised to intervals.

The generalisation of the calculations on exact values captured by the rules above to calculations on intervals is straightforwardly defined as follows:

Definition 13.1 $[min_1, max_1]$ op $[min_2, max_2] = [min_1 \ op \ min_2, max_1 \ op \ max_2]$, where op is one of $+$ (addition), $-$ (subtraction) and $\cdot$ (multiplication).

Notice that subtraction is only used in the context of probabilities. Notice also that we can combine an exact value with an interval by using the exact value as both the minimum and the maximum value of one of the intervals.

Remark Probabilities can never exceed the interval [0, 1], and the applicability of the rules in this chapter relies on valid input values. Moreover, for mutual exclusiveness, the sum of the input values can never exceed [0, 1]. When operating with less precise probability estimates in terms of intervals, it may be that the sum of the maximum input values is higher than 1. If more precise input values cannot be estimated, the sum of the maximum input values must be replaced by 1 in the output.

When using intervals that are understood only as maximum and minimum values, we may end up with large intervals that provide little information. Assume, for example, that threat scenario e will occur if at least one of the threat scenarios e_1, e_2, and e_3 occurs, and that e_1, e_2, and e_3 are mutually exclusive. Furthermore, assume that the probability interval [0.1, 0.3] is assigned to each of these three threat scenarios. By adding up the minimum values and maximum values, respectively, for e_1, e_2, and e_3, we obtain the probability interval [0.3, 0.9] for e. This means that the probability of e to occur is somewhere between 0.3 and 0.9, which does not say very much.

However, by using distributions rather than maximum and minimum values, we may obtain more specific information about likelihoods. For example, assume that we in the above example regard the intervals assigned to e_1, e_2, and e_3 as distributions with mean value in the middle of the interval, in other words 0.2. Examples of such distributions include the uniform distribution within the interval or the triangle distribution with its mode (in other words the most likely value) in the middle of the interval. As the mean of the sum of two distributions is equal to the sum of their means, we can then conclude that the mean value for the probability of e is 0.6. Intuitively, this would mean that the probability of e is most likely close to 0.6, but it can be as low as 0.3 or as high as 0.9.

Assigning distributions of likelihood may be particularly suitable when obtaining estimates from many different experts for the same threat scenarios, unwanted incidents or relations. In such cases, rather than insisting that the experts agree on a value or interval, a distribution could be used to capture the differing opinions of the experts. Obviously, the use of distributions could go much further than calculating mean values and extreme values in the way illustrated by the above example. For example, we could also study the variance, which would say something about *how likely* it is that the value is close to 0.6. However, calculations on distributions is a large field and will not be pursued further in this book. For more on the use of distributions in risk analysis, see, for example, [81].

13.2 Using CORAS Diagrams to Check Consistency

We now show how to check that the likelihood assignments are meaningful, in other words that they make sense given the semantics of CORAS diagrams. More specifically, we show how to check that the likelihoods assigned to the various parts of a CORAS diagram are consistent with each other. Unlike in the above sections, we now assume that likelihoods have been estimated for all the relevant threat scenarios, unwanted incidents, initiates relations and leads-to relations. Therefore, rather than calculating missing values from values assigned to other elements of the diagram, the task is now to check whether the already obtained estimates are consistent with each other.

For a given threat scenario or unwanted incident $e(l)$, where l is the assigned likelihood, we use the calculation rules to deduce $e(l')$ based on the likelihoods assigned elsewhere in the diagram. Checking consistency therefore amounts to comparing the two likelihoods l and l'. The requirements to consistency depend on whether the likelihoods are exact or given as intervals, and they depend on whether the diagrams are complete or not. In the following, we first address consistency of complete diagrams, and subsequently we address consistency of incomplete diagrams.

For exact likelihoods in complete diagrams, the consistency requirement is that the assigned likelihood l equals the likelihood l' that can be deduced from the other parts of the diagram. In case $l \neq l'$, there is an inconsistency that must be resolved. This can be done either by changing the assigned value l to the deduced value l', or by changing the likelihoods of relations, threat scenarios and/or unwanted incidents elsewhere in the diagram such that the deduced likelihood l' equals the already assigned likelihood l. The choice of which values to change and which ones to keep in order to ensure consistency depends on which values we have strongest confidence in and evidence for.

For likelihood intervals in complete diagrams, the consistency requirement is that the assigned interval $[l_i, l_j]$ of a threat scenario or unwanted incident e is wider than the interval $[l'_i, l'_j]$ that can be deduced from the parts that precede e in the diagram. This means that $[l'_i, l'_j] \subseteq [l_i, l_j]$, or equivalently that $l_i \leq l'_i$ and $l_j \geq l'_j$. As before, we resolve inconsistencies either by changing the assigned likelihood interval $[l_1, l_j]$, or by changing the likelihood assignments elsewhere in the diagram.

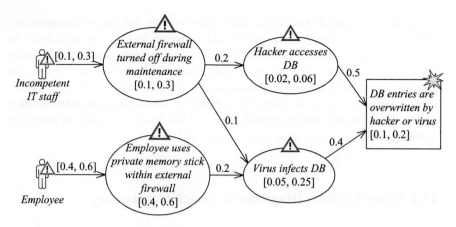

Fig. 13.7 A diagram used for checking the consistency of likelihood assignments

Example 13.11 Figure 13.7 shows the diagram that we use to demonstrate consistency checking. Except for the likelihood estimates, the diagram is identical to the one in Fig. 13.2, and we also use the same shorthand notation and make the same assumptions as in Example 13.1. In particular, we assume that the diagram is complete.

In practice, likelihood estimates are often given as intervals, or qualitative values corresponding to intervals, rather than exact values. We have therefore assigned probability intervals to the initiates relations, the threat scenarios and the unwanted incident in Fig. 13.7. For the leads-to relations, we have chosen to assign a single conditional likelihood. If desired, intervals could also have been assigned to these.

> **Remark** Both the likelihood annotations of threat scenarios and unwanted incidents in CORAS diagrams as well as the probability intervals are written in square brackets. When strictly adhering to the graphical notation of the CORAS language, these annotations should therefore appear in double square brackets in the diagrams. For reasons of readability, however, we omit the outermost brackets.

We start the consistency checking on the left-hand side of the diagram. For the threat scenarios ef and eu, we see that the probability intervals that have been assigned are the same as for the incoming initiates relations from the corresponding threats. This makes perfect sense, as we have assumed that $ef = ii \sqcap ef$ and $eu = em \sqcap eu$, in other words that these threat scenarios can only be initiated by the threats shown in the diagram.

The next step is to check the probability assigned to ha. This is done by calculating the likelihood of $ef \sqcap ha$, which by assumption is equal to ha, from the likelihood of ef and the conditional likelihood on the leads-to relation from ef to ha. We subsequently check whether the obtained result corresponds to the likelihood already assigned to ha. We use Rule 13.2 in the same way as in Sect. 13.1.2. The

13.2 Using CORAS Diagrams to Check Consistency

only difference is that we now use the generalisation of operations to intervals, as defined in Definition 13.1. Hence, we obtain the minimum value of $0.1 \cdot 0.2 = 0.02$ and the maximum value of $0.3 \cdot 0.2 = 0.06$, which gives the interval $[0.02, 0.06]$. This is in fact identical to the interval already assigned to ha, which means that the values are fully consistent.

A similar approach is taken for calculating the likelihood of vi. That is, we compare the likelihood already assigned to vi with the value we obtain from applying the appropriate rules from Sect. 13.1.2 generalised to intervals as defined by Definition 13.1. We use Rule 13.4 to calculate the likelihood of vi, which by assumption is identical to $(ef \sqcap vi) \sqcup (eu \sqcap vi)$, from the likelihood of ef, eu, and the leads-to relations from each of these threat scenarios to vi. By Rule 13.2, the probability of $ef \sqcap vi$ is $0.1 \cdot [0.1, 0.3] = [0.01, 0.03]$, and the probability of $eu \sqcap vi$ is $0.2 \cdot [0.4, 0.6] = [0.08, 0.12]$. We then use Rule 13.4 to calculate the probability of $(ef \sqcap vi) \sqcup (eu \sqcap vi)$, which is $([0.01, 0.03] + [0.08, 0.12]) - ([0.01, 0.03] \cdot [0.08, 0.12])$. This yields $[0.09, 0.15] - [0.0008, 0.0036]$, which finally equals the probability interval $[0.0892, 0.1464]$. This probability interval is a subset of the interval $[0.05, 0.25]$ that has already been assigned to the threat scenario vi. The likelihood is therefore consistent with the other likelihoods of the diagram.

Although not inconsistent, it is worth reconsidering assigned intervals that are not exactly equal to the likelihood interval that can be deduced from the diagram. The more narrow an interval, the more precise the likelihood estimate. We should therefore choose the most narrow interval when the correctness of this interval can be justified by the expert judgements or other evidence.

In the diagram of Fig. 13.7, the deduced likelihood of vi is much more precise than the already assigned likelihood. If there in this example is a very strong justification for the already assigned probability interval to vi, then the likelihoods of the preceding threats and threat scenarios should be reconsidered. If it is the other way around, the probability estimate of vi should be altered in order to make the likelihood estimates as precise as possible. For the remainder of the example, however, we keep the initial estimate $[0.05, 0.25]$ since our task is to check consistency of the already assigned estimates.

Finally, we check whether the likelihood interval assigned to de is consistent with the rest of the diagram. As discussed at the end of Sect. 13.1.2, we have no rule for calculating the likelihood of de from the likelihoods of ha, vi and the corresponding leads-to relations as ha and vi are not mutually exclusive or statistically independent. We therefore calculate maximum and minimum values for de from the likelihoods of $(ha \sqcap de)$ and $(vi \sqcap de)$ in a similar way as in Sect. 13.1.2 by using Rule 13.2 and the generalisation to intervals defined by Definition 13.1. Thus, we obtain the interval $[0.01, 0.03]$ for $(ha \sqcap de)$ and $[0.02, 0.1]$ for $(vi \sqcap de)$. The maximum likelihood for de is then the sum of the maximum values for these intervals, that is, 0.13. The minimum likelihood of de can clearly not be lower than the highest of the minimum likelihoods of $(ha \sqcap de)$ and $(vi \sqcap de)$. Therefore, the minimum likelihood for de is obtained by choosing the highest of the minimum likelihood values for $(ha \sqcap de)$ and $(vi \sqcap de)$, that is, 0.02.

According to these calculations, the likelihood assigned to *de* should be [0.02, 0.13]. This is not consistent with the likelihood assigned in the diagram, which says [0.1, 0.2]; the minimum likelihood 0.1 is too high.

The consistency checking in this example is based on the assumption that the threat diagram in question is complete. Such completeness assumptions allow us to check the consistency of both minimum and maximum likelihood values since we know that there are no other sources of the threat scenarios and unwanted incidents than those documented in the diagram. In cases of incompleteness, there are one or more threat scenarios and/or unwanted incidents that can occur as a result of threats and threat scenarios that are not described in the diagram. The likelihoods that have been assigned can therefore possibly be higher than what the diagram accounts for. However, since the actual likelihoods can never be smaller than what is contributed by the threats and threat scenarios that are described in the diagram, we can consistency check diagrams in that respect.

For exact likelihoods in incomplete diagrams, the consistency requirement is that the assigned likelihood l to a threat scenario or unwanted incident e is equal to or higher than the likelihood l' that can be deduced from the other parts of the diagram. In case $l < l'$, there is an inconsistency that must be resolved.

For likelihood intervals in incomplete diagrams, the consistency requirement is that the highest value l_j of the assigned interval $[l_i, l_j]$ of e is equal to or higher than the highest value l'_j of the interval $[l'_i, l'_j]$ that can be deduced from the parts that precede e in the diagram. The two intervals are therefore mutually inconsistent in case $l_j < l'_j$.

Example 13.12 If we throw out the completeness assumption of the previous example that *de* can only result from *ha* or *vi*, in other words discard that $de = (ha \sqcap de) \sqcup (vi \sqcap de)$, then the probability interval [0.1, 0.2] that has been assigned to *de* is consistent with the deduced interval [0.02, 0.13]. The reason is that in this case there may be additional threat scenarios that are not shown in the diagram and that may lead to *de*, so that the likelihood of *de* can indeed be higher than the likelihood of $(ha \sqcap de) \sqcup (vi \sqcap de)$. These additional threat scenarios could then account for the higher minimum likelihood assigned to *de*.

In Table 13.6, we give guidelines for how to consistency check likelihoods in CORAS diagrams. Notice that the consistency requirements are based on the assumption that the calculated values are deduced by reasoning from left to right in the diagrams. If we instead reason from right to left, the consistency requirements are the converse. For example, if we have an incomplete diagram with the specification $e_1(l_1)$, $e_2(l_2)$ and $e_1 \xrightarrow{l} e_2$, we can check the assigned likelihood l_1 against the deduced probability $l'_1 = \frac{l_2}{l}$. In that case the consistency requirement is that $l_1 \leq l'_1$.

Table 13.6 Guidelines for consistency checking likelihoods

How to check consistency of likelihoods in CORAS diagrams
Exact values in complete diagrams   Assigned value: $e(l)$ Calculated value: $e(l')$ Consistency check: $l = l'$
Exact values in incomplete diagrams   Assigned value: $e(l)$ Calculated value: $e(l')$ Consistency check: $l \geq l'$
Intervals in complete diagrams   Assigned interval: $e([l_i, l_j])$ Calculated interval: $e([l'_i, l'_j])$ Consistency check: $[l'_i, l'_j] \subseteq [l_i, l_j]$ or, equivalently $l_i \leq l'_i$ and $l_j \geq l'_j$
Intervals in incomplete diagrams   Assigned interval: $e([l_i, l_j])$ Calculated interval: $e([l'_i, l'_j])$ Consistency check: $l_j \geq l'_j$

13.3 Using CORAS to Analyse Scenarios with Logical Connectives

Scenarios and incidents described in natural language are often expressed with the logical connectives for conjunction ("and") and for disjunction ("or"), or at least they can be rephrased and understood in terms of these connectives. For example, the description "server A is down and server B is up or vice versa" can be understood as "either server A is down and server B is up or server A is up and server B is down". When modelling scenarios, it is important to structure the diagram in order to capture the aspects of relevance for the analysis. The combination of textual descriptions of scenarios and incidents and the CORAS language constructs gives a high degree of flexibility with respect to structuring diagrams. In this section, we show how different kinds of scenarios expressed in terms of logical connectives may be represented.

13.3.1 Using CORAS to Analyse Scenarios with Logical Conjunction

Logical conjunction is sometimes used to express that one event follows another, as in the phrase "there was a production flaw in the server and the server went down". Reversing the order of the conjuncts here would yield a slightly odd sentence. From

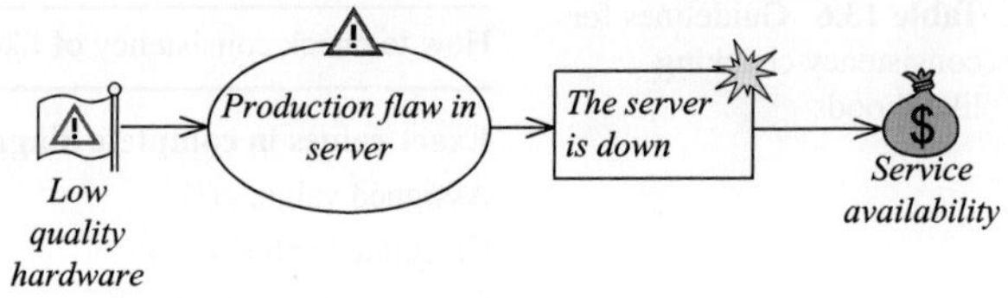

Fig. 13.8 A temporal interpretation of logical conjunction represented by a leads-to relation from *Production flaw in server* to *The server is down*

Fig. 13.9 The case where both servers go down is represented by a single unwanted incident

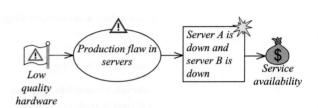

the meaning of the two conjuncts, it is reasonable to assume that the production flaw lead to the server going down. Such cases can be modelled with the use of a leads-to relation, as illustrated by Fig. 13.8.

A possibly more common use of logical conjunction is to express that the scenarios described by the conjuncts occur together, as in the phrase "server A is down and server B is down". Reversing the order of conjuncts would not change the meaning in this case. In CORAS, we may use a single threat scenario or unwanted incident to represent this as explained in the following example.

Example 13.13 Consider a service that should be available at all times. In order to ensure this, the service is deployed on two identical servers A and B. The two servers are made by the same manufacturer and share the same production flaws. A production flaw may lead to server A going down, or server B going down, or both. If one server goes down, all traffic will be directed to the other server while the problem is taken care of. The service will be unavailable only if both servers are down at the same time. No noticeable loss of response time or other problems is expected as long as at least one server is up. Figure 13.9 shows one way of capturing this scenario. The case where both servers go down is represented by a single unwanted incident *Server A is down and server B is down*. That only one server is down is not captured, as this will not by itself have any consequence for the service availability.

The approach taken in Example 13.13 is suitable in situations where we are only interested in the case where all the scenarios described by the conjuncts occur together. However, in practice, we often want to represent also cases where a conjunct occurs alone. This is the situation, for example, if a conjunct occurring alone will also impact an asset, or if we want a more fine-grained likelihood analysis. Example 13.14 shows how this can be represented.

Example 13.14 Assume that a service is deployed on two servers A and B in order to achieve redundancy as well as to handle a high amount of traffic. Server A is

13.3 Using CORAS to Analyse Scenarios with Logical Connectives

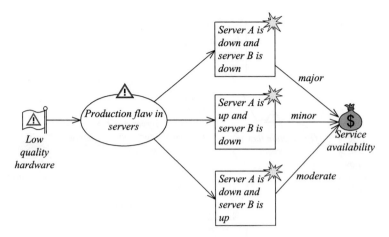

Fig. 13.10 Separate unwanted incidents have been included to represent the cases where exactly one server goes down

dedicated to running only the service in question, while server B also runs a number of other applications that require much of its resources. Therefore, the load balancer is configured so that server A handles most of the service requests as long as both servers are up. If only server A goes down, this will lead to a major increase in response time (which is seen as a moderate consequence for the service availability), while only server B going down will just lead to a minor increase. Figure 13.10 shows how this scenario can be captured. Separate unwanted incidents have been included for the cases where only server A or only server B goes down.

In relation to Example 13.14, we did not discuss whether the servers go down at exactly the same point in time or one after the other. Given that the threat scenario leading to the unwanted incidents is *Production flaw in servers*, it seems reasonable to assume the latter. This has not been represented in the diagram, which is perfectly acceptable if this aspect is not considered relevant for the analysis. However, there are cases where we want an explicit representation of the order, as illustrated by Example 13.15.

Example 13.15 Assume that the servers A and B from Example 13.14 are located at different sites. Server A is located at a site without easy access to spare components, while server B is located at a site where faulty hardware can be replaced swiftly. This means that if server B goes down first, it is more likely that the problem will be fixed before also server A goes down than the other way around. Figure 13.11 shows how this scenario can be captured. The two unwanted incidents *Server A is down and server B is up* and *Server A is up and server B is down* have leads-to relations to *Server A is down and server B is down* with different probabilities.

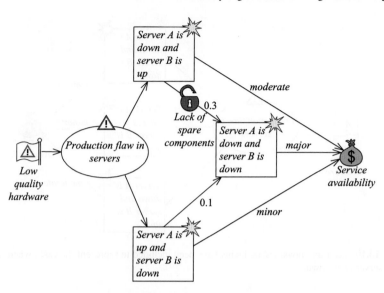

Fig. 13.11 The order of unwanted incidents have been explicitly represented

13.3.2 Using CORAS to Analyse Scenarios with Logical Disjunction

Outgoing relations from a threat scenario or an unwanted incident can be understood as logical disjunction between the diagram elements to which the relations lead. For example, the outgoing leads-to relations from the threat scenario *Production flaw in servers* in Example 13.14 can be understood as logical disjunction between the three unwanted incidents; the threat scenario may lead to *Server A is down and server B is down*, or to *Server A is up and server B is down*, or to *Server A is down and server B is up*.

In Fig. 13.10, we included separate unwanted incidents for all the possible combinations of servers A and B being up or down. However, there are cases where this level of detail is not needed. Typically, this will be the case of threat scenarios or unwanted incidents mirror each other, as will be the case when we have identical redundant components. In such cases, there is rarely any point in distinguishing between which of the components that are down. To reduce the number of diagram elements and obtain a simpler risk picture, we can then merge elements by the use of "or" in the description, as illustrated by Example 13.16.

Example 13.16 Assume that a service has been deployed on two identical servers A and B in order to achieve redundancy, as well as to handle a high amount of traffic while maintaining an acceptable response time. The load is shared equally between the two servers as long as both servers are up. If one server goes down then all traffic will be directed to the other server. This may increase the response time for the service, which is seen as a reduction of availability. The diagram in Fig. 13.12

13.4 How to Structure a Threat Diagram to Exploit the Potential for Likelihood Analysis

shows how this scenario may be captured. The case where either server A is up and server B is down or server A is down and server B is up is represented by a single unwanted incident. No distinction is made between the cases where server A is down and server B is down, as the two servers are identical.

13.4 How to Structure a Threat Diagram to Exploit the Potential for Likelihood Analysis

As illustrated by Example 13.1, we do not have rules for calculating likelihoods in cases where threat scenarios and unwanted incidents are neither statistically independent nor mutually exclusive, although we may calculate upper and lower bounds. In this section, we demonstrate two ways of restructuring the diagram to avoid such cases.

13.4.1 Enabling Application of Rules by Composition

For the diagram in Fig. 13.2, we are not able to calculate the likelihood of the unwanted incident *de* because the two threat scenarios *ha* and *vi* that may lead to *de* are neither statistically independent nor mutually exclusive. The easiest way of restructuring such a diagram is to simply compose the two threat scenarios into a single threat scenario. Figure 13.13 shows a diagram where we have composed the threat scenarios *ha* and *vi* into a single threat scenario.

Notice that we have combined the text for the original scenarios with *or*. In the following, we use the shorthand notation $ha \vee vi$ for the composed threat scenario *Hacker accesses DB or virus infects DB*. Relations to or from any of the original threat scenarios now lead to or from the composed scenario.

In cases where we need to restructure a diagram, the simplest approach is clearly to do this before we estimate and assign likelihoods to the diagram. This ensures that

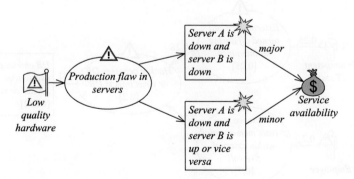

Fig. 13.12 A single unwanted incident has been included for the case where exactly one server is down, irrespective of whether it is server A or server B

the estimates apply to the restructured diagram rather than the original one. However, in the following we show how likelihoods in the original diagram, if already assigned, can be used to guide the assignment of likelihoods to the restructured diagram. In our case, the original diagram is the one from Fig. 13.2. We furthermore make the same assumptions and use the same shorthand notation as in Example 13.1.

The likelihoods of ef and eu in the composed diagram are the same as in the original diagram as these are not affected by the restructuring. Furthermore, the conditional likelihood for the leads-to relation from eu to $ha \vee vi$ is the same as for the relation from eu to vi in the original diagram in Fig. 13.2. The reason is that there is no relation from eu to ha in the original diagram, so the conditional likelihood that eu will lead to $ha \vee vi$ is the same as the conditional likelihood that eu will lead to vi.

In order to determine the conditional likelihood for the leads-to relation from ef to $ha \vee vi$, we need to take into consideration both leads-to relations from ef. Clearly, the conditional likelihood for the relation from ef to $ha \vee vi$ should lie somewhere between 0.2, which is the highest of the conditional likelihoods on leads-to relations from ef in the original diagram, and 0.3, which is the sum of the conditional likelihoods on leads-to relations from ef. Assuming the conditional likelihoods on the leads-to relations from ef to ha and vi in Fig. 13.2 are statistically independent, we apply Rule 13.6 to calculate the conditional likelihood $0.10 + 0.20 - 0.10 \cdot 0.20 = 0.28$ on the relation from ef to $ha \vee vi$.

Having established likelihoods for the leads-to relations from ef to $ha \vee vi$ and from eu to $ha \vee vi$, we may now calculate the likelihood of $ef \sqcap (ha \vee vi)$ and $eu \sqcap (ha \vee vi)$. By applying Rule 13.2, we obtain $0.25 \cdot 0.28 = 0.07$ for $ef \sqcap (ha \vee vi)$ and $0.7 \cdot 0.2 = 0.14$ for $eu \sqcap (ha \vee vi)$.

Based on the assumption that $ef \sqcap (ha \vee vi)$ and $eu \sqcap (ha \vee vi)$ are statistically independent, we may now apply Rule 13.4 to calculate the likelihood of $(ef \sqcap (ha \vee vi)) \sqcup (eu \sqcap (ha \vee vi))$, thus obtaining $0.07 + 0.14 - 0.07 \cdot 0.14 = 0.2002 \approx 0.2$. Maintaining the assumption that the diagram is complete, which means that $ha \vee vi = (ef \sqcap (ha \vee vi)) \sqcup (eu \sqcap (ha \vee vi))$, we therefore assign the likelihood 0.2 to $ha \vee vi$.

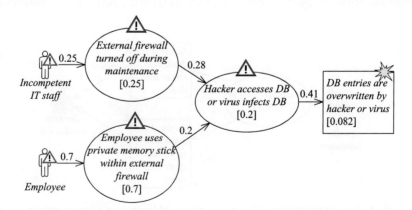

Fig. 13.13 Composition of threat scenarios in order to allow calculation of likelihood

13.4 How to Structure a Threat Diagram to Exploit the Potential for Likelihood Analysis

Table 13.7 Guidelines for composition of diagram elements

How to compose threat scenarios or unwanted incidents

Input: Scenarios/incidents e_1 and e_2
Output: The composition $e_1 \vee e_2$ (i.e. e_1 *or* e_2)
Procedure:

- Replace the scenarios/incidents e_1 and e_2 by the composed scenario/incident $e_1 \vee e_2$
- Replace each relation $e \to e_1$ by $e \to e_1 \vee e_2$
- Replace each relation $e \to e_2$ by $e \to e_1 \vee e_2$
- Note: For each e such that $e \to e_1$ and $e \to e_2$, the relation $e \to e_1 \vee e_2$ is the composed relation that replaces both of the former
- Replace each relation $e_1 \to e$ by $e_1 \vee e_2 \to e$
- Replace each relation $e_2 \to e$ by $e_1 \vee e_2 \to e$
- Note: For each e such that $e_1 \to e$ and $e_2 \to e$, the relation $e_1 \vee e_2 \to e$ is the composed relation that replaces both of the former

This demonstrates how we can extract additional information by composition of scenarios or incidents; in Example 13.1 we could not do any further reasoning about *ha* and *vi* since they are neither mutually exclusive nor statistically independent. In particular, Rule 13.3 and Rule 13.4 are not applicable. However, by the composition of these threat scenarios and the relations to them, we were able to deduce the likelihood of the composition $ha \vee vi$.

The leads-to relation $ha \vee vi \to de$ is the composition of the leads-to relations $ha \to de$ and $vi \to de$ from Fig. 13.2. The problem of calculating the probability of *de* remains, namely that *ha* and *vi* are neither mutually exclusive nor statistically independent, which are the assumptions required for the applicability of Rule 13.7 and Rule 13.8. We will therefore proceed with the analysis of the example by utilising decomposition techniques. Thereby, we will explain how we have obtained the conditional likelihood 0.41 for the leads-to relation and the probability 0.082 for the unwanted incident *de* as annotated in the diagram of Fig. 13.13. For now, the best approximation we have is still the interval $[0.065, 0.090]$ that we obtained in Example 13.1.

In Table 13.7, we give guidelines for how to conduct de composition of diagram elements.

13.4.2 Enabling Application of Rules by Decomposition

In some cases, the composition of threat scenarios, unwanted incidents and relations as demonstrated above will not be sufficient. For example, the resulting diagram may be considered to give a too course-grained picture. In such cases, we may take the alternative approach of decomposing the threat scenarios or unwanted incidents

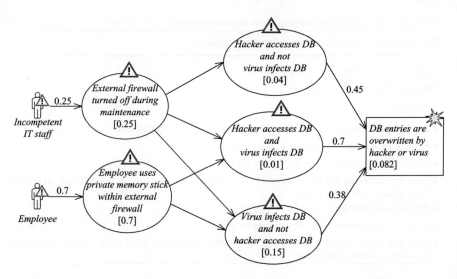

Fig. 13.14 Decomposition of threat scenarios in order to allow calculation of likelihood

in question into mutually exclusive scenarios or incidents, respectively. This can be achieved by combining the original scenarios or incidents with use of *and* and *not*. Figure 13.14 illustrates how the diagram in Fig. 13.2 can be restructured in this way. Here, the two overlapping threat scenarios *Hacker accesses DB* and *Virus infects DB* have been restructured into the mutually exclusive threat scenarios *Hacker accesses DB and not virus infects DB*, *Hacker accesses DB and virus infects DB* and *Virus infects DB and not hacker accesses DB*.

We now show how the likelihoods assigned to the original diagram in Fig. 13.2 have been used to guide the assignment of likelihoods to the diagram Fig. 13.14. We use the following shorthand notations for the annotated names and descriptions in Fig. 13.14:

$$ii = \textit{Incompetent IT staff}$$
$$em = \textit{Employee}$$
$$ef = \textit{External firewall turned off during maintenance}$$
$$eu = \textit{Employee uses private memory stick within external firewall}$$
$$ha \wedge \neg vi = \textit{Hacker accesses DB and not virus infects DB}$$
$$ha \wedge vi = \textit{Hacker accesses DB and virus infects DB}$$
$$vi \wedge \neg ha = \textit{Virus infects DB and not hacker accesses DB}$$
$$de = \textit{DB entries are overwritten by hacker or virus}$$

We first observe that the likelihoods that have been assigned to ha and vi in Fig. 13.2 give the following restrictions for the likelihoods of $ha \wedge \neg vi$, $ha \wedge vi$ and $vi \wedge \neg ha$:

13.4 How to Structure a Threat Diagram to Exploit the Potential for Likelihood Analysis 241

- The threat scenarios that include ha are $ha \land \neg vi$ and $ha \land vi$. As these threat scenarios are mutually exclusive, it is clear that the sum of their likelihood should equal the likelihood of ha, which according to Fig. 13.2 is 0.05.
- Similarly, the sum of likelihoods for $ha \land vi$ and $vi \land \neg ha$ should equal the likelihood of vi, which according to Fig. 13.2 is 0.16.

Now, we need to assign likelihoods to $ha \land \neg vi$, $ha \land vi$ and $vi \land \neg ha$ so that these restrictions are respected. One possible approach is to start by determining the sum of the probabilities for these scenarios. In the previous subsection, we showed how this can be done by composition of scenarios. There we obtained the value 0.2 for the likelihood of *Hacker accesses DB or virus infects DB*. We choose to use the same value now, that is, we set the sum of probabilities for $ha \land \neg vi$, $ha \land vi$ and $vi \land \neg ha$ to 0.2. Overloading the notation by using the names of the threat scenarios as variables for their probabilities, we get the following set of equations:

$$ha \land \neg vi + ha \land vi = 0.05$$
$$ha \land vi + vi \land \neg ha = 0.16$$
$$ha \land \neg vi + ha \land vi + vi \land \neg ha = 0.2$$

Solving these equations yields the following results, which have been inserted in the diagram in Fig. 13.14:

$$ha \land \neg vi = 0.04$$
$$ha \land vi = 0.01$$
$$vi \land \neg ha = 0.15$$

The next step is to assign conditional likelihoods to the leads-to relations from each of $ha \land \neg vi$, $ha \land vi$ and $vi \land \neg ha$ to the unwanted incident de. We start with the relation from $ha \land vi$.

In the original diagram of Fig. 13.2, the conditional likelihood on the leads-to relation from ha to de is 0.5, while the conditional likelihood on the leads-to relation from vi to de is 0.4. When considering the composition $ha \land vi$, we need to aggregate the conditional likelihood of these two leads-to relations. We assume these conditional likelihoods are statistically independent, and therefore assign the conditional likelihood $0.5 + 0.4 - 0.5 \cdot 0.4 = 0.7$ to the leads-to relation from $ha \land vi$ to de.

Next, we want to find the conditional likelihood for the leads-to relation from $ha \land \neg vi$ to de. In Fig. 13.14, the threat scenario ha from Fig. 13.2 has been decomposed into $ha \land \neg vi$ and $ha \land vi$. As these two scenarios are mutually exclusive, we can now use Rule 13.7 to "reverse engineer" this conditional likelihood by instantiating the rule as follows:

$$e_1 = ha \wedge \neg vi$$
$$p_1 = 0.04$$
$$l_1 = ?$$
$$e_2 = ha \wedge vi$$
$$p_2 = 0.01$$
$$l_2 = 0.7$$
$$e = de$$
$$(e_1 \sqcup e_2) = (ha \wedge \neg vi) \sqcup (ha \wedge vi) = ha$$

Here the question mark represents the conditional likelihood for the leads-to relation from $ha \wedge \neg vi$ to de that we want to find. From Fig. 13.2, we know that the conditional likelihood on the relation from ha to de is 0.5. From Rule 13.7, we then get

$$\frac{p_1 \cdot l_1 + p_2 \cdot l_2}{p_1 + p_2} = 0.5$$

This means that

$$l_1 = \frac{0.5 \cdot (p_1 + p_2) - p_2 \cdot l_2}{p_1}$$

By inserting the above values, we get $l_1 = 0.45$ as annotated on the leads-to relation from $ha \wedge \neg vi$ to de in Fig. 13.14.

We use the same approach to find the conditional likelihood for the leads-to relation from $vi \wedge \neg ha$ to de, this time instantiating Rule 13.7 as follows:

$$e_1 = vi \wedge \neg ha$$
$$p_1 = 0.15$$
$$l_1 = ?$$
$$e_2 = ha \wedge vi$$
$$p_2 = 0.01$$
$$l_2 = 0.7$$
$$e = de$$
$$(e_1 \sqcup e_2) = (vi \wedge \neg ha) \sqcup (ha \wedge vi) = vi$$

From Fig. 13.2, we know that the conditional likelihood on the relation from vi to de is 0.4. From Rule 13.7, we then get

$$\frac{p_1 \cdot l_1 + p_2 \cdot l_2}{p_1 + p_2} = 0.40$$

13.5 Summary

This means that

$$l_1 = \frac{0.40 \cdot (p_1 + p_2) - p_2 \cdot l_2}{p_1}$$

By inserting the above values, we get $l_1 = 0.38$ as annotated on the leads-to relation from $vi \wedge \neg ha$ to de in Fig. 13.14.

We now have probabilities for $ha \wedge \neg vi$, $ha \wedge vi$ and $vi \wedge \neg ha$, as well as conditional likelihoods for the leads-to relations from each of these threat scenarios to de. We can therefore apply Rule 13.2 to calculate the probabilities for $(ha \wedge \neg vi) \sqcap de$, $(ha \wedge vi) \sqcap de$ and $(vi \wedge \neg ha) \sqcap de$. This gives $0.04 \cdot 0.45 = 0.018$ for $(ha \wedge \neg vi) \sqcap de$, $0.01 \cdot 0.7 = 0.007$ for $(ha \wedge vi) \sqcap de$ and $0.15 \cdot 0.38 = 0.057$ for $(vi \wedge \neg ha) \sqcap de$.

As these three threat scenarios are mutually exclusive, we can calculate the probability of their composition $((ha \wedge \neg vi) \sqcap de) \sqcup ((ha \wedge vi) \sqcap de) \sqcup ((vi \wedge \neg ha) \sqcap de)$ by taking the sum of their probabilities, thus obtaining the probability 0.082. Technically, this is done by applying Rule 13.3 twice. Furthermore, the completeness assumption means that

$$de = \big((ha \wedge \neg vi) \sqcap de\big) \sqcup \big((ha \wedge vi) \sqcap de\big) \sqcup \big((vi \wedge \neg ha) \sqcap de\big)$$

Hence, the probability 0.082 has been assigned to de in Fig. 13.14.

Note that we have not assigned conditional likelihoods to the leads-to relations from ef and eu in Fig. 13.14. The reason is that we were able to obtain likelihoods for the threat scenarios and unwanted incident without doing this.

Having obtained the likelihood for the unwanted incident de, we can revisit the diagram in Fig. 13.13 and assign the value 0.082 to de also here. Since we already have obtained the likelihood 0.2 for the composed threat scenario $ha \vee vi$, we can use Rule 13.2 to deduce the conditional likelihood of the leads-to relation from $ha \vee vi$ to de. This conditional likelihood equals the likelihood of de divided by the likelihood of $ha \vee vi$, which is 0.41 as annotated in Fig. 13.13.

In Table 13.8, we give guidelines for how to conduct decomposition of diagram elements.

13.5 Summary

Likelihood estimation and documentation is a crucial part of risk analysis as the likelihoods of unwanted incidents, together with their consequences, serve as the basis for the risk estimation. In this chapter, we have explained and demonstrated how to analyse and reason about likelihoods using CORAS diagrams.

The CORAS diagrams modelling risks can be complete or incomplete. The likelihoods can furthermore be exact or given as intervals, and they can be in terms of probabilities or in terms of frequencies. In any case, the CORAS diagrams provide support for calculating likelihoods of unwanted incidents and threat scenarios and for checking the consistency of likelihood estimates. This chapter gives precise rules

Table 13.8 Guidelines for decomposition of diagram elements

How to decompose threat scenarios or unwanted incidents

Input: Two overlapping scenarios/incidents e_1 and e_2
Output: Three mutually exclusive scenarios/incidents:

- $e_1 \wedge \neg e_2$ (e_1 *and not* e_2)
- $e_1 \wedge e_2$ (e_1 *and* e_2)
- $e_2 \wedge \neg e_1$ (e_2 *and not* e_1)

Procedure:

- Replace the scenarios/incidents e_1 and e_2 by the decomposition $e_1 \wedge \neg e_2$, $e_1 \wedge e_2$ and $e_2 \wedge \neg e_1$
- Replace each relation $e \to e_1$ by the decomposition $e \to e_1 \wedge e_2$ and $e \to e_1 \wedge \neg e_2$
- Replace each relation $e \to e_2$ by the decomposition $e \to e_1 \wedge e_2$ and $e \to e_2 \wedge \neg e_1$
- Replace each relation $e_1 \to e$ by the decomposition $e_1 \wedge e_2 \to e$ and $e_1 \wedge \neg e_2 \to e$
- Replace each relation $e_2 \to e$ by the decomposition $e_1 \wedge e_2 \to e$ and $e_2 \wedge \neg e_1 \to e$

Note: If the input e_1 and e_2 are not overlapping, but rather mutually exclusive, the scenario/incident $e_1 \wedge e_2$ of the output is empty. In that case the output of the decomposition is equivalent with the input.

and guidelines for how to do such likelihood analysis. The chapter furthermore describes how to restructure CORAS diagrams in order to enable further likelihood analysis, and to facilitate an improved understanding of likelihoods.

Chapter 14
The High-level CORAS Language

The CORAS diagrams that we have presented so far in this book are "flat" in the sense that every element is at the same level of abstraction and there is no hierarchy. There are, however, situations where it would be helpful to differentiate between different levels of abstraction or layers within the same diagram, that is, situations where hierarchy in the information would be useful.

The first of these situations is when we want to detail risk documentation already at hand. We may, for example, have identified a high-level threat scenario and see the need to decompose the scenario into a series of subscenarios in order to better understand how the overall scenario may happen and what its likelihood is. Or we may have identified a high-level unwanted incident that is actually a class of incidents, and we want to differentiate the incidents that make up this class to better understand what assets are affected and what the consequence of the unwanted incident is.

The other situation is when we need to abstract the risk information. In a risk analysis, we often identify several threat scenarios and several unwanted incidents that are almost the same, but that still must be documented separately in order to ensure a precise description of the risk picture. Sometimes there may be single threat scenarios that each leads to many different unwanted incidents, or single unwanted incidents that each follows from many different threat scenarios. Or it might be that a single treatment scenario provides treatment to several risks. In such cases, the risk documentation may be complex and difficult to follow, and we may see the need to group related things together.

Both detailing and abstraction might be needed in order to present the risk documentation in a scalable way. We may also need to document risk at different levels or layers at the same time; the technician that is going to implement a treatment, for example, needs different information than the decision maker that determines which treatments to implement. The technician needs the details, while the decision maker needs the big picture.

Using only flat diagrams, we would have to make separate CORAS diagrams for each level of abstraction. In this chapter, we introduce two mechanisms, namely *referenced diagram* and *referring element*, that mitigate this problem by allowing modelling at different levels of abstraction in the same diagram. Basically, a referring element is a reference to a referenced diagram. Because detailing and abstraction are kind of inverse operations, referenced diagrams and referring elements can be applied both when we need to detail and when we need to abstract. In Sect. 14.1, we introduce referenced diagrams and referring elements by means of examples.

Referring elements, besides being references to referenced diagrams, are also modelling elements in their own respect, and can be assigned likelihoods. Similar to the rules for analysing likelihood in basic CORAS diagrams, we have rules analysing the likelihood in high-level CORAS. This topic is dealt with in Sect. 14.2. Furthermore, high-level diagrams can also specify consequences and risk levels, which we explain in Sect. 14.3 and Sect. 14.4, respectively.

High-level CORAS introduces a number of new constructs, which means that the translation of CORAS diagrams to English prose provided in Chap. 4 is not sufficient. In Sect. 14.5, we explain how the translation scheme is extended and how diagrams in high-level CORAS can be translated into English. The formal syntax and semantics of the high-level CORAS language are provided in Appendices A and B.

The examples we use to introduce the high-level mechanisms in Sect. 14.1 are devised just for that purpose, and they are therefore too simple to really show how high-level CORAS can be utilised to handle large and complex threat and risk models. For this reason, we demonstrate high-level CORAS on a more challenging example in Sect. 14.6. In Sect. 14.7, we summarise.

14.1 Referring Elements and Referenced Diagrams

A referenced diagram can be seen as a fragment of a CORAS diagram that is contained inside a CORAS language element. A referenced diagram is referred to from another diagram by a referring element. In a way, the referenced diagram represents a view inside the referring element. A diagram containing referring elements that refer to referenced diagrams are called *high-level diagrams*. A diagram that is not high-level is referred to as a *basic diagram*.

By "referring", we mean that a referring element in a diagram can be syntactically replaced by its referenced diagrams without changing the meaning the diagram containing the reference. We refer to this operation of syntactical replacement as *expanding* a referring element.

Threat scenarios, unwanted incidents, risks and treatment scenarios can all be referring elements. They refer to referenced diagrams of the same kind as the reference in question, that is, a referring threat scenario refers to a referenced threat scenario, a referring unwanted incident refers to a referenced unwanted incident, and so forth. Somewhat different rules apply for the different kinds of referring elements, and in the following we go through them one by one.

The various kinds of referenced diagrams and referring elements have several features in common. We introduce and explain these common features when introducing referring and referenced threat scenarios and do not repeat them when presenting the other kinds of diagrams.

14.1 Referring Elements and Referenced Diagrams

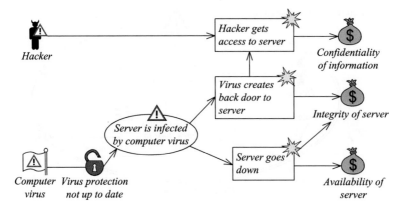

Fig. 14.1 Basic CORAS diagram from Chap. 4

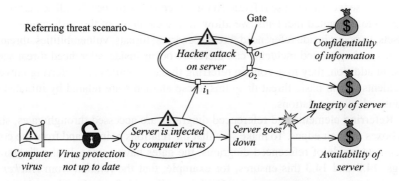

Fig. 14.2 Diagram with referring threat scenario, matching the diagram in Fig. 14.1

14.1.1 Threat Scenarios

We take the basic CORAS diagram of Fig. 14.1 (originally Fig. 4.7 from Chap. 4) as starting point for explaining referenced threat scenarios. As highlighted above, one of the aims of high-level CORAS is to facilitate abstraction. Assume that we want to group the elements of the diagram in Fig. 14.1 that are related to hacker attack and present these as one scenario. In the diagram of Fig. 14.2, we have substituted these elements with the referring threat scenario *Hacker attack on server*. As we see, referring threat scenarios look like ordinary threat scenarios, only with double instead of single border.

A referring element refers to a referenced diagram of the same name (identifier) and kind as the referring element. Hence, the referring threat scenario *Hacker attack on server* refers to a referenced threat scenario named *Hacker attack on server*. This referenced threat scenario is shown in Fig. 14.3. Notice that the name of referenced diagrams is placed immediately below the border.

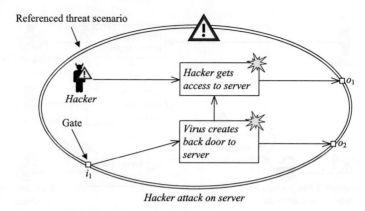

Fig. 14.3 Referenced threat scenario *Hacker attack on server*

Because we want to use referenced threat scenarios to specify full scenarios leading up to unwanted incidents, we allow all the elements of threat diagrams, except assets, to occur inside referenced threat scenarios; threats, vulnerabilities, threat scenarios and unwanted incidents can therefore occur inside referenced threat scenarios. In addition, they may contain referring threat scenarios and referring unwanted incidents. As in basic threat diagrams, these elements are related by initiates relations or leads-to relations.

Referring elements and referenced diagrams are accessed through gates, shown as boxes on their edges. To make sure that the relations point to and from the correct elements inside of referenced diagrams, the gates have names. In the diagrams of Figs. 14.2 and 14.3 this ensures, for example, that the relation from *Hacker gets access to server* points to the asset *Confidentiality of information* (through gate o_1), while the relation from *Virus creates back door to server* points to the asset *Integrity of server* (through gate o_2). In cases where the paths are unambiguously given (for example if a referenced diagram has only one gate), we can choose to not show the gate names in the diagram, that is we can let them be implicit.

The gates have direction, which means we distinguish between in-gates and out-gates. Relations to an element inside a referenced diagram must go through an in-gate, and relations from an element inside a referenced diagram must go through an out-gate. In the diagrams shown in this section, we have adopted the convention that in-gates are named i_x and out-gates o_y, and further that every gate has a unique name. These are not strict rules; the only restriction on the assignment of names to gates is that all the gates on the border of the same referenced diagram or referring element must have different names.

Within the scope of a referenced diagram, we do not necessarily know what kind of elements the gates are related to on the outside. In particular, this is the case when we reuse referring elements in different high-level diagrams. Consider, for example, the gate o_1 in the high-level diagram in Fig. 14.2. The relation from o_1 to the asset *Confidentiality of information* is clearly a harms relation. However, if we reuse the referring threat scenario in a setting where o_1 is related to a threat scenario,

14.1 Referring Elements and Referenced Diagrams

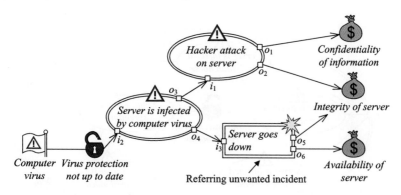

Fig. 14.4 Diagram with referring threat scenario and referring unwanted incident

an unwanted incident or a risk, the relation originating from o_1 must be a leads-to relation. The reader is referred to Fig. 14.10 for an example.

Relations inside a referenced threat scenario that have gates as their source or target are therefore formalised as leads-to relations, with the exception of relations from threats to gates, which are initiates relations. Still, these relations are given a special treatment. When expanding, the relations on the outside will decide the kind of the combined relation. In the semantics, we therefore say that an element "is initiated via" a gate and "has an effect via" a gate, even though the relations are leads-to relations.

In the above example, we looked at abstraction—how to make a high-level CORAS diagram from a "flat" diagram. As stated in the introduction, we can also apply the high-level constructs for the purpose of detailing elements. We now use the diagram of Fig. 14.2 as starting point, noticing that the threat scenario *Server is infected by computer virus* is rather high-level and underspecified as there are many different ways in which a server may be infected by computer virus.

In the diagram of Fig. 14.4, we have changed the threat scenario in question to the referring threat scenario *Server is infected by computer virus*. This allows us to detail the threat scenario in a separate referenced threat scenario as shown in Fig. 14.5. As illustrated by the diagram, we can do both sequential and parallel decompositions of referring threat scenarios; the referenced diagram details the series of events compromising the scenario *Server is infected by computer virus*, but also specifies several parallel subscenarios.

Observe that while the high-level diagram of Fig. 14.4 shows that computer virus on the server may lead to both hacker attack on the server and the server going down, it does not show how. In the more detailed picture shown in the referenced threat scenario of Fig. 14.5, however, we explicitly see that it is Trojan horse that may lead to hacker attack, whereas email worm or SQL worm may lead to the server going down.

As can be observed from the diagram of Fig. 14.5, we can have several relations using the same gate as long as they all go in the same direction. We have, for example, relations to *Server is infected by Trojan horse*, *Server is infected by email worm*, and *Server is infected by SQL worm* coming in via the gate i_2.

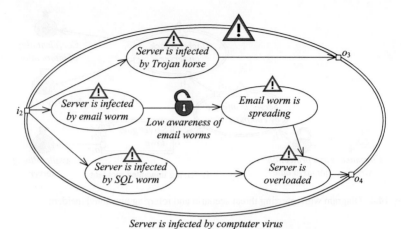

Fig. 14.5 Referenced threat scenario *Server is infected by computer virus*

Each path that we can follow through a high-level diagram via its referenced diagrams is valid. This means, for example, that the threat *Computer virus* in Fig. 14.4 initiates the treat scenarios *Server is infected by Trojan horse*, *Server is infected by email worm*, and *Server is infected by SQL worm* in Fig. 14.5.

14.1.2 Unwanted Incidents

Referenced unwanted incidents are used for specifying different ways in which an unwanted incident may manifest itself. In other words, we use referenced unwanted incidents for specifying subincidents that produce the effect of the high-level incident. Such a decomposition allows us to specify and study various more specific instances of the same more general event. Conversely, we can use referenced unwanted incidents to make an abstract and general representation of several specific events that can be understood as different instances of the same more general unwanted incident.

In Fig. 14.6, the referenced unwanted incident *Server goes down*, that is referred to from Fig. 14.4, is shown. We see from the diagram that the incident is decomposed into the unwanted incidents *Server crashes* and *Server does not respond*, which are two different ways in which the server may go down. The initial unwanted incident *Server goes down* is rather high-level, and therefore hides details. From the decomposition in Fig. 14.6 we can, for example, explicitly see that it is only *Server crashes* that harms the asset *Integrity of server*.

Unwanted incidents are by definition events and, contrary to threat scenarios, we only allow parallel decomposition of unwanted incidents. The reason is, as stated above, that the purpose of the decomposition is to specify the different ways that an unwanted incident may happen, and not to specify the events leading up to the

14.1 Referring Elements and Referenced Diagrams

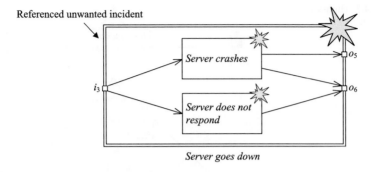

Fig. 14.6 Referenced unwanted incident *Server goes down*

incident, which is the purpose of the threat scenarios. As a consequence of this, referenced unwanted incidents are only allowed to contain other unwanted incidents and referring unwanted incidents. Further, the internal unwanted incidents should not be related to each other by leads-to relations.

The diagram of Fig. 14.6 is a further example of relations that use the same gate, which is allowed provided the relations all go in the same direction. Here, we have relations from both *Server crashes* and *Server does not respond* going out via the gate o_6 in *Server goes down*.

Following the paths of the high-level diagram in Fig. 14.4 via the referenced threat scenario *Server is infected by computer virus* and the referenced unwanted incident *Server goes down*, we see, for example, that the threat *Computer virus* initiates *Server is infected by SQL worm*, which leads to *Server is overloaded*. The latter threat scenario leads to *Server does not respond* which impacts the asset *Availability of server*.

14.1.3 Risks

In CORAS, we specify risks by assigning likelihoods and consequences to unwanted incidents. This is reflected in the CORAS language, as explained in Chap. 4, in the way that the unwanted incidents of threat diagrams are the basis for the risks in risk diagrams.

In high-level CORAS diagrams, we follow the same procedure, which means that referring unwanted incidents in high-level threat diagrams are the basis for referring risks in high-level risk diagrams, and the referenced unwanted incidents are the basis for the referenced risks. Figure 14.7 shows the high-level risk diagram based on the high-level threat diagram of Fig. 14.4. This illustrates how referring risks can be specified in high-level risk diagrams in the same way as risks are specified in basic CORAS risk diagrams. We use the same conventions with respect to risk names as we did in Chap. 4, namely that the description of the unwanted incident is prefixed by a unique risk identifier.

Fig. 14.7 High-level risk diagram

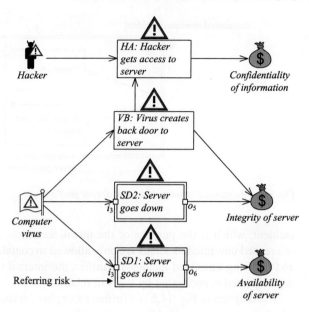

Fig. 14.8 Referenced risk *SD1: Server goes down*

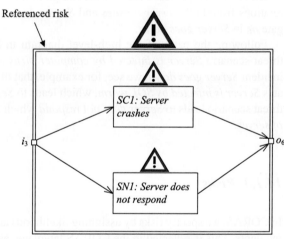

As with a risk in basic CORAS, a referring risk of high-level CORAS can be related to one asset only. Further, we require that it can be related to the asset with only one harms relation. This means that the referring unwanted incident *Server goes down* the diagram of Fig. 14.4 results in two referring risks, which are given the identifiers *SD1* and *SD2*. Note that the high-level risk diagram of Fig. 14.7 has exactly the same structure as the risk diagram of Fig. 4.10 in Chap. 4.

The referenced risks *SD1* and *SD2* are shown in Figs. 14.8 and 14.9, respectively. As with the other referenced elements, they have double border, and are accessed through gates. Note that the referring risks only have one out-gate each, even though

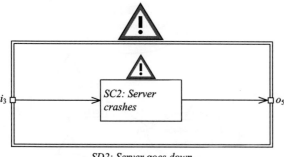

Fig. 14.9 Referenced risk *SD2: Server goes down*

the referenced unwanted incident *Server goes down* has two out-gates. The reason for this is that, as stated above, each referring risk can only be related to one asset with one relation. Because the unwanted incident in this example relates to each of the assets through only one gate, the referring risks need only to keep that gate. In order to illustrate how this works, we let the referring risks use the same gate names as the unwanted incident.

We see that this also influences the structure of the referenced risks; a referenced risk only "inherits" the risks that is related to the particular gates that it "inherits" from the unwanted incident. As a result of this they only contain the risks that are relevant for the asset that they are related to. The referenced risk *SD1* contains two risks, while *SD2* contains only one risk. This is because only one unwanted incident in *Server goes down*, and hence only one risk, is relevant for the asset *Integrity of server*. In other words, *SD1* contains two risks because two unwanted incidents inside *Server goes down* is related to gate o_6, while *SD2* only contains one risk because only one unwanted incident is related to gate o_5.

14.1.4 Treatment Scenarios

A high-level treatment diagram is, as a basic CORAS treatment diagram, a threat diagram where the risks from the risk diagram are inserted at their appropriate places, and where treatment scenarios are specified and related to the relevant diagram elements. In Fig. 14.10, we show a high-level treatment diagram. The diagram is the result of combining the high-level threat diagram in Fig. 14.4 with the high-level risk diagram in Fig. 14.7. In addition, the treatment diagram specifies both basic and high-level treatment scenarios.

Referring treatment scenarios have, as the other high-level elements, the appearance of basic treatment scenarios with a double border. In the diagram of Fig. 14.10, we have specified one basic treatment scenario and two referring treatment scenarios. These illustrate the three possibilities we have for using high-level elements in

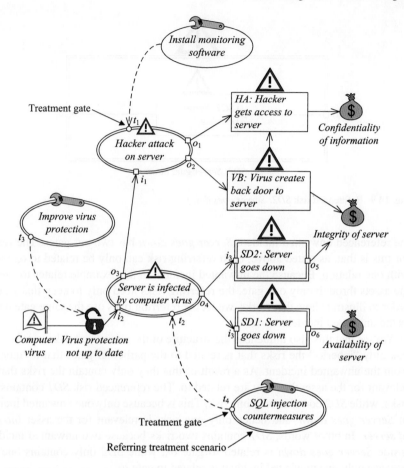

Fig. 14.10 High-level treatment diagram

treatment diagrams: A basic treatment scenario treating a referring element, a referring treatment scenario treating a basic element, and a referring treatment scenario treating a referring element. We only look at the latter case in detail, as this will also explain other cases.

A referring treatment scenario refers to a referenced treatment scenario, and a referenced treatment scenario is a collection of treatment scenarios. In Fig. 14.11, we can see the referenced treatment scenario *SQL injection countermeasures* as a collection of the two treatment scenarios *Improve routines for patching SQL server* and *Improve input validation*. The treats relations use gates in the same way as the other relations. These treatment gates are shown as white circles on the border and, as with the other kinds of gates, are named.

In Fig. 14.12 an updated version of the referenced threat scenario *Server is infected by computer virus* is shown. Here, we can see a treats relation from the treatment gate t_2 to the threat scenario *Server is infected by SQL worm*. In this way, we

Fig. 14.11 Referenced treatment scenario *SQL injection countermeasures*

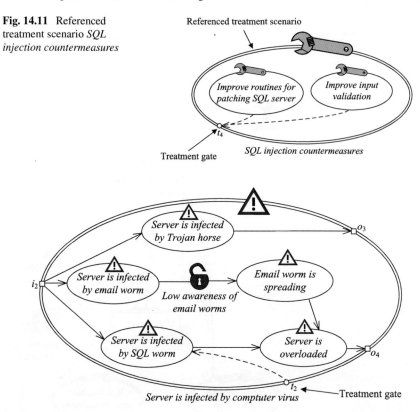

Fig. 14.12 Referenced threat scenario *Server is infected by computer virus* updated with treatment

specify that *SQL injection countermeasures* provides treatment to this specific threat scenario.

We can also use the referring treatment scenarios in high-level treatment overview diagrams. Figure 14.13 shows the risk diagram of Fig. 14.7 with the treatment scenarios from the diagram of Fig. 14.10. As in basic treatment overview diagrams, the treatment scenarios are directed towards the relevant risks.

When we decide what the relevant risks of a treatment scenario in high-level treatment overview diagrams are, we have to take special care. This can be illustrated with the referring treatment scenario *SQL injection countermeasures*. Apparently, it is relevant for all the risks since in the treatment diagram of Fig. 14.10 it is related to *Server is infected by computer virus*, which again is related to *Hacker attack on server*, and thus indirectly to all risks. However, if we inspect the referenced threat scenario in Fig. 14.12 we can see that *SQL injection countermeasures* is related (through the treatment gate t_2) only to the gate out-gate o_4 and not to o_3. Hence, it is clearly relevant for risks *SD1* and *SD2*, but has no effect on the threat scenario *Hacker attack on server* and is therefore not relevant for the risks *HA* and *VB*. In the treatment overview diagram of Fig. 14.13, it is therefore related to *SD1* and *SD2*, but not *HA* and *VB*.

Fig. 14.13 High-level treatment overview diagram

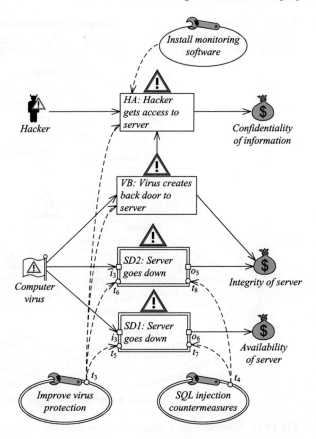

Fig. 14.14 Referenced risk *SD2: Server goes down* updated with treatments

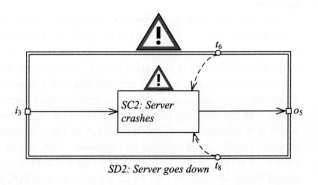

Finally, in Fig. 14.14, we show an updated version of the referenced risk *SD2: Server goes down* where the treats relations from the treatment gates to the risks inside the referenced diagram is shown.

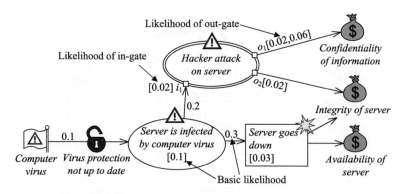

Fig. 14.15 Likelihood specification in referring element

14.2 Likelihoods in High-level CORAS

So far, the high-level diagrams we have presented in this chapter have not been assigned likelihoods. In this section, we explain how to specify and reason about likelihoods in high-level CORAS.

In high-level CORAS, the basic elements and the relations can be assigned likelihoods in the same manner as in basic CORAS, while the referring elements are treated differently. Instead of assigning a likelihood directly to a referring element, we annotate the gates of the referring elements with likelihoods as exemplified by the diagram in Fig. 14.15. As for basic CORAS, we can use qualitative or quantitative values, we can use probabilities or frequencies, and we can use exact values or likelihood intervals.

It is furthermore only the in-gates and out-gates of referring threat scenarios and referring unwanted incidents that are assigned likelihood. The gates of referring risks are not assigned likelihoods, since the likelihood value of the corresponding referring unwanted incident is embedded in the risk level assigned to the referring risk.

In the diagram of Fig. 14.15, the in-gate i_1 is annotated with the likelihood 0.02. This likelihood documents the contribution from *Server is infected by computer virus* via gate i_1 to the likelihood of the referring threat scenario *Hacker attack on server*. Because a referring element is an abstraction of a set of elements there is generally no unique likelihood that can be assigned to a referring threat scenario or a referring unwanted incident. Likelihoods are therefore annotated on the out-gates to document the contribution of referring elements to the likelihood of the other diagram elements that they may lead to. In Fig. 14.15, the out-gate o_1 is annotated with the likelihood interval [0.02, 0.06] to document the likelihood of *Hacker attack on server* to have an effect via this gate. This likelihood annotation on the in-gates and out-gates of referring elements is used in the reasoning about likelihoods in high-level diagrams, but also in the reasoning about the relation between referring elements and corresponding referenced elements.

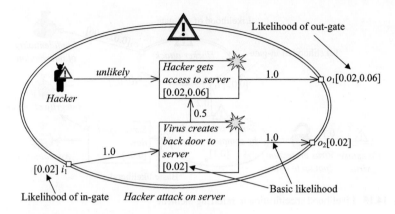

Fig. 14.16 Likelihood specification in referenced diagram

In referenced diagrams, we can likewise place likelihoods on basic elements and relations in the same way as in basic CORAS diagrams. In addition, we can assign likelihoods to the gates at the border of the referenced threat scenarios and referenced unwanted incidents as exemplified by the diagram in Fig. 14.16. We can then use the likelihood 0.02 of the in-gate i_1 when reasoning about the likelihoods of the elements within the referenced threat scenario *Hacker attack on server*. The likelihoods of the out-gates document the contribution of the referenced elements to the likelihood of other elements via these gates. The likelihood [0.02, 0.06] of o_1, for example, shows the likelihood of *Hacker attack on server* when we are considering its effect via the gate o_1. From the diagram, we see that this likelihood partly stems from the threat *Hacker*, and partly via the in-gate i_1.

In practice, this approach to specifying likelihoods for referring elements and referenced diagrams means that we treat the gates, both in high-level diagrams and in referenced diagrams, as a kind of pseudo-elements. These pseudo-elements serve as placeholders for likelihoods, and allows us to reason about of the likelihoods in a high-level diagram in a modular way; we can reason about the likelihoods of a high-level diagram separately from reasoning about the likelihoods in the referenced diagrams to which it refers.

When we are expanding the referring elements of a high-level diagram to flatten it out and make an equivalent basic diagram, the gates with their likelihoods disappear. The likelihood information is then instead represented as annotations on the diagram elements and relations. The basic threat diagram of Fig. 14.17 shows the result of expanding the high-level diagram of Fig. 14.15.

In this case, each of the likelihoods of the gates of the referring threat scenario *Hacker attack on server* equals the likelihoods of the corresponding gates of the referenced threat scenario, and determining which likelihoods to use on the expanded diagram of Fig. 14.17 is therefore straightforward. The gate i_1, for example, serves as a pseudo-element that represents the occurrences of *Server is infected by computer virus* that lead to *Virus creates back door to server*. The likelihood 0.02 of i_2 is given by Rule 13.2 (see Chap. 13) from the likelihood 0.1 of the threat scenario

14.2 Likelihoods in High-level CORAS

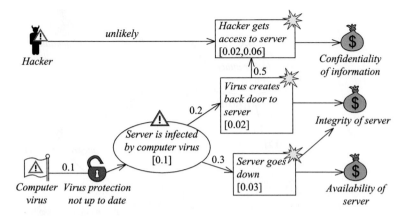

Fig. 14.17 Likelihoods in expanded diagram

and the conditional likelihood 0.2 on the leads-to relation. In the expanded diagram, the likelihood information on the i_1 is redundant since it is now directly specified on the elements and relations.

In high-level CORAS, we can specify and reason about likelihoods at each layer of abstraction independent from the other layers. In particular, we can reason about the likelihoods in a referenced diagram independent of the corresponding referring elements, and vice versa. This may lead to likelihoods on the gates of a referenced diagram that are not equal to the likelihoods on the corresponding gates of the respective referring elements; the likelihoods may even be mutually inconsistent. Detecting and resolving such conflicts is therefore a separate issue when expanding high-level diagrams with likelihoods.

In general, reasoning about likelihoods in CORAS diagrams serves two purposes: We can calculate likelihoods for elements (without likelihood estimates) from the likelihoods assigned to other elements and relations, and we can check the mutual consistency of the different likelihoods assigned to a diagram. And as already explained, when we introduce likelihoods in high-level CORAS, we furthermore add a new dimension to the reasoning, namely reasoning across levels. Reasoning about likelihoods in high-level CORAS can therefore be divided into three issues:

- Reasoning about the likelihoods in a high-level diagram.
- Reasoning about the likelihoods in a referenced diagram.
- Analysing the relation between the likelihoods of a referring element and the likelihoods in the referenced diagrams.

With respect to the two first issues we can use the rules and techniques from Chap. 13 quite straightforwardly both for calculating likelihoods and for consistency checking likelihood values. As to the third issue, we introduce some further principles.

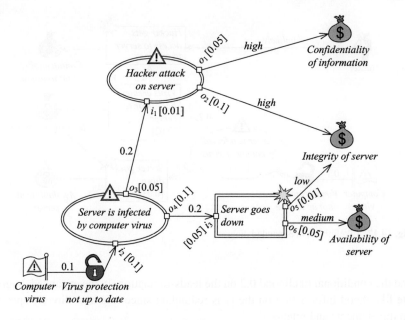

Fig. 14.18 High-level threat diagram with likelihoods

14.2.1 Reasoning About the Likelihoods in a High-level Diagram

When reasoning about likelihoods in high-level diagrams, we make use of the same techniques and rules as prescribed for basic CORAS diagrams, but there is one important difference. As explained above, referring elements are not annotated with likelihoods directly; instead the gates of the elements are assigned likelihoods. Analysing the likelihoods of a high-level diagram therefore includes analysing the likelihoods of the gates of the referring elements. This is, however, quite straightforward as we simply treat the gates as basic elements, in other words as threat scenarios or unwanted incidents, and apply the techniques and rules presented in Chap. 13.

Example 14.1 Consider the diagram in Fig. 14.18. This is the high-level threat diagram from Fig. 14.4 with all relations and gates decorated with likelihoods. Now we want to check the consistency of the likelihood assignments, assuming that the diagram is complete. We use the following shorthand notation:

$$cv = Computer\ virus$$
$$vn = Virus\ protection\ not\ up\ to\ date$$
$$si = Server\ is\ infected\ by\ computer\ virus$$
$$ha = Hacker\ attack\ on\ server$$
$$sd = Server\ goes\ down$$

14.2 Likelihoods in High-level CORAS

We furthermore use "dot-notation" to refer to gates so that, for example, $ha.i_1$ refers to the gate i_1 on the border of *Hacker attack on server*.

First, we use Rule 13.1 for initiates relations to check the likelihood of the in-gate $si.i_2$, which we treat as a basic diagram element with a likelihood. Since the diagram is complete, we have that $si.i_2 = cv \sqcap si.i_2$. From $cv \xrightarrow{0.1} si.i_2$, we get the likelihood 0.1 for $cv \sqcap si.i_2$, which is consistent with the annotation in the diagram.

Second, we use Rule 13.2 to check the likelihoods of $ha.i_1$ and $sd.i_3$. By diagram completeness, we have that $ha.i_1 = si.o_3 \sqcap ha.i_1$. By Rule 13.2, the likelihood of $ha.i_1$ is then $0.05 \cdot 0.2 = 0.01$, which is consistent with the diagram. Similarly, we calculate the likelihood 0.02 for $sd.i_3$ which is not consistent with the already assigned likelihood.

This inconsistency must be resolved either by changing one or more of the likelihoods of the diagram, or by reconsidering the assumptions that we made. In order to express the uncertainty concerning the likelihood estimates, we choose to change the likelihood of $si.o_4$ to the interval $[0.1, 0.25]$. This uncertainty then propagates to $sd.i_3$, the likelihood of which we change to the interval $[0.02, 0.05]$.

One way of decreasing this uncertainty through further analysis is to decompose *Server is infected by computer virus* and *Server goes down* into their respective referenced threat scenario and referenced unwanted incident, and then estimate the likelihoods at a more detailed level.

14.2.2 Reasoning About the Likelihoods in a Referenced Diagram

The reasoning about likelihoods in referenced diagrams is similar to that of referring elements; we make use of the techniques and rules from Chap. 13, and we reason about the gates at the border of the diagrams as if they were ordinary diagram elements with likelihoods.

Example 14.2 In Fig. 14.19, we present a decomposition of the threat scenario *Server is infected by computer virus* from Fig. 14.18 with likelihoods assigned to the relations and to some of the elements. We now want to deduce the missing likelihoods by analysing the diagram under the assumption that the diagram is complete. We use the following shorthand notation:

$$st = \text{Server is infected by Trojan horse}$$
$$se = \text{Server is infected by email worm}$$
$$es = \text{Email worm is spreading}$$
$$ss = \text{Server is infected by SQL worm}$$
$$so = \text{Server is overloaded}$$

Notice that we refer to the gates inside this referenced threat scenario by their names only, that is, by i_2, o_3 and o_4.

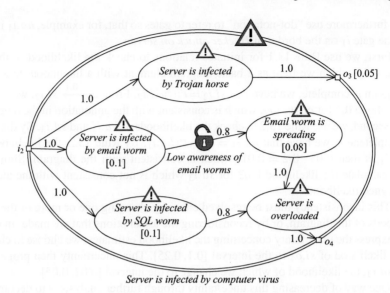

Fig. 14.19 Referenced threat scenario *Server is infected by computer virus* with likelihoods

We start with *Server is infected by Trojan horse*. Since we have the likelihood of o_3 and the conditional likelihood of the leads-to relation from st to o_3 we can use Rule 13.2 backwards and deduce the likelihood 0.05 for st.

Next, we wish to deduce the likelihood for the gate i_2. By the completeness assumption, we actually have that the likelihoods of each of the three threat scenarios st, se and ss should equal each other, since they all are initiated only via the in-gate i_2. Considering the diagram, this assumption should be reconsidered; it does not seem plausible that the likelihood of these three scenarios are exactly the same. For the purpose of this example, however, we maintain the assumption and rather change the likelihood of st from 0.05 to the interval $[0.05, 0.1]$. This change then propagates to o_3, and we therefore assign the same interval also to this gate. We then choose to assign the likelihood 0.1 to the in-gate i_2. This likelihood can be deduced from either se or ss by applying Rule 13.2. Since this value is contained in the interval $[0.05, 0.1]$ assigned to st, the two are mutually consistent.

Finally, we want to find a value for the threat scenario so and the gate o_4. Since the diagram is complete, we have that $so = (ss \sqcap so) \sqcup (es \sqcap so)$. By Rule 13.2, the likelihood of $ss \sqcap so$ is $0.1 \cdot 0.8 = 0.08$, and the likelihood of $es \sqcap so$ is $0.08 \cdot 1.0 = 0.08$. The likelihood of so is at least the lowest value of $ss \sqcap so$ and $se \sqcap so$, which is 0.08. The likelihood is furthermore at most the sum of the two, which is 0.16. According to the likelihoods already assigned to the diagram, the likelihood of so is therefore within the interval $[0.08, 0.16]$. Assuming that $ss \sqcap so$ and $es \sqcap so$ are statistically independent, we can use Rule 13.4 and narrow this rough approximation to the exact likelihood $0.08 + 0.08 - 0.08 \cdot 0.08 = 0.1536 \approx 0.15$. By Rule 13.2, we can also use the value 0.15 for the gate o_4.

14.2.3 Analysing the Relation Between the Likelihoods of a Referring Element and the Likelihoods in the Referenced Diagrams

A referring element and the referenced diagram it refers to are required to have matching gates. It should be obvious that reasoning about and analysing the relation between likelihoods of a referring element and the likelihoods of the referenced diagram involves comparing the likelihoods of matching gates.

When we do a detailing, we get a more precise picture of the situation. The same holds for detailing likelihoods. When detailing likelihoods, we increase the precision and reduce the uncertainty in the forward direction. In this case, reducing the uncertainty means narrowing the likelihood intervals on the out-gates. The likelihood of an out-gate in a referenced diagram should therefore be included in the likelihood of the matching out-gate on the referring element.

At the same time, when we detail the likelihoods, it might be that the assumptions made at the higher level (that is, in the high-level diagram) are too strong and must be loosened. The assumptions in this setting are the likelihoods of the in-gates. We therefore allow wider likelihood intervals on the in-gates in referenced diagram. In other words, the likelihood of an in-gate on the border of a referring element should be included in the likelihood of the matching in-gate in the referenced diagram.

All likelihoods on gates can be represented by intervals; if a likelihood is an exact value l, the interval is $[l, l]$. Let re be a referring element and rd be the referenced diagram to which it refers. The rules for analysing the relation between the likelihoods of a referring element in a high-level diagram and the likelihoods of the referenced diagram to which it refers may be summarised by the following:

- For each out-gate $re.og([l_i, l_j])$ on the border of re and its matching out-gate $og([l'_i, l'_j])$ on the border of rd, we must have that $[l'_i, l'_j] \subseteq [l_i, l_j]$.
- For each in-gate $re.ig([l_i, l_j])$ on the border of re and its matching in-gate $ig([l'_i, l'_j])$ on the border of rd, we must have that $[l_i, l_j] \subseteq [l'_i, l'_j]$.

Example 14.3 In the above examples, we analysed the threat scenario *Server is infected by computer virus*, first as a referring threat scenario in Fig. 14.18, and then as a referenced threat scenario in Fig. 14.19. In both cases, the analysis involved assigning likelihoods to the gates of this scenario. We now wish to analyse their mutual consistency, that is, to check the likelihood of each gate of the referring threat scenario against the likelihood of the matching gate of the referenced threat scenario.

Using the updated likelihoods that we obtained in the previous examples, we have for the high-level diagram the likelihoods 0.1 for i_2, 0.05 for o_3 and $[0.1, 0.25]$ for o_4. The respective likelihoods for the matching gates of the referenced diagram are 0.1, $[0.05, 0.1]$ and 0.15.

For the in-gate i_2, the likelihoods are obviously consistent as they are equal. For the out-gate o_3, we have the interval $[0.05, 0.05]$ for the referring element and $[0.05, 0.1]$ for the referenced element. As the latter is not a subset of the former, the likelihoods are mutually inconsistent.

This inconsistency is resolved by changing one or both of the already obtained values for o_3. For each such change, we then need to take a step back and ensure that the new value is consistent with the rest of the diagram in which we do the change. For example, if we choose to change the likelihood of $si.o_3$ in the high-level diagram of Fig. 14.18 we need to reconsider the likelihoods of the gates of the referring threat scenario ha in the same diagram.

Finally, for the out-gate o_4 we have the interval $[0.1, 0.25]$ for the referring element and $[0.15, 0.15]$ for the referenced element. Since the latter is a subset of the former, these likelihoods are mutually consistent.

14.3 Consequences in High-level CORAS

In high-level CORAS, assets and harms relations can only occur in high-level diagrams, and not in referenced diagrams. Consequence estimation and specification are therefore issues concerning the impacts relations ending in the assets in the high-level diagrams, and not the referenced diagrams. There is therefore no need for consistency checking consequences across the different levels of high-level CORAS diagrams.

However, it does make sense to abstract consequences in cases where specific harms relations are combined to form a more general and abstract harms relation. The combined consequence is then a single value that represents each of the consequences of the harms relations that are combined. Roughly speaking, the abstract consequence should be the average of the consequences that are combined. However, we need to take into account also the likelihoods of the respective unwanted incidents and give corresponding weights to the various consequences.

The unwanted incidents shown in Fig. 12.4 of Chap. 12 may serve as an example. The incidents *Thief steals jewelry* and *Thief steals car* both harm the asset *Savings*. The former has the likelihood 0.2 and the monetary consequence 12,000, whereas the latter has the likelihood 0.1 and the consequence 80,000. If desired, we can combine these two incidents into the more abstract and general incident *Theft*. This requires that also the two harms relations are combined into one. The consequence of this general incident is then either 12,000 or 80,000 since *Theft* means that one of the initial incidents occurs. However, we need to assign a single value that represents both of the specific incidents. One option is of course to use the interval [12000, 80000]. Another option is to calculate the average, where we count *Thief steals jewelry* twice since it is twice as likely as *Thief steals car*. This yields the combined value 34,667 that represents the average consequence of theft.

In the practical setting of a risk analysis, we often operate with consequence values that cannot be straightforwardly aggregated. This is particularly the case for qualitative consequence values or intangible assets like reputation or trust. The general principle nevertheless remains, namely that we need to estimate one abstract consequence value that represents each of the more concrete consequence values of the harms relations that are combined.

14.3 Consequences in High-level CORAS

Fig. 14.20 Fragment of threat scenario with consequence values

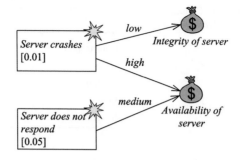

Fig. 14.21 Abstracted unwanted incident with unknown consequence value

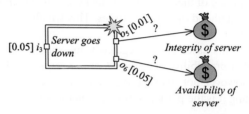

Fig. 14.22 Referenced unwanted incident

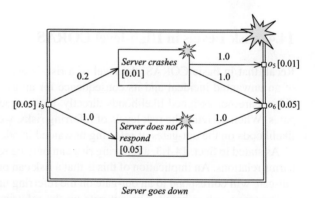

Example 14.4 Consider the diagram fragment in Fig. 14.20 for which we have made the abstraction shown in Fig. 14.21. The referenced diagram is shown in Fig. 14.22. Based on the consequence annotations in Fig. 14.20, we now wish to deduce the consequences of the combined harms relations from the out-gates o_5 and o_6 in Fig. 14.21, where for now there are only question marks.

The harms relation from o_5 only represents the harms relation from the unwanted incident *Server crashes* to *Integrity of server* and should therefore have the same consequence, namely *low*. The harms relation from o_6 to *Availability of server*, on the other hand, is the combination of two harms relations and we need to find a suitable aggregate.

Assume that we operate with the consequence scale shown in Table 14.1. The aggregate of the two consequence values in question, namely *high* and *medium* lies in their intersection. The fact that the interval *high* is much wider than *medium* pulls the aggregate in that direction. However, we also need to take into account that the

Table 14.1 Consequence scale for *Availability of server*

Consequence	Description
Catastrophic	Downtime in range [1 week, ∞)
High	Downtime in range [1 day, 1 week)
Medium	Downtime in range [1 hour, 1 day)
Low	Downtime in range [1 minute, 1 hour)
Insignificant	Downtime in range [0, 1 minute)

likelihood of the unwanted incident *Server does not respond* is five times as high as the likelihood of *Server crashes*. This weight pulls the aggregate in the direction of *medium*.

In this case, the aggregated value should probably be *medium* because of the higher probability of *Server does not respond*. Choosing the appropriate aggregate is, however, not an exact science when we use qualitative values, and in the practical setting it is the participants of the analysis that must make the final decision.

14.4 Risk Levels in High-level CORAS

Recall that in basic CORAS, the level of a risk is determined from the likelihood of an unwanted incident and its consequence for an asset. Referring unwanted incidents are not assigned likelihoods directly, but instead have likelihoods on their gates. When deriving the risk levels of referring risks, we therefore make use of the likelihoods on the out-gates of referring unwanted incidents.

As stated in Sect. 14.1.3, a referring risk can only be related to one asset with one harms relations. An implication of this is that a risk can only have one out-gate. This out-gate will correspond to an out-gate on the referring unwanted incident on which the referring risk is based. The out-gate on the referring unwanted incident has a likelihood value and is the source of a harms relation annotated with a consequence value; these are the values from which we derive the risk level of the referring risk.

Example 14.5 For the purpose of the example, we assume that likelihoods in the interval $[0, 0.01)$ together with the consequence *low* yield the risk level *acceptable*, while likelihoods in the interval $[0.01, 0.1)$ together with consequence *medium* yield risk level *unacceptable*.

Figure 14.23 shows the risk diagram obtained from the threat diagram in Fig. 14.18. The risk diagram is identical to the risk diagram in Fig. 14.7 except that the risks are given risk levels. The two basic risks are based on basic unwanted incidents, so we are not concerned with them here.

The referring risks *SD1* and *SD2* are based on the referring unwanted incident *Server goes down* in the threat diagram in Fig. 14.18. The gate o_6 has the likelihood 0.05 and the harms relation from the gate to the asset has the consequence *medium*. The referring risk *SD1* therefore has the risk level *unacceptable*. The gate o_5 has

Fig. 14.23 Risk levels in high-level CORAS

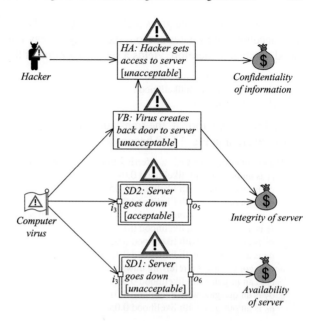

likelihood 0.01 in the threat diagram and the harms relation from the gate to the asset has the consequence *low*. Hence, the referring risk *SD2* has the risk level *acceptable*.

14.5 How to Schematically Translate High-level CORAS Diagrams into English Prose

High-level CORAS introduces to the CORAS language the notions of referring elements, referenced diagrams and gates. When translating diagrams to English text, we need to be able to translate these new elements as well. To this end, we extend the translation scheme presented in Chap. 4 to include the additions to the CORAS language. We first address the referring elements, and then proceed with the translation of referenced diagrams.

14.5.1 Referring Elements

High-level CORAS introduces four referring elements, namely referring threat scenarios, referring unwanted incidents, referring risks and referring treatment scenarios. When we translate the referring elements, we also list their gates. The translations of referring threat scenarios and unwanted incidents therefore have the form:

...is a referring threat scenario with gate(s) ....
...is a referring unwanted incident with gate(s) ....

The translation of the gates includes their kind. The translation furthermore takes into account that in-gates and out-gates may or may not be annotated with a likelihood. Each gate therefore yields a sentence of one of the following forms:

... is an in-gate (with likelihood ...).
... is an out-gate (with likelihood ...).
... is a treatment gate.

Following this format, the translation of the referring elements of the high-level threat diagram in Fig. 14.18 becomes:

Hacker attack on server is a referring threat scenario with gates i_1, o_1 and o_2.
  i_1 is an in-gate with likelihood 0.01.
  o_1 is an out-gate with likelihood 0.05.
  o_2 is an out-gate with likelihood 0.1.
Server is infected by computer virus is a referring threat scenario with gates i_2, o_3 and o_4.
  i_2 is an in-gate with likelihood 0.1.
  o_3 is an out-gate with likelihood 0.05.
  o_4 is an out-gate with likelihood 0.1.
Server goes down is a referring unwanted incident with gates i_3, o_5 and o_6.
  i_3 is an in-gate with likelihood 0.05.
  o_5 is an out-gate with likelihood 0.01.
  o_6 is an out-gate with likelihood 0.05.

Referring risks may have a risk level in addition to the gates. The format of the translation depends on whether or not the risk level has been specified:

Referring risk ..., with gate(s) ..., occurs with risk level ....
Referring risk ..., with gate(s) ..., occurs with undefined risk level.

The two referring risks in the high-level risk diagram in Fig. 14.23 are therefore translated as:

Referring risk *SD1: Server goes down*, with gates i_3 and o_6, occurs with risk level *unacceptable*.
  i_3 is an in-gate.
  o_6 is an out-gate.
Referring risk *SD2: Server goes down*, with gates i_3 and o_5, occurs with risk level *acceptable*.
  i_3 is an in-gate.
  o_5 is an out-gate.

The translation of referring treatment scenarios follows the same format as referring threat scenarios and unwanted incidents. The format for these elements is:

... is a referring treatment scenario with gate(s) ....

If we now consider the high-level treatment diagram in Fig. 14.10, the translation of the referring treatment scenarios yields the following:

Improve virus protection is a referring treatment scenario with gate t_3.
  t_3 is a treatment gate.
SQL injection countermeasures is a referring treatment scenario with gate t_4.
  t_4 is a treatment gate.

In high-level CORAS, the relations may have the gates on the border of the referring elements as source or target. In order to capture this when we translate the

14.5 How to Schematically Translate High-level CORAS Diagrams into English Prose

relations in a high-level diagram, we need to extend the translation scheme of the relations.

Initiates relations have four new forms that are used for translating them into sentences. The reason for the number of different forms is that we still need to consider all possibilities of the relations being annotated with likelihoods and vulnerabilities. The four forms are:

...initiates ... via gate ....
...initiates ... via gate ... with likelihood ....
...exploits vulnerability ... to initiate ... via gate ....
...exploits vulnerability ... to initiate ... via gate ... with likelihood ....

As an example, consider the initiates relation in the diagram in Fig. 14.18. In the translation, the relation produces the following sentence:

Computer virus exploits vulnerability *Virus protection not up to date* to initiate *Server is infected by computer virus* via gate i_2 with likelihood 0.1.

The new forms of sentences that leads-to relations may produce are similar to the translations of the initiates relations in high-level CORAS. Additionally, we need to take into account that the relations may have gates both as source and as target. The sentences are therefore of the following forms:

...leads to ... via gate(s) ....
...leads to ... via gate(s) ... with conditional likelihood ....
...leads to ... via gate(s) ..., due to vulnerability ....
...leads to ... via gate(s) ... with conditional likelihood ..., due to vulnerability ....

The two leads-to relations in Fig. 14.18 are therefore translated into:

Server is infected by computer virus leads to *Hacker attack on server* via gates o_3 and i_1 with conditional likelihood 0.2.
Server is infected by computer virus leads to *Server goes down* via gates o_4 and i_3 with conditional likelihood 0.2.

Harms relations can have gates as source, but not as target. We therefore get two new forms of sentences for these relations:

...impacts ... via gate ....
...impacts ... via gate ... with consequence ....

Consider again the high-level diagram in Fig. 14.18. The harms relations in this diagram are translated to:

Hacker attack on server impacts *Confidentiality of information* via gate o_1 with consequence *high*.
Hacker attack on server impacts *Integrity of server* via gate o_2 with consequence *high*.
Server goes down impacts *Integrity of server* via gate o_5 with consequence *low*.
Server goes down impacts *Availability of server* via gate o_6 with consequence *medium*.

Finally, we extend the translation of treats relations with a new sentence form to handle the treatment gates:

...treats ... via gate(s) ....

Using this new form, the treats relations in the high-level treatment diagram in Fig. 14.10 is translated into:

Install monitoring software treats *Hacker attack on server* via gate t_1.
Improve virus protection treats *Virus protection not up to date* via gate t_3.
SQL injection countermeasures treats *Server is infected by computer virus* via gates t_4 and t_2.

14.5.2 Referenced Diagrams

When we translate a referenced diagram into English prose, there are three differences compared to translating a basic CORAS diagram. The first difference is that we start by producing a sentence that serves as a heading of the diagram translation. These sentences have the form:

...is a referenced diagram with gate(s) ...:
...is a referenced treatment scenario with gate(s) ...:

Under the heading, we then proceed by producing sentences for the gates, the elements and the relations of the diagram according to the general translation scheme. For example, the referenced diagram shown in Fig. 14.19 are translated into the text:

Server is infected by computer virus is a referenced diagram with gates i_2, o_3 and o_4:
i_2 is an in-gate.
o_3 is an out-gate with likelihood 0.05.
o_4 is an out-gate.
Threat scenario *Server is infected by email worm* occurs with likelihood 0.1.
etc.

The referenced treatment scenario in Fig. 14.11 is translated into:

SQL injection countermeasures is a referenced treatment scenario with gate t_4:
t_4 is a treatment gate.
Improve routines for patching SQL server is a treatment scenario.
etc.

Because relations inside a referenced diagram can have the gates at the border of the diagram as source or target, we have to do additional extensions to the translation scheme. Initiates relations and leads-to relations with gates as targets have the same translation, given by the following form:

...has an effect via gate ....
...has an effect via gate ..., due to vulnerability ....

Leads-to relations with gates as source are translated using the following sentence forms:

...is initiated via gate ....
...is initiated via gate ... with conditional likelihood ....
...is initiated via gate ..., due to vulnerability ....
...is initiated via gate ... with conditional likelihood ..., due to vulnerability ....

The reason for not using the expression "leads to" in any of these translations is that gates are not proper elements, and it therefore does not make sense to say, for example, that a gate leads to an incident or that a scenario leads to a gate.

Continuing with the referenced diagram in Fig. 14.19, the relations with gates as source or target are translated into the following sentences:

Server is infected by Trojan horse is initiated via gate i_2 with conditional likelihood 1.0.
Server is infected by email worm is initiated via gate i_2 with conditional likelihood 1.0.
Server is infected by SQL worm is initiated via gate i_2 with conditional likelihood 1.0.
Server is infected by Trojan horse has an effect via gate o_3 with conditional likelihood 1.0.
Server is overloaded has an effect via gate o_4 with conditional likelihood 1.0.

In addition to the new sentence forms of initiates and leads-to relations, we need new sentence forms for treats relations inside referenced diagrams. We start with the treats relations inside referenced treatment scenarios. These relations are translated into sentences using the following format:

...provide treatment via gate ....

As an example, the treats relations inside the referenced treatment scenario of Fig. 14.11 are translated into:

Improve routines for patching SQL server provides treatment via gate t_4.
Improve input validation provides treatment via gate t_4.

The treats relations inside referenced threat scenarios, unwanted incidents and risks are translated with the following sentence form:

...is treated via gate ....

Examples of such a treats relations are found in the referenced risk in Fig. 14.14. The treats relations in the diagram are translated into the sentences:

SC2: Server crashes is treated via gate t_6.
SC2: Server crashes is treated via gate t_8.

14.6 Example Case in High-level CORAS

The examples of the use of high-level CORAS presented in the preceding sections of this chapter are not very realistic; the examples are too small and simple for justifying the use high-level CORAS in risk analysis. However, if we look back at the AutoParts example presented in Part II of this book, we needed five full pages just to show the completed threat diagrams (see Figs. 9.6–9.10 on pp. 137–141). These diagrams, which are similar to diagrams we often see in real-life analyses, have reached the point where the complexity of the diagrams begins to make them difficult to work with. These example diagrams are moreover only a selection of the diagrams that are typically made during a risk analysis of that kind and size. In this section, we will apply the AutoParts example as a case for demonstrating the use of high-level CORAS.

14.6.1 Threat Diagram

Figure 14.24 shows a high-level CORAS diagram made from the threat diagrams of the AutoParts example. When logically grouping the scenarios of these threat diagrams, each of the five diagrams gives rise to two referenced threat scenarios, resulting in ten referenced threat scenarios in all. The high-level diagram in Fig. 14.24 therefore has ten referring threat scenarios, each referring to one of these referenced threat scenarios. We do not show all of the referenced threat scenarios. Instead, we show three of them as examples, namely *Hacker attack on customer DB* in Fig. 14.25, *Employee mistake* in Fig. 14.26, and *Hacker attack on online store* in Fig. 14.27.

These diagrams show how we can use referring elements and referenced diagrams to hide much of the complexity of threat diagrams in order to make nice high-level overviews of threat scenarios, but still keep the detailed information about the scenarios. In order to achieve this, it was necessary to duplicate some of the information in the referenced threat scenarios, for example, the threats *Hacker* and *Employee*, but at the same time we get the benefit of having the scenarios presented separately without numerous relations going in every direction between the them.

In a similar way, the unwanted incidents specified in the AutoParts threat diagrams are logically grouped together in referenced unwanted incidents. In this example, the grouping is according to the assets to which the incidents are related. Because only one unwanted incident is related to the asset *Inventory DB* in the original threat diagrams, the high-level diagram in Fig. 14.24 has three referring unwanted incidents. As examples we show two of the corresponding referenced unwanted incidents, namely *Leakage from customer database* in Fig. 14.28 and *Online store down* in Fig. 14.29. We see here how we can use referenced diagrams to create classes of unwanted incidents that keep the essence but hide the details of the incidents. In the example, the essence of the incidents is leakage from the database and that the online store goes down, respectively. What data is leaked or the way the online store goes down are details that we do not necessarily need in order to give an overview of the risk picture.

The likelihoods of the basic elements in these diagrams are obtained from diagrams in Chaps. 10 and 12 (specifically Figs. 10.2–10.4 on pp. 154–157 and Fig. 12.2 on p. 193). The out-gates of the referring elements in the high-level diagram are also assigned likelihoods. Because the high-level diagram is an abstraction of the original AutoParts threat diagrams, these likelihoods are derived from the likelihoods on the basic elements inside the referenced diagrams. In other words, we have done bottom-up reasoning with respect to the likelihoods. The reader is referred to Example 11.2 on p. 170 for an explanation of how the likelihood *possible* on the out-gate o_1 of *Online store down* can be deduced.

Further, the combined harms relations in the high-level diagram are assigned consequence values. As explained in Sect. 14.3, we combine consequences by determining an aggregated consequence that appropriately represents each of them.

The occurrence of the high-level referring unwanted incident *Leakage from customer database* of Fig. 14.24, for example, is more specifically the occurrence of

14.6 Example Case in High-level CORAS   273

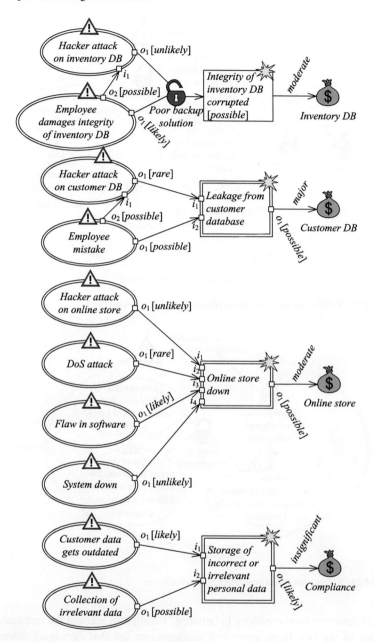

Fig. 14.24 High-level threat diagram for AutoParts

either *Payment card data leaks to third party* or *Personally identifiable information leaks to third party*, as shown in Fig. 14.28. The former has the consequence *catastrophic*, and the latter has the consequence *moderate*. The combined consequence

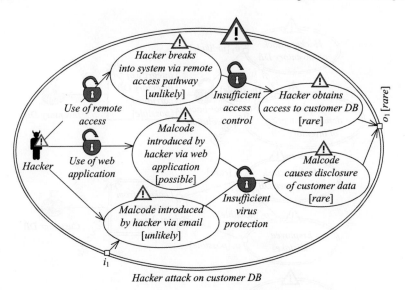

Fig. 14.25 Referenced threat scenario *Hacker attack on customer DB*

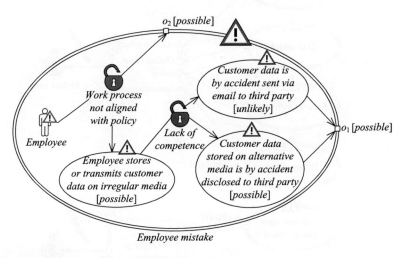

Fig. 14.26 Referenced threat scenario *Employee mistake*

should therefore be somewhere in between. Taking into account the respective likelihoods of the two basic unwanted incidents, we see that *Personally identifiable information leaks to third party* is by far the most likely of the two, which means that the combined consequence gravitates to *moderate*. We have, however, chosen to use the consequence *major* in the high-level threat diagram in order to allow for the more severe consequence of *Payment card data leaks to third party*. The eventual choice depends on the customer or parties of the analysis, and in other cases it could be more reasonable to use the worst case estimates. Abstraction always im-

14.6 Example Case in High-level CORAS

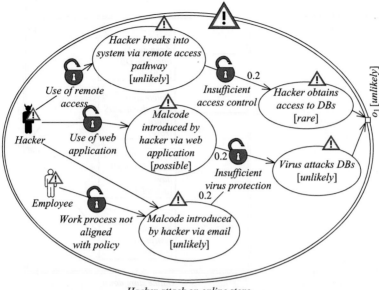

Fig. 14.27 Referenced threat scenario *Hacker attack on online store*

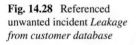

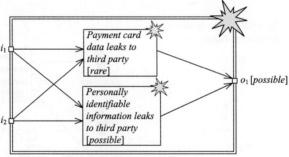

Fig. 14.28 Referenced unwanted incident *Leakage from customer database*

plies loss of information, and it is the purposes and objectives of the analysis that should determine which information we choose to document at the higher levels.

The reader is referred to Example 11.2 for an explanation of how the consequence *moderate* on the leads-to relation from o_1 of *Online store down* can be deduced.

14.6.2 Risk Diagram

Figure 14.30 shows the high-level risk diagram based on the high-level threat diagram of Fig. 14.24. The diagram shows both threats and risks. In the high-level

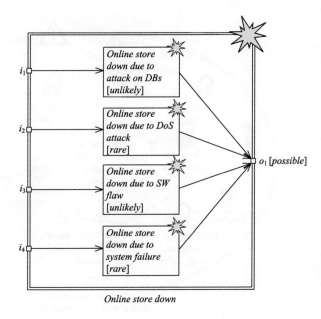

Fig. 14.29 Referenced unwanted incident *Online store goes down*

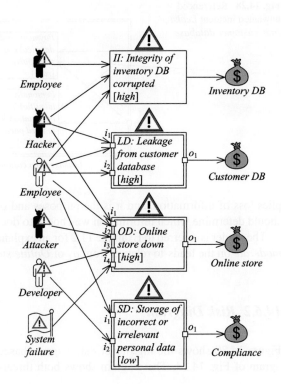

Fig. 14.30 High-level risk diagram for AutoParts

Fig. 14.31 Referenced risk
LD: Leakage from customer database

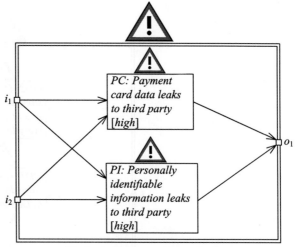

threat diagram, these threats do not show as they are inside the referenced threat diagrams, but in the high-level risk diagram we have chosen to show them as this is customary in risk diagrams.

Figures 14.31 and 14.32 show the referenced risks *LD: Leakage from customer database* and *OD: Online store down*, respectively. Similar to the referenced unwanted incidents on which they are based, they provide a classification of the risks where the essence of the risks are shown in the high-level overview and the details are shown in the referenced diagrams.

The basic risks inside the referenced diagrams have the risks levels as assigned in the basic risk diagrams in Fig. 10.5 on p. 159 and Fig. 11.1 on p. 168. The risk levels of the referring risks in the high-level risk diagram, however, are calculated directly from the likelihood and consequence values in the high-level threat diagram in Fig. 14.24, using the risk functions for the AutoParts example documented in Tables 8.12–8.14 on p. 120.

14.6.3 Treatment Diagram

In Fig. 14.33, we show a high-level treatment diagram based on the threat diagram in Fig. 14.24 and the risk diagram in Fig. 14.30. In Chap. 12, five treatment scenarios were identified for the AutoParts example. Two of them, namely *Implement new, secure backup solution* and *Increase awareness of security risks*, are shown in the high-level treatment diagram and are related to the appropriate threat scenarios. The remaining three treatment scenarios have been grouped together in a referenced treatment scenario *Protection software*. This referenced treatment scenario is shown in Fig. 14.34, and a reference to it can be seen in the high-level treatment diagram as the referring treatment scenario *Protection software*.

Fig. 14.32 Referenced risk
OD: Online store down

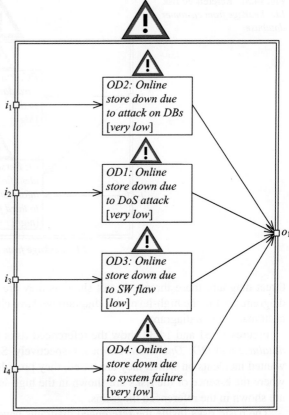

OD: Online store down

In Figs. 14.35, 14.36 and 14.37, we show updated versions of the referenced threat scenarios from Figs. 14.25, 14.26 and 14.27. Here, the referenced threat scenarios have been updated with treatment gates (given to the corresponding referring threat scenarios in the high-level treatment diagram) and with treats relations showing which of the inside elements that are given treatments via the gates.

The treatment diagram for AutoParts shown in Fig. 14.33 admittedly contains a lot of treats relations. These are necessary in order to keep track of which elements inside the referenced threat scenarios in the diagram are treated by the different treatment scenarios inside the referenced treatment scenario *Protection software*. Although we would like to have this information documented, it might be that we do not have to show it in all situations. In Fig. 14.38, we therefore offer a simplified version of the treatment diagram in Fig. 14.33. In this simplified diagram, the details of the treats relations are hidden by removing the treatment gates and by merging treats relations that have the same source and target element. When the purpose of a treatment diagram is to provide a high-level overview of the treatments available, such simplified versions of treatment diagrams, with some of the details hidden, can be a good alternative.

14.7 Summary

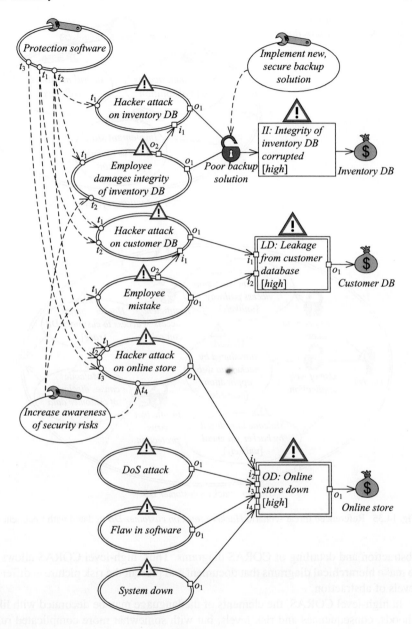

Fig. 14.33 High-level treatment diagram for AutoParts

14.7 Summary

In this chapter, we have presented high-level CORAS. High-level CORAS introduces referring elements and referenced diagrams as two mechanisms for doing

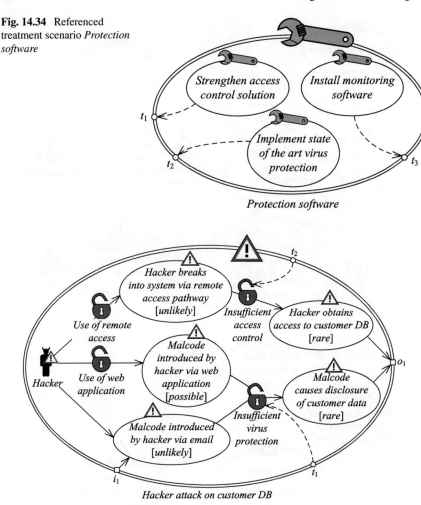

Fig. 14.34 Referenced treatment scenario *Protection software*

Fig. 14.35 Referenced threat scenario *Hacker attack on customer DB* updated with treatments

abstraction and detailing of CORAS diagrams. Thus, high-level CORAS allows us to make hierarchical diagrams that document and present the risk picture at different levels of abstraction.

In high-level CORAS, the elements of the language may be decorated with likelihoods, consequences and risk levels, but with somewhat more complicated rules than for basic CORAS. The rules and techniques for analysing likelihood as defined for basic CORAS nevertheless apply. However, since high-level CORAS is hierarchical, we have enhanced the approach with techniques for analysing likelihoods across the various levels of abstraction.

High-level CORAS is designed to handle large and complex risk models, and to allow detailed analysis of selected aspects. In order to show the use of high-level CORAS beyond mere toy examples, we have included in this chapter also a larger

14.7 Summary

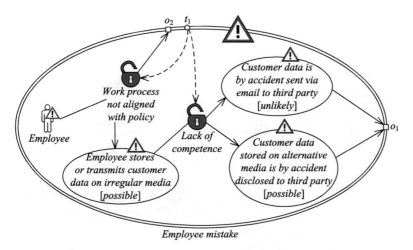

Fig. 14.36 Referenced threat scenario *Employee mistake* updated with treatments

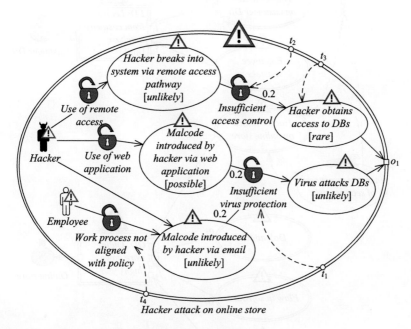

Fig. 14.37 Referenced threat scenario *Hacker attack on online store* with treatments

example case. In this example case, we have applied the mechanisms of high-level CORAS to the AutoParts example developed in Part II of this book.

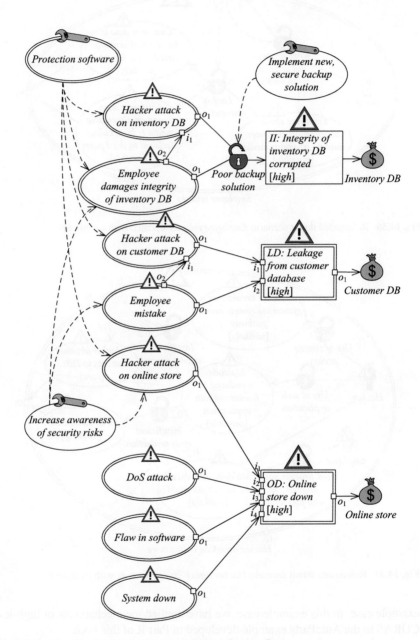

Fig. 14.38 Simplified high-level treatment diagram for AutoParts

Chapter 15
Using CORAS to Support Change Management

A risk analysis typically focuses on a particular configuration of the target at a particular point in time, and is valid under the assumptions made in the analysis. However, the target, its configuration, as well as the assumptions that may be made will change and evolve over time. We therefore need methods and techniques for how to deal with change and evolution in relation to risk analysis. Such methods and techniques are the topic of this chapter.

How we manage change in a risk analysis depends to a large degree on the context and the kinds of changes we are dealing with: Are the changes the result of maintenance or of bigger, planned changes? Are the changes a transition from one stable state of the target to another or the continuous evolution of a target designed to change over time? Do the changes occur in the target or in the environment of the target? The answers to such questions decide how we manage the changes.

We start by classifying the various kinds of changes of relevance to risk analyses in Sect. 15.1. In Sect. 15.2, we explain how we manage change in a practical risk analysis setting. We distinguish between three main scenarios, referred to as perspectives, in which change is an issue in relation to risk analysis. In Sect. 15.3, we provide a summary.

15.1 Classification of Changes

In a risk analysis, information is collected and organised to describe the target of analysis and its environment. The scope and focus of the analysis is furthermore characterised, defining the parts of the system that are most relevant to the analysis. Together, such information serve as the input to the analysis. Any change in this information may cause the risk analysis to produce a different outcome. We refer collectively to the information that forms the input to a risk analysis as the target description. The elements of the target description are described in Chap. 7. In summary, the target description documents:

- The target of analysis.
- The scope and focus of the analysis.
- The environment of the target.
- The assumptions of the analysis.
- The parties and assets of the analysis.
- The context of the analysis.

In addition to managing the effect of changes in the target description, we must also consider updating the risk picture because we have gained new or better knowledge about the target and its environment.

In the following, we explain how changes to each element of the target description as well as changes in our knowledge may impact the risk picture.

15.1.1 Target of Analysis

The target of analysis is the system, organisation, enterprise, and so forth, or parts thereof, that is the subject of the analysis. Changes to the target must be expected, even in what we would consider a stable target. Consider for example bug-fixes distributed from third party software vendors. Another obvious example of changes to the target is implementation of treatments identified in a risk analysis. But changes may also be more extensive, such as introduction of new functionality in a software system or replacement of software components, work processes or hardware components. We allow full generality when defining the target, that is, we impose no restrictions on what we can define as target, and allow the same generality with respect to changes to the target. It is therefore necessary to characterise in more detail what changes to the target may constitute. In all, we distinguish between four main kinds of changes to the target of analysis:

1. *Changes to the functions/functionality of the target*: This represents changes to all physical or logical parts of the target that exhibit relevant behaviour. This may be computer hardware or software, but also mechanical or moving parts.
2. *Changes to the non-functional properties of the target*: This includes, among other things, changes to security mechanisms and safety systems, and introduction of barriers.
3. *Changes to the processes of the target*: There are often work processes associated with the target. These may be of equal importance to the risk analysis as the components of the target, and changes to the processes must be considered changes to the target. Such changes may also include organisational changes.
4. *Changes in policies associated with the target*: Policies restrict functionality and processes. This means that changes in policies may be of equal relevance to the risk analysis as changes to the functionality or the processes themselves.

15.1.2 Scope and Focus

The scope of the analysis defines the extent or range of the target of the analysis. The scope defines the border of the analysis, that is, what is held inside and what is held outside of the analysis, or in other words, what is considered to be part of the target of analysis and what is considered to be part of the target's environment. The focus, on the other hand, defines the main issue or central area of attention in the risk analysis. It is clear that the focus is defined within the scope of the analysis.

Any change in the focus or scope will almost inevitably result in changes in the risk analysis and thereby outdate existing risk analysis results and documentation. Change of scope or focus may obviously be relevant also for cases in which the system, organisation or enterprise in question has not undergone changes, since a client or other stakeholders may wish to gain knowledge of the risks in a wider setting or with a different focus. However, change of focus or scope is particularly relevant when the system, organisation or enterprise in question is changing or evolving. If functionality is changed, services are substituted, new groups of end-users emerge, other interfaces are introduced, and so forth, it may often be that the focus or the scope of the analysis must be reconsidered in order to derive a more relevant and up to date risk picture.

15.1.3 Environment

It is not only changes to the target of analysis itself that may affect or outdate risk analysis documentation and results. There can be changes to the world outside the boundaries of the target that might be of equal or even greater relevance for the risk picture.

The environment of the target is anything in the surroundings of the target that is of relevance or that may affect or interact with the target; in the most general case the environment is the rest of the world. One specific change of the environment is that a new kind of threat emerges or that a threat disappears or is no longer relevant for the risk analysis. Obvious examples of new threats (in a computer security setting) are the invention of new kinds of computer viruses or hacker attacks. The emergence of electronic warfare and cyber crime are other examples.

Another kind of change in the environment is changes in the likelihood of threat scenarios due to changes in external factors. An example of this is threat scenarios involving blackouts. The likelihood of such threat scenarios may be dependent on the stability of external power supply, so if there are changes in the reliability of the external power supply, the likelihood of the threat scenarios might change.

15.1.4 Assumptions

Sometimes it is not changes to the target or its environment that triggers the need to reconsider the risk analysis documentation and results, but changes to the assumptions made in the analysis. There are many reasons why we might want to change the assumptions after the completion of a risk analysis, and most often changes in the assumptions mean that we also do changes to the scope of the analysis. It might be that parts of a system were assumed to be secure and for that reason kept outside the target, but that we later obtain evidence for the contrary, or for other reason start to doubt the validity of this assumption, and therefore want to include these parts in the target. Changes in assumptions are particularly relevant for changing and evolving systems since certain assumptions may have to be discarded after the introduction or replacement of functionality, services, end-users, and so forth.

15.1.5 Parties and Assets

A party of a risk analysis is an organisation, company, person, group or other stakeholder on whose behalf the risk analysis is conducted. Every risk that is identified and assessed is associated with a party. More specifically, each risk is associated with an asset, and an asset is something to which a party assigns value and hence for which the party requires protection. This means that any change of party will directly affect the outcome of the risk analysis.

There are two additional ways in which changes in parties may be relevant in a risk analysis. First, there may be organisational changes with respect to the customer of an analysis as a result of changes in parties. An example might be that the company for which a risk analysis was conducted is bought by another company, and the new owners have different priorities. Second, we may want to use a risk analysis conducted in the past as a template or pattern for a new risk analysis. This may be the case if we are analysing a target similar (or even identical) to a target we have analysed in the past, or if the previous analysis was within the same kind of domain. In this case, we may think of the existing risk documentation as a template or a pattern that is parameterised with respect to party.

Since an asset always belongs to a party, any changes in parties will result in changes of assets. Irrespective of changes in parties, however, it may also be that the value of an asset is reassessed and hence modified, that an asset is completely removed from the target (for example, because it is transferred to another party, or because the new asset value equals zero), or that new assets are introduced.

15.1.6 Context

The context of the analysis is the premises for and the background of the analysis, including the purposes of the analysis and to whom the analysis is addressed. Moreover, changes in the context may require changes to other parts of the target description such as the parties, assets, assumptions, scope and focus. Therefore, the context must also be taken into account when dealing with change in risk analysis.

The context of a particular analysis might, for example, be that the customer wants a high-level analysis of the target with respect to the security concerns of integrity, confidentiality and availability, and that the intended audience for the analysis results are decision makers. If the context later changes to a technical analysis with respect to integrity, and with developers as the intended audience, then clearly this affects the risk analysis and the analysis results.

15.1.7 Changes in our Knowledge

Another kind of change that may affect the outcome of a risk analysis, and that we therefore must consider, is the change in knowledge. If we obtain new or better knowledge, for example through monitoring, we may want to change the already documented estimates in order to ensure correspondence with this updated knowledge. Changes in our knowledge may also reveal new threats or threat scenarios.

When conducting a risk analysis, we choose the focus and scope, we make the assumptions, and we make decisions about the parties and the context of the analysis. These issues are therefore not subject to observations and empirical study in the same way as the target and its environment. Changes in our knowledge about the target and its environment may nevertheless substantially affect the decisions and assumptions upon which a risk analysis is based. The complete target description must therefore be reconsidered for possible changes and updates when we obtain new knowledge of relevance for the risk analysis in question.

We may think of a risk picture as our best attempt at representing the reality at a suitable level of abstraction, or as the best approximation of the reality that we can establish with the available knowledge sources. Changes in our knowledge can therefore be treated as real changes for the purpose of updating the risk picture. One important difference, however, is that changes in our knowledge usually increase our confidence in the correctness of the risk analysis documentation and results.

15.2 Managing Change

In this section, we look at how we manage change in relation to risk analysis from a practical point of view. To this end, we distinguish between three concrete risk analysis scenarios, referred to as perspectives, in which change is a major issue.

1. *The maintenance perspective*: Sometimes the target evolves over time, changes accumulate unnoticed, and risk analysis documentation and results may become outdated. An outdated risk analysis may give a false picture of the risks associated with the target and we may need to conduct a new risk analysis. Conducting a risk analysis from scratch is expensive and time-consuming, and we would rather like to update the documentation from the risk analysis that we have already conducted.

2. *The before-after perspective*: We often plan and anticipate changes, and major changes to the target may even be the main motivation for a risk analysis. Such planned changes require special treatment for two reasons. First, it is important to have a clear understanding of what characterises the target "as-is" and what characterises the target "to-be", and of what are the differences between these two. Second, the process of change may itself be a source of risks.
3. *The continuous evolution perspective*: There may be cases where we plan for the target to evolve over time or where we can anticipate gradual changes. What is common to such cases is that the target can be described as a function of time. Obviously then, it would be a benefit if we could also provide a risk picture as a function of time.

In general, when dealing with changes, we can, with advantage, view the target and the other elements of the target description as a collection of distinguishable, but possibly related, parts. In the following, we refer to such parts as *entities*. We should furthermore map the elements of the risk documentation to these entities, which we in the remainder of this chapter refer to as the *mapping model*. The entities of the target description can include, but are not restricted to, physical and logical components such as hardware and software. In many analyses, human and organisational issues are also relevant, so the entities may include such things as human actors and organisational units.

Mapping the elements of the risk documentation to the entities of the target description means, for example, for each threat determining whether it constitutes an entity of the target or belongs to the environment, for each threat scenario and unwanted incident determining which entities are invoked, for each vulnerability determining to which entity it belongs, and so forth.

15.2.1 Maintenance Perspective

The scenario corresponding to the maintenance perspective may be described as follows: We as risk analysts have conducted a risk analysis in the past, say three years ago, and are now requested by the same customer to reassess and update the risk picture to reflect recent changes to the target or to the environment and thereby restore the validity of the risk analysis documentation.

The changes we address from this perspective are changes that accumulate more or less unnoticed over time, and that are not really radical. Such changes can, for example, be bug-fixes and security patches, increase in network traffic or increase in the number of attacks. In this case, the risk picture remains more or less the same, but risk values may have changed such that previously acceptable risks may now be unacceptable or vice versa. The objective is then to maintain the documentation of the previous risk analysis to reflect the current situation of the target.

Figure 15.1 illustrates the principle by which such a reassessment is conducted. Assume we have available a description of the old target and either the updated target description or a description of the updates. We then use the mapping model

15.2 Managing Change

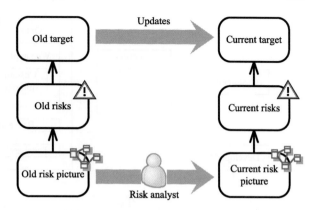

Fig. 15.1 The maintenance perspective

to identify the elements of the risk picture that are affected by the changes. When the elements of the risk picture affected by the changes are identified, we proceed by identifying the paths through the threat diagrams that contain affected elements. These paths require attention, and a walk-through of these paths should be made. They should be analysed with respect to whether or not they are still relevant and whether we need to make changes to their likelihood and consequence estimates.

Risk levels are determined by functions over likelihoods and consequences. Hence, we must update the levels of identified risks in accordance with the new likelihoods and consequences. Moreover, we must take into account that identified risks that earlier were judged to be acceptable, now may be unacceptable.

The procedure can also be applied to update a risk analysis after treatments have been implemented. In this special case, we also get some help from the specification of treatment scenarios. If we can associate a change with a specified treatment scenario, we can use the treatment scenario to track the elements affected by the change. This approach also works for introduction of new entities in the special case where the new entities can be associated with a treatment scenario (for example the implementation of barriers or the introduction of policies).

In the maintenance perspective, changes to assets are typically restricted to changes in asset values. In general, the impact of an unwanted incident on an asset depends on the value of the asset; the greatest harm that can be inflicted on an asset is equal to the value of the asset. In other words, the maximal possible reduction in asset value is the value of the asset. This means that if the value of an asset is changed, we need to do a walk-through of all the unwanted incidents related to this asset and reassess the consequence of each of the unwanted incidents. As explained above, this may result in changes to levels of risks, which hence need to be updated as well.

Example 15.1 Figure 15.2 maps elements of a threat diagram to a target description. (Both are based on the AutoParts example in Part II.) The mapping is illustrated by the dashed lines, and represents what we refer to as the mapping model.

From the mapping, we can deduce that changes to the online store may affect the path through the threat diagram, while changes to the gateway will probably not.

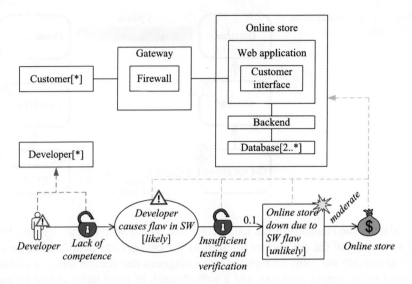

Fig. 15.2 Maintenance perspective example

Determining the latter depends of course to which extent the target description, the risk models and the mapping models are complete.

The mapping model implies that if the online store is maintained, the threat diagram must be reassessed and possibly updated. However, if only the gateway was updated there is in most cases no need to update the threat diagram.

From the figure, we can also see that the asset is related to the online store. If the value of the online store has changed since the old risk picture was established, this may affect the consequence of the unwanted incident. If, for example, the online store has increased in value because it produces more revenue than before, it may be that the consequence of downtime has increased as well.

15.2.2 Before-after Perspective

The scenario corresponding to the before-after perspective may be described as follows: We as risk analysts are asked to conduct a risk analysis with the aim to predict the effect on the risk picture of implementing certain changes in the target. In other words, we are asked to predict the future risk picture based on the current risk picture and a specified process of change.

The changes we address in this perspective are planned and anticipated, and may furthermore be radical. Such changes can, for example, be the rollout of a new system or major organisational changes such as the implementation of a merger agreement between two companies.

Figure 15.3 illustrates the principle by which we conduct a risk analysis from the before-after perspective. From the current target and the intended change process,

15.2 Managing Change

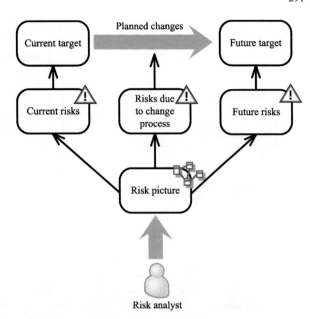

Fig. 15.3 The before-after perspective

we establish the future target. The current and the future target, as well as the change process itself, are input to the risk analysis where the goal is to establish a coherent risk picture that unambiguously describes the current risks, the future risks, as well as the impact of the risks associated with the change process.

The before-after perspective often involves the introduction of new entities in the form of processes, components, applications, and so forth, or the replacement or removal of such entities. In the former case, we must first determine how the introduction affects other entities, in other words how the changes propagate. Second, we must go through the risk identification step of the CORAS process with respect to the new and affected entities in order to determine whether new threats, vulnerabilities, threat scenarios and unwanted incidents have been introduced.

With respect to replacement of entities, we have to go through much the same process, although we may to a larger degree build on the current risk picture.

Removal of entities from the target is more straightforward, because it usually entails removal of elements from the risk model. However, removals must still be analysed carefully as also removal of entities can result in new risks; consider, for example, removal of safety barriers or security mechanisms. For this reason, the procedure for handling introduction of entities should also be applied in the case of removal of entities.

New threats may emerge either because the environment changes independently of the target, because we adopt a broader view of the environment, or because changes in the target result in the target changing its interface towards the environment. We can think of the latter case as the target getting in touch with new parts of the environment, with the result that the assumptions about the environment must be changed.

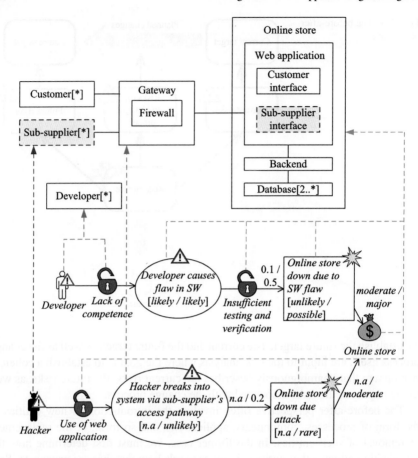

Fig. 15.4 Before-after perspective example

The emergence of new threats and related threat scenarios requires additional risk analysis. It is, however, often possible to get some help from existing risk documentation. As part of the additional risk analysis, we should identify the entities that might be affected by the new threats and examine whether these threats may exploit any of the vulnerabilities that have already been identified. We must additionally look for new vulnerabilities that are specific for the new threats.

Example 15.2 Assume the diagrams in Fig. 15.2 capture our current target and the current risk picture. Further, assume our client plans to introduce new entities to the target, in particular an interface through which sub-suppliers can interact with the online store. As result of this in Fig. 15.4, we extend the target description with a new entity, namely sub-supplier, as well as a sub-supplier interface. We use dashed frames and shading to highlight the changes to the target description.

We express the anticipated changes to the risk picture by assigning pairs of values to the elements of the threat diagram. The first value of the pair represents the

15.2 Managing Change

current value, while the second is the anticipated future value. Hence, with respect to Fig. 15.4, the likelihood of *Developer causes flaw in SW* is the same in both the current and the future risk picture. The *n.a.* (not applicable) decoration is used when the current risk picture has features not relevant for the future risk picture, and the other way around. For example, with respect to Fig. 15.4, *Hacker breaks into system via sub-supplier's access pathway* is not relevant in the current risk picture, and the first value of its likelihood pair is therefore *n.a.*

Through propagation, the online store is affected by the introduction of the sub-supplier interface, and the mappings show us that the (single) path through the original threat diagram is relevant for the changes. This path is therefore reassessed and some of the values are changed: Likelihoods are increased because the added functionality to the online store increased the likelihood of software flaws, and the consequence is increased because downtime after the change will affect not only users, but also sub-suppliers. In an additional risk analysis with respect to the new entities, we identify a threat *Hacker* since the new interface provided to sub-supplier might be exploited by this threat.

In the before-after perspective, the process of change may in itself be a source of risk, especially when dealing with larger changes. When doing a risk analysis in the before-after perspective, we should therefore also consider the process of change to assess what may happen during the transition from the current to the future state, and how this may impact the future risk picture.

In the identification of risks related to the change process, we should pay particularly attention to the following. First, there might be vulnerabilities that are not present neither in the current nor in the future state of the target, but which open for exploitations during the transition from the current to the future target. Second, parts of the change itself may fail, causing unwanted incidents. It is of course important to also document the risks originating specifically from the change/transition of the target.

In the before-after perspective, we must also consider the possibilities of removal of assets and introduction of new assets. Removal of an asset can be seen as assigning the asset a new value zero, and can therefore be treated as special case of change in asset value. The introduction of a new asset, however, is more complicated.

In the case where the new asset is an indirect asset, we proceed by simply applying the procedure for dealing with indirect assets as described in Chap. 11. In the case of a new direct asset, we should make a walk-through of all identified unwanted incidents and assess whether they have impact on the new asset. But we must also remember that the new asset may be harmed by unwanted incidents not identified in the original analysis. We must therefore do additional risk analysis with respect to the new asset. In this additional analysis, we should, similar to what we do in the case of changes to the target, try to associate the asset with entities of the target.

A change of party will first and foremost result in changes to the assets. The new party must decide for each of the assets of the original analysis whether the asset is relevant and, if so, assign it a value. The party must also decide whether new assets are needed. In addition, we must remember that the risk evaluation criteria are

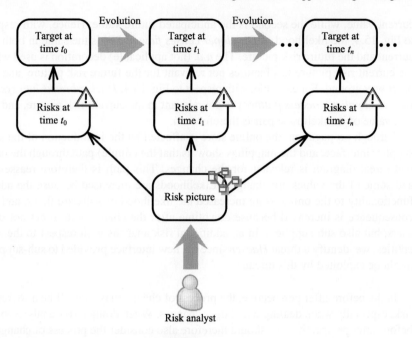

Fig. 15.5 The continuous evolution perspective

chosen by (or with respect to) the parties. Changing the party of the analysis may therefore also result in modification of the risk evaluation criteria, or new evaluation criteria in the case of new assets. If the risk evaluation criteria are changed, all risks must be reevaluated with respect to the new criteria.

15.2.3 Continuous Evolution Perspective

The scenario corresponding to the continuous evolution perspective may be described as follows: We as risk analysts are requested to conduct a risk analysis with the aim of establishing a dynamic risk picture reflecting the expected evolution of the target as a function of time.

The changes we address from the continuous evolution perspective are predictable, gradual evolvements that can be described as functions of time. The predictions can be based on well-founded forecasts or planned developments. Examples of planned continuous evolutions are the gradual increase of the number of components working in parallel and the gradual inclusion of more sites. Examples of well-founded forecasts can be the expected gradual increase of end-users, adversary attacks and annual turnover.

Figure 15.5 illustrates the principle by which we conduct a risk analysis from the continuous evolution perspective. Assuming that we have a description of the target as a function of time such that the target at any point in time can be derived, we

15.2 Managing Change

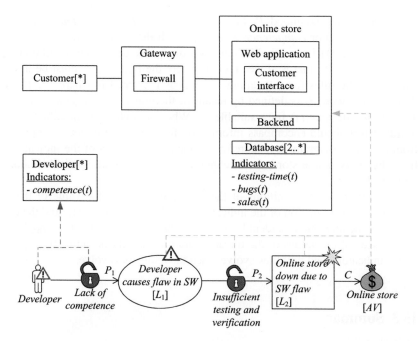

Fig. 15.6 Continuous evolution perspective example

use this as input to the risk analysis. Knowing how the target and its environment evolve over time, the objective is then to work out a risk picture also represented as a function of time that describes how the risks evolve.

This means that we are concerned with the situation in which we plan for the target to evolve over time or we in other ways are able to anticipate gradual changes over time. In this perspective, the target description is not updated or changed as with the other perspectives; instead, evolution is explicitly represented in the target description. Evolution will therefore also need to be explicitly represented in the risk model.

We represent evolution in the target description by defining properties of the entities as functions over time. We refer to these functions as *indicators*. An indicator may characterise any property of an entity.

In a risk model, we represent evolutions by defining the likelihoods, consequences, and other values, as functions of time. These functions may, for example, be defined with the help of the indicators in the target description.

Example 15.3 Figure 15.6 shows the target description and threat diagram from Fig. 15.2 with two modifications: We have specified indicators, with the time variable t as argument, for the entities developer and online store, and we have replaced all values in the threat diagram with variables. In order to obtain the evolving risk picture, we define the values of the threat diagram by means of functions.

We define the likelihood P_1 (associated with *Lack of competence*) as a function of the indicator *competence* associated with developer. Hence, we get $P_1 = f_1(competence(t))$ for a suitable definition of f_1. We define the likelihood L_1 as a function over P_1, that is, $L_1 = f_2(P_1)$, where the function f_2 must be in accordance with the rules for calculating likelihoods presented in Chap. 13. The conditional likelihood P_2 is defined by means of the indicator *testing-time*, so we get $P_2 = f_3(testing\text{-}time(t))$ for a suitable f_3. When we define the likelihood L_2, we decide that it should encompass information from both the target description and from the threat diagram. It is therefore defined as a function of the indicator *bugs* from the entity online store, the likelihood L_1 and the conditional likelihood P_2: $L_2 = f_4(bugs(t), L_1, P_2)$.

In addition to the definitions of the likelihoods, we let the value of the asset *Online store* be a function of the indicator *sales*, $AV = f_5(sales(t))$, and the consequence of the unwanted incident be a function of this asset value, $C = f_6(AV)$. As we can see, all the values of the threat diagram is directly or indirectly dependent on the indicators, and will thus evolve when the indicators evolve.

15.3 Summary

A risk analysis typically focuses on a particular configuration of the target at a particular point in time, and is valid under the assumptions made in the analysis. However, both the risk analysis target and its environment change over time, and such changes must be managed.

Change management in relation to risk analysis involves changes to the target description, as well as changes to our knowledge of the target and its environment. Central to change management in relation to risk is the definition of a mapping model that maps elements of the risk model to entities of the target description. This mapping model provides the necessary traceability to track how changes in the target description affects the risk model.

The context of the changes, as well as the setting of the analysis, is highly relevant for the handling of change in relation to risk analysis. We distinguish between three main scenarios, referred to as perspectives, when providing our guidelines: The maintenance perspective, the before-after perspective and the continuous evolution perspective. In the maintenance perspective we are concerned with updating risk analysis documentation and results after the target has been changed. In the before-after perspective we focus on predicting future risks after large, planned changes, as well as documenting the risks associated with the process of change. Finally, in the continuous evolution perspective we address the issue of forecasting the continuous evolution of risks as a result of planned and anticipated evolution of the target and its environment.

Chapter 16
The Dependent CORAS Language

The environment of the target is the surrounding things of relevance that may affect or interact with the target; in the most general case, the environment is the rest of the world. The CORAS language as defined so far in this book offers no support for the explicit documentation of environment assumptions, or assumptions in general. This may be unfortunate since the validity of the diagrams we make during a risk analysis, and thereby the validity of the risk analysis results, may depend on assumptions. If we only operate with informal descriptions of the assumptions, they may easily be forgotten with the effect that the conclusions that we draw from the risk analysis are believed to be more general than they really are.

Assumptions are important in relation to risk analysis for several reasons:

- In a risk analysis, we often make assumptions to simplify or focus the analysis. It may for example be that the customer has great trust in a particular aspect of the target, like the encryption algorithms or the safety system, and we are therefore asked to assume that there will be no incidents originating from that aspect, thus keeping it outside the target of the analysis. Alternatively, it may be that the customer is only worried about risks in relation to the daily management of a business process and we are asked to ignore situations caused by more extreme environmental events like fire or flooding.
- Most artefacts are only expected to work properly when used in their intended environment. There are, for example, all kinds of expectations and requirements to the behaviour of a car when driven on a motorway, but less so when it is dropped from a ship in the middle of the ocean. When doing a risk analysis, it is often useful to make assumptions about the environment in which artefacts of the target are used in order to avoid having to take into consideration risks that are of no real relevance for the target of analysis.
- Environment assumptions are also often useful in order to facilitate reuse of risk analysis results. We may, for example, be hired by a service provider to conduct a

---

This chapter was coauthored by Gyrd Brændeland, SINTEF ICT.
Email: Gyrd.Brandeland@sintef.no

risk analysis of its service from the point of view of a potential service consumer. The service provider wants to use the risk analysis to convince potential service consumers that the service is trustworthy. Environment assumptions can then be used to characterise what we may assume about the infrastructure of these potential service consumers. In fact, the correct formulation of these assumptions may be essential for a successful outcome. If the environment assumptions are too weak, it may be that we identify risks that are irrelevant for the practical use of the service and thereby give the impression that the service is less trustworthy than it really is. If the environment assumptions are too strong, it may be that we frighten away potential service consumers because we without any real reason impose assumptions that their infrastructure does not fulfil.
- Explicit documentation of assumptions is also an important means to facilitate reasoning about risk analysis documentation. Assume, for example, that a potential service consumer already has completed a risk analysis with respect to its own infrastructure. When this service consumer is presented with the risk analysis, we have conducted on behalf of the service provider, it should in principle be possible from these two analyses to deduce an overall risk picture for the combined infrastructure. This requires, however, a careful inspection of the assumptions under which these two analyses have been carried out.

The assumptions of a risk analysis are what we take as granted or accept as true, although they actually may not be so. An assumption is obviously not something that comes out of the blue, but is usually something that we have strong evidence for or high confidence in, for example, that there are no malicious insiders within a particular organisation or that passwords fulfilling certain criteria are resistant to brute force attacks. We may furthermore use assumptions as a means to choose the desired or appropriate focus and scope of the analysis. For example, even though there is strong reason to believe that there are no malicious insiders, we may still assume the contrary in order to ensure an acceptable risk level even in case of disloyal servants. Or we may, for example, assume that power failure, fire and flood do not occur while knowing that this is not so, in order to understand the general risk picture and risk level when such incidents are ignored.

In this chapter, we extend the CORAS language to facilitate the documentation of and reasoning about risk analysis assumptions. We refer to this extended language as dependent CORAS since we use it to document dependencies on assumptions. The assumptions are mainly of relevance in relation to threat scenarios and unwanted incidents that document the potential impact of events. Dependent CORAS is therefore only concerned with the two kinds of CORAS diagrams that can express these, namely threat diagrams and treatment diagrams.

16.1 Modelling Dependencies Using the CORAS Language

In this section, we first introduce the dependent CORAS language and explain how we construct dependent CORAS diagrams. Thereafter, we explain how such di-

16.1 Modelling Dependencies Using the CORAS Language

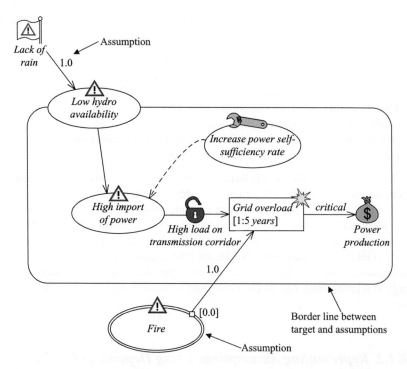

Fig. 16.1 Dependent treatment diagram

agrams should be interpreted by describing how to unambiguously translate any dependent CORAS diagram into English prose.

16.1.1 Dependent CORAS Diagrams

Figure 16.1 exemplifies a dependent treatment diagram. The only thing that is really new in this diagram is the border line separating the target from the assumptions. Everything inside the border line belongs to the target; everything on the border line, like the threat scenario *Low hydro availability*, and every relation crossing the border line, like the leads-to relation from *Fire* to *Grid overload*, also belong to the target. The remaining elements, that is everything completely outside the border line like the high-level threat scenario *Fire*, the threat *Lack of rain*, and the initiates relation from *Lack of rain* to *Low hydro availability*, are the assumptions that we make for the sake of the analysis.

As already pointed out, the only difference between a dependent diagram on the one hand and an ordinary or high-level threat diagram or treatment diagram as defined in Chaps. 4 and 14 on the other hand is the border line. Hence, if we remove the border line, the diagrams are required to be syntactically correct according to the definitions given in Chaps. 4 and 14.

With respect to the border line, we impose the following constraints for both kinds of dependent CORAS diagrams:

- Relations cannot cross the border line from the inside to the outside.
- Assets can occur only on the inside of the border line.
- Impacts relations can occur only on the inside of the border line.
- Threats cannot occur on the border line. They may, however, occur both on the inside and on the outside of the border line.
- Initiates relations cannot cross the border line. This means that initiates relations are either fully inside or fully outside the border line.

From the above, it follows that threat scenarios as well as unwanted incidents may occur on the border line. They may also occur on the outside and on the inside of the border line. Leads-to relations are the only relations that may cross the border line. They may also occur on the outside and on the inside of the border line. There are no special restrictions on the treats relation. Whether conditional likelihoods on leads-to relations that cross the border are placed inside or outside the border line is furthermore irrelevant. In Fig. 16.1, for example, the conditional likelihood on the leads-to relation from *Fire* to *Grid overload* is incidentally placed within the border line.

16.1.2 Representing Assumptions Using Dependent CORAS Diagrams

In the following, we present some small, illustrative examples in order to explain how assumptions can be represented and dealt with using dependent CORAS. The examples correspond to the motivation presented in the introduction of this chapter.

The dependent diagram of Fig. 16.1 exemplifies how we can explicitly utilise assumptions to simplify and focus a risk analysis. For example, by assuming that fires do not occur, and hence assigning the probability 0.0 to this threat scenario, this potential source of risk is ignored in the analysis. This assumption is of course generally not true, but if the customer of the risk analysis in question is not responsible for maintaining power supply and emergency planning in case of disasters such as fire, the assumption makes sense; given the assumption, the customer will get an evaluation of the risks that are within the customer's area of responsibility. If, on the other hand, the customer is required to maintain an acceptable risk level even when fires are considered, this particular assumption must be assigned a realistic likelihood estimate. In any case, dependent diagrams allow the explicit documentation of such assumptions, and thereby also the documentation of the assumptions under which the results of the risk analysis are valid.

Providers of software and services commonly specify hardware (HW) requirements such as processor (CPU) power, memory size and hard drive size. In order to guarantee, for example, a certain level of software dependability and service availability, the provider will then typically assume that the HW of the end-users fulfils the requirements.

16.1 Modelling Dependencies Using the CORAS Language

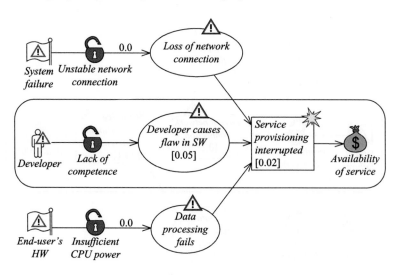

Fig. 16.2 Assumptions of service provider

The dependent diagram of Fig. 16.2 exemplifies a fragment of a risk analysis conducted by a service provider in order to ensure that the risk of loss of service availability for the end-users is acceptable. Since the service is expected to work properly only when the network connection is up and when the end-user's HW fulfils the requirements to CPU power, the fulfilment of these requirements are explicitly specified as assumptions. In particular, the likelihoods for the threats *System failure* and *End-user's HW* to initiate, respectively, the threat scenarios *Loss of network connection* and *Data processing fails* are set to 0.0. These likelihoods are assumptions that may not hold, but by making such assumptions the service provider can guarantee a certain level of service availability provided that the end-users deploy the service in the intended environment.

A risk analysis with explicit assumptions as illustrated in Fig. 16.2 can furthermore conveniently be reused for different purposes and in different analyses. For example, if different end-users have different needs, the service provisioning can be divided into different categories such as standard version and light version. If the different categories have different HW requirements, the analysis can be conducted separately for each category, and thereby for each kind of end-user. The dependent diagram of Fig. 16.2 could, for example, be reused in a setting in which availability is less critical for the end-user and where the HW requirements therefore can be eased. Such a change in the assumptions could be the replacement of the likelihood 0.0 on the initiates relation from the threat *End-user's HW* by the likelihood 0.2. The likelihood of the unwanted incident must then be updated to reflect the new assumption.

The diagram of Fig. 16.3 illustrates the risk analysis of the same service provisioning. The difference from Fig. 16.2 is that the analysis is conducted for the service consumer, that is for the end-user, instead of for the service provider. From

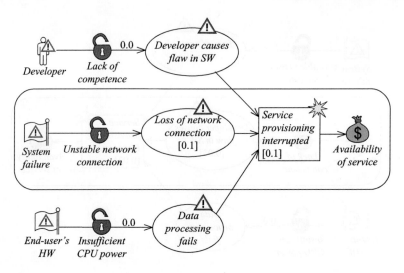

Fig. 16.3 Assumptions of service consumer

the likelihood 0.0 of the initiates relation pointing to the threat scenario *Data processing fails* of the assumptions, we see that also this analysis takes into account the HW requirements for deploying the service. Again, it may not be true that the end-user will always deploy the service using HW with sufficient CPU power. However, by making this assumption the analysis yields the risk level under this assumption.

The dependent diagrams of Figs. 16.2 and 16.3 furthermore show that assumptions in one analysis can be part of the target in another analysis, and vice versa. The assumption *Loss of network connection*, for example, is in Fig. 16.2 an assumption about the end-user made by the service provider. In Fig. 16.3, the same threat scenario is part of the target. The service consumer similarly makes the assumption *Developer causes flaw in SW* in Fig. 16.3, which is part of the target in Fig. 16.2.

The two separate analyses with different assumptions also show how separate analysis results can be combined into an overall risk picture by combining diagrams. For example, if the service consumer have conducted a risk analysis that is partly documented by the dependent diagram of Fig. 16.3 and is then presented the analysis results of the service provider, as partly documented by the dependent diagram of Fig. 16.2, she can combine the diagrams into the dependent diagram shown in Fig. 16.4.

The initial assumption *Developer causes flaw in SW* in Fig. 16.3 must then be assigned the likelihood value 0.05 as estimated in the target of Fig. 16.2. This particular threat scenario then no longer represents an assumption in the combined picture. Similarly, the assumption *Loss of network connection* in Fig. 16.2 is assigned the likelihood value 0.1. Once the assumptions in the one diagram is harmonised with the target in the other diagram and vice versa, the diagrams can be combined into one dependent diagram. As shown in the resulting diagram of Fig. 16.4, the elements that are assumptions in both diagrams, as the threat scenario *Data processing fails*, remain assumptions in the combined diagram.

16.1 Modelling Dependencies Using the CORAS Language

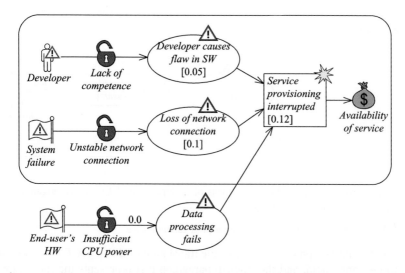

Fig. 16.4 Combining assumptions of service provider and service consumer

16.1.3 How to Schematically Translate Dependent CORAS Diagrams into English Prose

In Appendix B, we define the semantics of the CORAS language as a schematic translation of any ordinary CORAS diagram, or a fragment thereof, into a paragraph in English. In the following, we describe the semantics of dependent CORAS in the same way as we presented the semantics of basic CORAS diagrams in Chap. 4 and high-level CORAS diagrams in Chap. 14. The reader is referred to Appendix B for a more formal description.

Using a shorthand notation, we denote any dependent CORAS diagram by $A \triangleright T$. T denotes the part of the diagram that belongs to the target, which is everything inside the border line, every leads-to relation crossing the border line, as well as every threat scenario and unwanted incident on the border line. A denotes the assumptions, which is the rest. In other words, A denotes everything outside the border line including relations pointing at threat scenarios or unwanted incidents on the border line.

Both T and A are fragments of CORAS threat diagrams or treatment diagrams, possibly high-level. We have already explained how to schematically translate ordinary and high-level threat diagrams and treatment diagrams into English. What remains is therefore to explain how to translate the border line that defines what belongs to the target and what belongs to the environment in a dependent CORAS diagram.

In the previous subsection, we precisely characterised which part of a dependent diagram that constitutes the target T and which part that constitutes the assumptions A given the border line. Given this unambiguous partition of any dependent

Fig. 16.5 Simple dependent threat diagram

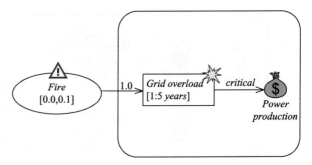

diagrams into two parts, the translation of a dependent CORAS diagram yields a paragraph of the following form:

...Assuming : ... To the extent there are explicit dependencies.

The English paragraph that represents the translation of the target T is inserted into the former open field, and the English paragraph that represents the translation of the assumptions A is inserted into the latter open field.

The suffix "To the extent there are explicit dependencies" is significant. It implies that if there are no relations explicitly connecting A to T, we do not gain anything from assuming A.

Example 16.1 Consider the dependent threat diagram in Fig. 16.5. In order to translate the diagram into English, we first determine for each element whether it is an assumption or belongs to the target. Since all elements inside or on the border line, as well as all relations that cross the border line, belong to the target, it is only the threat scenario *Fire* that constitutes an assumption. The rest belongs to the target.

By the description in Chap. 4 of how to translate threat scenarios into English, the translation of the assumption yields the following sentence:

Threat scenario *Fire* occurs with likelihood $[0.0, 0.1]$.

The part of the diagram that belongs to the target makes a threat diagram that is translated into the following:

Unwanted incident *Grid overload* occurs with likelihood $1 : 5$ *years*.
Power production is a direct asset.
Fire leads to *Grid overload* with conditional likelihood 1.0.
Grid overload impacts *Power production* with consequence *critical*.

Using the above schematic procedure for translating the full dependent diagram of Fig. 16.5, we then get the following paragraph in English:

Unwanted incident *Grid overload* occurs with likelihood $1 : 5$ *years*.
Power production is a direct asset.
Fire leads to *Grid overload* with conditional likelihood 1.0.
Grid overload impacts *Power production* with consequence *critical*.
Assuming:
   Threat scenario *Fire* occurs with likelihood $[0.0, 0.1]$.
To the extent there are explicit dependencies.

16.2 Reasoning and Analysis Using Dependent CORAS Diagrams

Table 16.1 Naming conventions

Variable	Diagram construct
D	Threat diagram or treatment diagram, possibly high-level
T	Fragment of threat diagram or treatment diagram, possibly high-level, that constitutes the target part of a dependent diagram
A	Fragment of threat diagram or treatment diagram, possibly high-level, that constitutes the assumption part of a dependent diagram
$A \triangleright T$	Dependent diagram of assumptions A and target T
e	Diagram element of threat, threat scenario, unwanted incident or asset
$e_1 \rightarrow e_2$	Initiates relation, leads-to relation or impacts relation from e_1 to e_2
P	Diagram path of connected relations

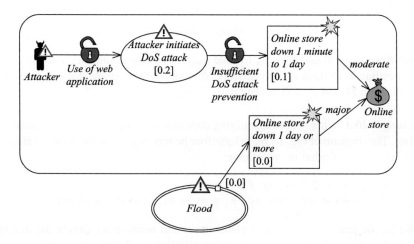

Fig. 16.6 Diagram paths and dependencies

16.2 Reasoning and Analysis Using Dependent CORAS Diagrams

In order to facilitate reasoning about and analysis of assumptions, we introduce four deduction rules. We start by defining some central concepts and helpful notation. For arbitrary diagrams, diagram elements and diagram relations of relevance for this chapter, we use the syntactic variables shown in Table 16.1.

Example 16.2 The dependent diagram of Fig. 16.6 will serve as an example to illustrate the new concepts. The diagram is based on the AutoParts example introduced in Chap. 5 and shows the explicit environmental assumption of flood and how it may lead to loss of availability of the online store.

For sake of readability, we use the following shorthand notations for the various elements of the diagram:

$$at = Attacker$$
$$aa = Attacker\ initiates\ DoS\ attack$$
$$oy = Online\ store\ down\ 1\ minute\ to\ 1\ day$$
$$fl = Flood$$
$$oe = Online\ store\ down\ 1\ day\ or\ more$$
$$os = Online\ store$$

When we are reasoning about assumptions we need to identify dependencies of target elements upon the assumptions. These dependencies are explicitly represented by diagram paths from the assumptions to the target. A path P is a set of connected, consecutive relations:

$$P = \{e_1 \to e_2, e_2 \to e_3, \ldots, e_{n-1} \to e_n\}$$

We write $e_1 \to P$ and $P \to e_n$ to state that P is a path commencing in element e_1 and ending in element e_n, respectively. We use $e \in D$ to denote that e is an element in diagram D, and $P \subseteq D$ to state that P is a path in diagram D.

Example 16.3 When we are identifying dependencies, we can ignore the vulnerabilities. The diagram of Fig. 16.6 can therefore be represented by the set es of elements and the set er of relations as follows:

$$es = \{at, aa, oy, fl, oe, os\}$$
$$rs = \{at \to aa, aa \to oy, oy \to os, fl \to oe, oe \to os\}$$

We let the pair $D = (es, rs)$ of elements es and relations rs denote the diagram. A path is a set of connected, consecutive relations, so the set $\{at \to aa, aa \to oy\}$ is an example of a path in D.

Let D be a CORAS diagram, D' be a fragment of D, and D'' the result of removing this fragment from D. An element $e \in D''$ is independent of D' if for any path $P \subseteq D$ and element $e' \in D$

$$e' \to P \wedge P \to e \quad \Rightarrow \quad e' \notin D'$$

Hence, e is independent of D' if there are no paths to e in D commencing from an element e' in D'.

We say that D'' is independent of D' if each element in D'' is independent of D', in which case we write $D' \ddagger D''$. By $D' \cup D''$, we denote the diagram fragment resulting from conjoining D' and D''. Hence, $D = D' \cup D''$.

Example 16.4 Let T denote the fragment of the dependent diagram D of Fig. 16.6 that consists of all the elements and relations inside the border, as well as the relation

16.2 Reasoning and Analysis Using Dependent CORAS Diagrams

crossing the border. Using our shorthand notation, we can represent this fragment by the following pair of elements and relations:

$$T = \big(\{at, aa, oy, oe, os\}, \{at \to aa, aa \to oy, oy \to os, fl \to oe, oe \to os\}\big)$$

Let, furthermore, A denote the result of removing T from D. The diagram fragment $A = (\{fl\}, \emptyset)$ then represents the assumptions.

In order to check whether T is independent of A, we need to check whether each of the elements of T is independent of A in D. Clearly, each of the elements at, aa and oy of T is independent of A since there is no path in the diagram D that leads from an element in A to any of at, aa, and oy. The unwanted incident oe, however, is not independent of A since there is a path commencing in fl that ends in oe. Hence, the fragment T of D is not independent of A.

Notice that when we are reasoning about dependencies in CORAS diagrams we are considering how threat scenarios and unwanted incidents as documented in one part of a diagram depend on threat scenarios and unwanted incidents as documented in another part. Since vulnerabilities are annotations on relations, we do not need to take them into account when identifying and analysing dependencies. In the setting of dependency analysis, we can likewise ignore treatment scenarios and treats relations since they can be understood as mere annotations on the dependency relevant information that is conveyed by a threat diagram. Therefore, as shown by Table 16.1, we do not include vulnerabilities, treatment scenarios or treats relations in the shorthand notation for this chapter.

Having introduced the notion of dependency, we now turn to the rules for reasoning about dependencies. These rules assume dependent CORAS diagrams D that are partitioned into the fragments T and A for target and assumptions, respectively, such that $T \cup A = D$. The rules are of the following form:

$$\frac{P_1 \quad P_2 \quad \ldots \quad P_n}{C}$$

We refer to $P_1, \ldots, P_n$ as the premises and to C as the conclusion. The interpretation is that if the premises are valid, so is the conclusion.

16.2.1 Assumption Independence

The following rule states that if we have deduced T assuming A, and T is independent of A, then we may deduce T.

Rule 16.1 (Assumption independence)

$$\frac{A \triangleright T \quad A \ddagger T}{\triangleright T}$$

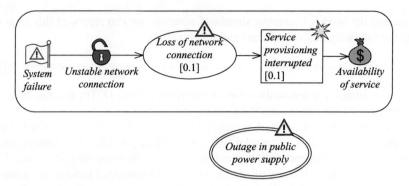

Fig. 16.7 Assumption independence

From the second premise, it follows that there is no path from A to an element in T. Since the first premise states T assuming A to the extent there are explicit dependencies, we may deduce T.

Example 16.5 The dependent diagram of Fig. 16.7 is a simple example of assumption independence. If, for example, redundant power supply is ensured within the target, the risk of loss of service availability is independent of the public power supply. Since there is no path from the assumption to the target, the target is independent of this assumption. By Rule 16.1, the assumption can be removed when we reason about the risks in this particular diagram.

16.2.2 Assumption Simplification

The following rule allows us to remove a fragment of the assumptions that is not connected to the rest.

> **Rule 16.2** (Assumption simplification)
>
> $$\frac{A \cup A' \triangleright T \quad A \ddagger A' \cup T}{A' \triangleright T}$$

The second premise implies that there are no paths from A to the rest of the diagram. Hence, the validity of the first premise does not depend upon A in which case the conclusion is also valid.

Example 16.6 The diagram of Fig. 16.8 exemplifies assumption simplification by the threat scenario *Outage in public power supply*. Since the target as well as the remaining assumptions are independent of this assumption, it can by Rule 16.2 be removed along with the leads-to relation ending in it when we are reasoning about the dependencies of the target on the assumptions in this diagram.

16.2 Reasoning and Analysis Using Dependent CORAS Diagrams

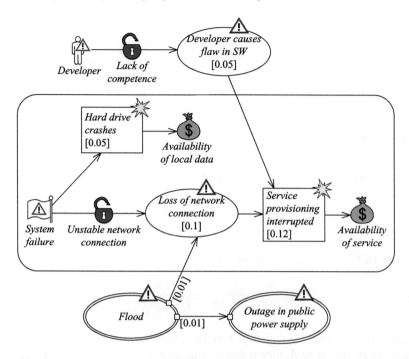

Fig. 16.8 Assumption and target simplification

16.2.3 Target Simplification

The following rule allows us to remove a fragment of the target as long as it is not situated in-between the assumption and the fragment of the target we want to keep.

Rule 16.3 (Target simplification)

$$\frac{A \triangleright T \cup T' \quad T' \ddagger T}{A \triangleright T}$$

The second premise implies that there is no path from A to T via T'. Hence, the validity of the first premise implies the validity of the conclusion.

Example 16.7 The dependent diagram of Fig. 16.8 exemplifies target simplification by the unwanted incident *Hard drive crashes* and the asset *Availability of local data*. If we want to reason about risks in relation to the asset *Availability of service* only, we can remove the parts of the target on which this asset does not depend. Since there is no path from the assumptions via *Hard drive crashes* or *Availability of local data* to any of the other target elements, the unwanted incident and asset in question can by Rule 16.3 be removed from the target.

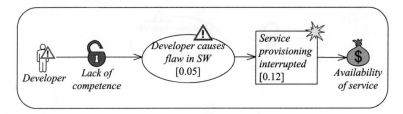

Fig. 16.9 Premise of consequence rule

16.2.4 Assumption Consequence

To make use of these rules when scenarios are composed, we also need a consequence rule.

Rule 16.4 (Assumption consequence)

$$\frac{A \cup A' \vartriangleright T \quad \vartriangleright A}{A' \vartriangleright T}$$

Hence, if T holds assuming $A \cup A'$ to the extent there are explicit dependencies, and we can also show A, then it follows that $A' \vartriangleright T$.

Given a dependent diagram $A \cup A' \vartriangleright T$ we can use Rule 16.4 to combine this diagram with a diagram D in which the validity of A has been shown. In particular, we can make a new diagram based on $A \cup A' \vartriangleright T$ by moving the fragment A from the assumptions to the target as described in the following example.

Example 16.8 Figure 16.9 shows a dependent diagram that documents the unwanted incident *Service provisioning interrupted* where the threat *Developer* and the threat scenario *Developer causes flaw in SW* are within the target. This is a separate analysis of elements that form part of the assumptions in Fig. 16.8, and we can therefore use this diagram as the second premise of Rule 16.4 to deduce the part of the diagram of Fig. 16.8 that depends on these assumptions.

The dependent diagram of Fig. 16.10 shows the result of applying Rule 16.4 with Figs. 16.8 and 16.9 as premises. In order to explicitly document that the elements that previously were mere assumptions are no longer so, as shown in the diagram of Fig. 16.9, these elements have in Fig. 16.10 been placed within the target.

The element of Fig. 16.8 that can be removed by Rule 16.2, namely *Outage in public power supply*, as well as the elements that can be removed by Rule 16.3, namely *Hard drive crashes* and *Availability of local data*, have also been removed in Fig. 16.10.

Notice that for the cases in which we can show the validity of all assumptions A in a dependent diagram $A \vartriangleright T$, possibly after assumption and target simplification, we can by Rule 16.4 conclude T under no further assumptions. This is conveniently captured by a specialisation of Rule 16.4 in which A' is empty.

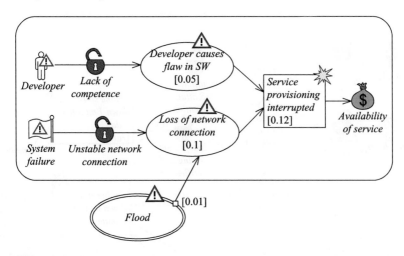

Fig. 16.10 Assumption consequence

Remark If we can show the validity of all assumptions A of the dependent diagram $A \triangleright T$, we can use the following specialisation of Rule 16.4:

$$\frac{A \triangleright T \quad \triangleright A}{\triangleright T}$$

16.3 Example Case in Dependent CORAS

In order to illustrate the usage of dependent CORAS diagrams, we continue with the AutoParts example introduced in Chap. 5. In the following, we use dependent CORAS diagrams to document assumptions about threat scenarios in the environment that may affect the risk level of the target. In this case, the documented assumptions serve two purposes: (1) Simplifying the analysis by allowing us to assume that certain events will not take place. (2) Facilitate reuse by describing the assumptions in a generic manner.

16.3.1 Creating Dependent Threat Diagrams

Figure 16.11 shows a dependent threat diagram for the planned online store of AutoParts. For the purpose of this example case, we assume that AutoParts has conducted a risk analysis of the old system a few years ago. The previous analysis focused on risks towards confidentiality of data stored in the inventory and customer databases, related to remote access. The CORAS diagrams documenting risks and

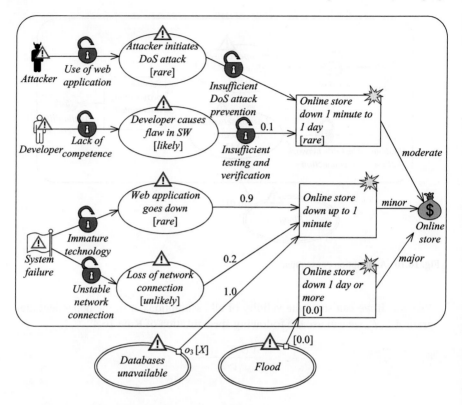

Fig. 16.11 Dependent CORAS diagram for the online store

treatments for the databases in Chaps. 9 through 12 document the previous analysis. AutoParts now wants to analyse any new risks that may be introduced in connection with the planned launch of the online store. The focus of this new analysis is the availability of the services provided by the online store to the customers and to the sub-suppliers. We use the consequence scale from Table 8.7 and the likelihood scale from Table 8.10.

During the initial steps of the analysis, the target team of the risk analysis agrees that the risk of flood should be held outside of the analysis. If the system is hit by a flood, it may render the online store unavailable for many days. However, we simply assume that the chance of a flood is zero in order to evaluate the overall risk level when this scenario is not taken into account. We state our assumptions about factors in the environment that may affect the risk level of the online store by placing them outside the border of the target as shown in the dependent diagram of Fig. 16.11. The relation from the assumption *Flood* to the target illustrates in what way this assumption is assumed to affect the target.

Since the scenarios outside the border line of the dependent diagram describe our assumptions they might not be verified with respect to the target, and may therefore be wrong. They simply describe the assumptions upon which the validity of

16.3 Example Case in Dependent CORAS

our analysis relies. We document the assumption that the chance of flood is zero by placing the referring threat scenario *Flood*, with likelihood zero, outside of the border.

The target team also agrees that the inventory and customer databases should be left out of the target of the analysis, as these have already been analysed in the previous study. We assume that some of the recommended treatments from the previous study have been implemented and that the threat diagrams have been updated to document the current risk level for the databases. The availability of the online store relies on the availability of customer information and the inventory database. Therefore, threats towards the availability of the databases may also pose threats towards the availability of the online store. The analysis team documents this dependency in the diagram by having a referring threat scenario *Databases unavailable* outside the target border that leads to the unwanted incident *Online store down up to 1 minute*.

For the purpose of identifying risks, the likelihood of this threat scenario is not important. We therefore parameterise the likelihood value of the assumption *Databases unavailable*. The possibility to generalise assumptions through parameterisation facilitates reuse of analysis documentation. The likelihood of the incident *Online store down up to 1 minute* depends upon the likelihood of all three threat scenarios leading to it, including the assumption *Databases unavailable*. Hence, we cannot compute the likelihood of this unwanted incident until we instantiate the likelihood of *Databases unavailable* with a specific value. However, we could of course characterise the likelihood of *Online store down up to 1 minute* by an expression referring to the parameter X.

To summarise, the referring threat scenarios *Flood* and *Databases unavailable* of Fig. 16.11 represent analysis assumptions, while everything else, including the leads-to relations from the assumptions, represents the target.

16.3.2 Combining Dependent Threat Diagrams

When the risks towards the online store have been identified and risk values have been estimated in workshops, it is up to the analysis team to combine the documentation from the old and the new analysis in order to obtain an overall risk picture of the system as a whole. To keep the combined analysis manageable, the analysis team first creates a high-level diagram from the dependent diagram in Fig. 16.11, the result of which is depicted in Fig. 16.12.

Figure 16.13 shows the high-level threat diagram with likelihoods for unwanted incidents regarding the databases. There is no element in this diagram corresponding directly to the referring threat scenario *Databases unavailable* in the dependent diagram in Fig. 16.12. However, both of the referring threat scenarios *Hacker attack on inventory DB* and *Hacker attack on customer DB* may lead to unavailability of the databases. The analysis team therefore creates a new high-level diagram, shown in Fig. 16.14 with the referring threat scenario *Databases unavailable* that contains the referring threat scenarios *Hacker attack on customer DB* and *Hacker attack on*

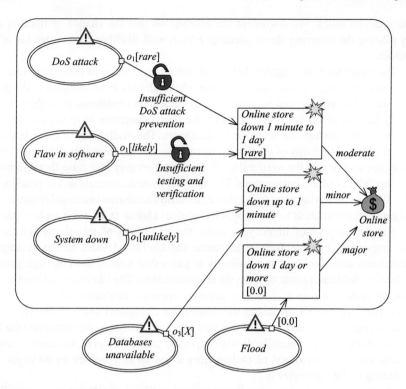

Fig. 16.12 High-level dependent CORAS diagram for the online store

inventory DB. Figure 16.15 shows the referenced threat scenario *Databases unavailable*. The referenced threat scenario *Hacker attack on customer DB* is shown in Fig. 14.25.

The referenced threat scenario *Databases unavailable* has five gates, two in-gates and three out-gates. For presentation purposes we show in the corresponding referring threat scenarios only the gates that are relevant for the diagram in which these referring threat scenarios occur. For example, in the diagrams in Figs. 16.11 and 16.12 we only show the out-gate o_3, whereas in the diagram in Fig. 16.14 this gate is not shown. In the referenced threat scenario of Fig. 16.15, however, all gates are shown. The in-gates i_1 and i_2 as well as the out-gates o_1 and o_2 are shown also on the corresponding referring threat scenario of Fig. 16.14. The out-gate o_3 explains the relation from the assumption *Databases unavailable* in the dependent diagram of Fig. 16.11.

Figure 16.16 presents the combined high-level threat diagram for the databases and the online store. The threat diagram of Fig. 16.14 allows the assumption *Databases unavailable* of the dependent diagram in Fig. 16.12 to be moved to the target, while the assumption about zero probability of flooding remains an assumption.

The dependencies between the two diagrams in Figs. 16.12 and 16.14 go only one way, that is to say the former diagram depends on elements in the latter, but not

16.3 Example Case in Dependent CORAS

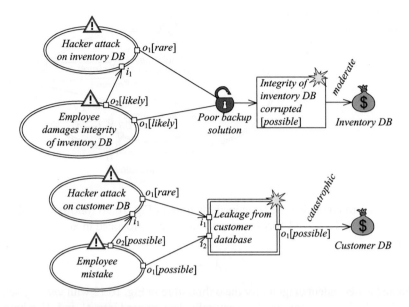

Fig. 16.13 High-level threat diagram for the databases

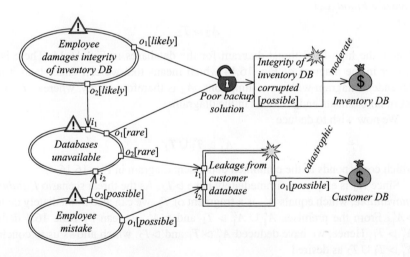

Fig. 16.14 High-level threat diagram for the databases combining threat scenarios

vice versa. It is therefore straightforward to combine the two diagrams as illustrated in Fig. 16.16, when substituting X with *unlikely* in Fig. 16.12. Initially, the unwanted incident *Online down up to 1 minute* was not assigned any likelihood because of the dependency on the assumption *Databases unavailable*. In the combined diagram, however, we use the additional information about this assumption to deduce the likelihood *possible* to this unwanted incident, as shown in Fig. 16.16.

Fig. 16.15 Referenced threat scenario *Databases unavailable*

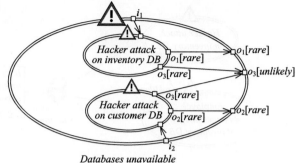

In order to argue that the combined diagram in Fig. 16.16 follows from the diagrams in Figs. 16.12 and 16.14, we apply the rules introduced in Sect. 16.2.

Let

$$A_1 \triangleright T_1$$

denote the dependent diagram for the online store in Fig. 16.12 with *unlikely* substituted for X. The assumptions A_1 is naturally decomposed into A_1' and A_1'' where A_1' is the referring threat scenario *Databases unavailable* and A_1'' is the referring threat scenario *Flood*. Let

$$A_2 \triangleright T_2$$

denote the high-level threat diagram for the databases in Fig. 16.14. There is no border in the diagram of Fig. 16.14 which means that we can understand it as a dependent diagram with no assumptions. A_2 is therefore empty, whereas T_2 is the set of all elements and relations in the diagram.

We now wish to deduce

$$A_1'' \triangleright T_1 \cup T_2$$

which corresponds to the combined dependent diagram in Fig. 16.16.

Since A_2 is empty, we immediately have $\triangleright T_2$. As the threat scenario *Databases unavailable*, which equals A_1', is a fragment of T_2, we can also immediately deduce $\triangleright A_1'$. From the premises $A_1' \cup A_1'' \triangleright T_1$ and $\triangleright A_1'$, we can by Rule 16.4 deduce $A_1'' \triangleright T_1$. Hence, we have deduced $A_1'' \triangleright T_1$ and $\triangleright T_2$, which allows us to conclude $A_1'' \triangleright T_1 \cup T_2$ as desired.

16.4 Summary

In this chapter, we have introduced the dependent CORAS language, which facilitates the documentation of and reasoning about risk analysis assumptions. Assumptions are used in risk analysis to simplify the analysis, to avoid having to consider risks of no practical relevance and to support reuse. The only syntactic difference

16.4 Summary

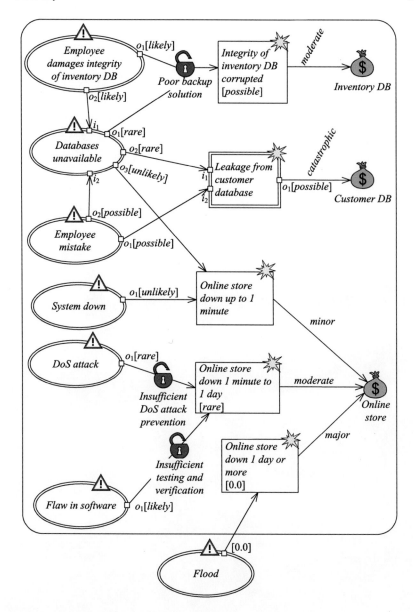

Fig. 16.16 Combined threat diagram for the data bases and online store

between a dependent CORAS diagram and an ordinary CORAS diagram is the border line that distinguishes the target from the assumptions. The reasoning about assumptions and dependencies as documented by dependent diagrams is supported by four deductions rules.

Fig. 16.16 Combined threat diagram for the data bases and online store

Chapter 17
Using CORAS to Analyse Legal Aspects

When we are conducting risk analyses, we often need to understand and to take into account legal aspects in order to properly present the risk picture. A company may, for example, have legally binding contracts or agreements with customers, suppliers or partners, and if something goes wrong we need to understand who is liable and whether there are any risks related to potential legal claims. Other examples are laws and regulations in relation to data protection and privacy that must be adhered to.

For cases in which legal aspects play a crucial role and may have a significant impact on the overall risk level, we must be prepared to take special measures in order to analyse and document these aspects properly. In this chapter, we explain and demonstrate the specific support for legal aspects in CORAS. In particular, we use the eight steps of the CORAS method to handle legal risks, and we introduce new language constructs to support the modelling and analysis of legal aspects.

In Sect. 17.1, we introduce and explain in the notion of legal risk, and relate this notion to the more general notion of risk that we have used so far. In Sect. 17.2, we explain how the analysis of legal aspects involves analysis along a legal dimension and a factual dimension, as well as how these dimensions are combined. In Sect. 17.3, we extend the standard CORAS language to provide support for the modelling of legal aspects. We refer to the resulting extension as the legal CORAS language. In Sect. 17.4, we go through the eight steps of the CORAS method to introduce and exemplify the specific support for analysing legal aspects. Finally, in Sect. 17.5, we conclude by summarising.

17.1 Legal Risk

We use the notion of legal risk to refer to risks that arise, completely or partially, because of legal norms. A legal risk is therefore a special case of risk. In order to understand and to deal with legal risks, we must therefore be able to identify, understand and reason about the legal norms that may be the origin of risks.

When we think of norms in general, we often think of social norms. These are rules that apply to individuals in a group and that define acceptable or expected behaviours and values. A norm is not descriptive, that is to say it does not describe how the world is. Rather, a norm describes how the world should be. Examples of norms are that people are expected to wear black in a funeral and that people are expected to be truthful. A given norm applies, that is, it is triggered, under given circumstances. In the circumstance of a funeral, for example, the norm stating that people should wear black is triggered. We refer to these circumstances as the antecedent of a norm. We refer to the requirements imposed by the norm, for example, to wear black, as the consequent of the norm.

For a norm to have effect, it must somehow be enforced. Social norms are usually enforced by individuals themselves, that is, the individuals of a group know and accept the norms, and they adhere to the norms without being told to do so. But the group also enforces norms by reacting to the behaviour of individuals. By adhering to norms an individual is accepted and appreciated, while by failing to adhere to norms an individual becomes unpopular and may be excluded from the group.

A legal norm is a norm that stems from a legal source such as domestic or international law, legal regulations, contracts, and legally binding agreements. Failure to adhere to a legal norm can lead to claims or legal actions. As an example of a legal norm, consider a mail order company that signs a legally binding agreement with its customers saying that the company is to be held liable in case the ordered goods are damaged when collected by the customer. More precisely, the legal norm may be formulated "If the goods are damaged when collected by the customer, the company is obliged to cover the damages". The legal source in this case is the agreement. The antecedent of the norm is "The goods are damaged when collected by the customer", and the consequent is "The company is obliged to cover the damages".

Definition 17.1 A *legal norm* is a norm that stems from a legal source, such as domestic or international law, contracts or legally binding agreements.

The UML class diagram of Fig. 17.1 gives a conceptual overview of the notion of legal norm. In case a legal norm applies, that is the antecedent is triggered, the legal norm prescribes that the consequent should follow. Both the antecedent and the consequent are described by or refer to some set of circumstances. These circumstances may be certain behaviours or certain scenarios, and they may be certain states of affairs.

Legal norms often define specific behaviour and scenarios as obliged, permitted or prohibited under given circumstances. But legal norms may also specify so-called legal qualifications [48]. A legal qualification refers to a certain state of affairs and prescribes that this state of affairs counts as, in other words is qualified as, something. For example, if two business partners are working out a document that regulates their partnership, it is only when certain requirements are fulfilled that this document is qualified as a legally binding contract. The relevant legal norm can then be understood as a legal definition of what is a valid contract. Another example is that of personal data, such as personally identifying information; whether or not certain pieces of information about a person should be held as personal data is a question of legal qualification as defined by legal norms.

Fig. 17.1 Elements of a legal norm

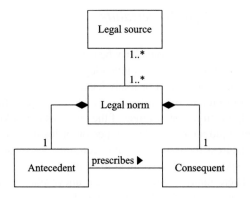

A legal norm is legally enforced by institutions, senior public servants, civil servants, and so forth. For example, if a legal claim is brought to court, the relevant legal norms are enforced by the court of justice and the judge. Legal norms can also be enforced by for example, police officers, for example, by fining someone for speeding.

When we are analysing a target of analysis with respect to legal aspects, the identification of legal norms that may be sources of risk is a key task. Furthermore, in order to understand the impact of a given legal norm we must deal with the problem of estimating two types of uncertainty that are associated with the legal norm, namely factual uncertainty and legal uncertainty.

17.2 Uncertainty of Legal Aspects

A legal norm applies if certain circumstances that satisfy the antecedent of the legal norm occur. When we are conducting a risk analysis that takes legal aspects into account, we must therefore identify such potential circumstances. Furthermore, in order to estimate the significance of a legal norm with respect to potential risks, we need to estimate the uncertainty of when the norm applies. This uncertainty is decomposed into two dimensions, namely a legal dimension and a factual dimension. We refer to the respective notions of uncertainty as legal uncertainty and factual uncertainty.

The legal uncertainty is the uncertainty of whether a legal norm actually applies to given circumstances, that is, the uncertainty of whether the circumstances satisfy the antecedent of the norm. The factual uncertainty, on the other hand, is the uncertainty of whether the given circumstances will actually occur and thereby trigger the legal norm. It is by combining the estimates for these two notions of uncertainty that we can estimate the significance of a legal norm.

In the following, we first separately address these two notions of uncertainty in more detail, and subsequently explain how they are combined.

17.2.1 Legal Uncertainty

The elements we consider when we address the issue of legal uncertainty are certain circumstances and a given legal norm. The legal uncertainty is the uncertainty regarding the correct or likely legal judgement of the legal norm in relation to these circumstances. Basically, we need to determine the likelihood of the legal norm to apply given the occurrence of the circumstances. Whether or not the norm applies is a legal judgement made by a person or institution that can enforce the norm. When we address the issue of legal uncertainty, we must therefore estimate the likelihood of a specific legal judgement to be made.

There many reasons why legal uncertainty may arise [8, 48]. One thing is that the law or other legal sources may be vague. A specific statute, for example, may be formulated in general terms and intended to cover several circumstances, and it may then be uncertain whether it applies to some specific circumstances. Another cause may be the variations of legal sources that are available to the legal decision maker such as a judge; since a legal judgement can be made on the basis of different sources, there may be an uncertainty of the outcome of the judgement. Legal uncertainty also arises because there may be differences regarding the interpretation of the legal sources and the legal norms. There may moreover be some necessary uncertainty in legal norms to intentionally allow legal decision makers to exercise discretion. A further issue that may be relevant is the decision maker's client or other loyalty that makes some legal judgements more attractive than other possible alternatives. The law's overall certainty may also be a cause of legal uncertainty, where increased certainty often precludes the possibility of considering the unique nature of the case. Finally, the "safety-valve" in law may allow the legal decision maker to make adjustments in case the consequence of the decision do not appear fair or reasonable according to extra-legal norms.

As an example of legal uncertainties, consider again a mail order company that signs a legally binding agreement with its customers in which it is stated that "If the goods are damaged when collected by the customer, the company is obliged to cover the damages". In order to identify and analyse the potential risks that may be caused by this legal norm, we must identify potential scenarios that may trigger the antecedent "the goods are damaged when collected by the customer". Assume that the following three scenarios are considered:

A The goods are damaged during manufacturing
B The goods are damaged during shipment
C The original packaging of the goods are damaged

For each of the scenarios, we must then estimate the legal uncertainty for the scenario to be held as a valid fulfilment of the antecedent of the legal norm. More precisely, we must estimate the likelihood of a legal judgement, for example by a judge in court, stating that the legal norm actually applies to the scenario. This likelihood estimate may, for example, be specified as a probability.

Scenario A above in which the goods have manufacturing defects is a typical case in which the customer can refer to the agreement and demand the mail order

17.2 Uncertainty of Legal Aspects

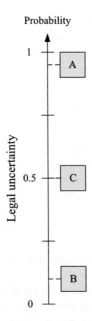

Fig. 17.2 Estimating legal uncertainty

company to cover the damages. Taking into account potential exceptional cases in which the legal norm does not apply to this scenario, we estimate the legal uncertainty to the probability 0.95. The scale depicted in Fig. 17.2 shows legal uncertainty as a probability ranging from 0 to 1, into which scenario A is plotted according to the given estimation.

At the other end of the scale, we find scenario B, that is, the scenario in which the goods are damaged during shipment. Assuming that the mail order company is not liable for damages caused by the postal services and other shipment and logistics services, the likelihood of the scenario to trigger the legal norm is low. In Fig. 17.2, it is therefore plotted in with likelihood 0.1.

The third scenario that may occur and that may be judged to trigger the legal norm is scenario C, namely that the original packaging of the goods are damaged. The scenario may be relevant for certain goods, such as collector's items the value of which depends on the condition of not only the goods themselves but also the packaging. Such cases may not have been thought of when the content of the customer agreement were worked out, and the outcome of a legal judgement may therefore be more open. As shown in Fig. 17.2, the likelihood estimate is 0.5.

17.2.2 Factual Uncertainty

Having estimated the legal uncertainty related to a given legal norm and some circumstances, we know the extent to which the legal norm is relevant under these circumstances. No matter how relevant the legal norm is under these circumstances,

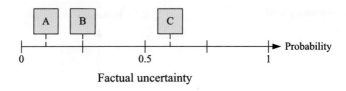

Fig. 17.3 Estimating factual uncertainty

however, there is little reason to bother if the circumstances never occur given our target of analysis. In order to estimate the significance of the legal norm, we must therefore also estimate the likelihood of the relevant circumstances to occur within the target of analysis. We refer to this likelihood as the factual uncertainty.

The estimation of factual uncertainties is precisely what we do in standard risk analyses when we are estimating the likelihood of scenarios and incidents; after having identified threat scenarios and unwanted incidents, we use expertise, statistics, experience, historical data, and so forth as a basis for assigning likelihood values to them.

As examples of factual uncertainties, we revisit the three scenarios from the previous subsection that were identified as possible ways of triggering the antecedent of the legal norm stating that "If the goods are damaged when collected by the customer, the company is obliged to cover the damages".

Scenario A is that the ordered goods are damaged during manufacturing. As shown by the scale depicted in Fig. 17.3, the probability of this scenario occurring is estimated to 0.1. Such an estimation is typically based on historical data and past experiences. As to scenario B, namely that the goods are damaged during shipment, the probability estimation describing the factual uncertainty is 0.25. Finally, scenario C, namely that the original packaging of the goods are damaged when collected by the customer, is estimated to the probability of 0.6.

Given these likelihood estimates, we know the probabilities or frequencies of relevant scenarios or circumstances to occur. However, in the same way as we cannot evaluate the estimates of legal uncertainties separately from the estimates of factual uncertainties, we can also not evaluate the estimates of the factual uncertainties separately from the legal ones.

17.2.3 Combining Legal and Factual Uncertainty

In order to understand the significance of a legal norm with respect to certain circumstances, we need to combine the legal uncertainty and the factual uncertainty. The diagram of Fig. 17.4 shows how the representation of legal uncertainties can be combined with the representation of factual uncertainties for a given legal norm. In particular, the diagram shows how the estimations with respect to the three aforementioned scenarios are decomposed along a legal axis and a factual axis. We can understand the axes as two orthogonal dimensions of a risk analysis that takes legal aspects into account, namely the legal dimension and the factual dimension.

17.2 Uncertainty of Legal Aspects

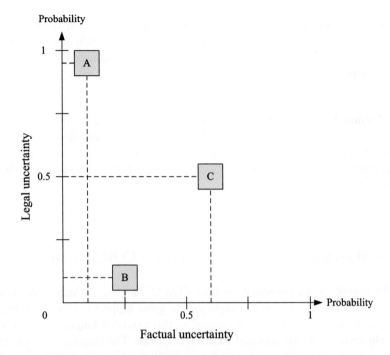

Fig. 17.4 Combining legal and factual uncertainty

When we are addressing legal aspects, we may initially keep the dimensions apart and do the estimations of the legal and factual uncertainties separately. For a given legal norm and some circumstances to which the norm may apply, we thereafter combine the two uncertainty estimations to derive the significance or impact of the legal norm.

Consider scenario A, for example, that is, the scenario in which the goods have manufacturing defects when collected by the customer. The estimation of the legal uncertainty is that there is a probability of 0.95 that the legal norm "If the goods are damaged when collected by the customer, the company is obliged to cover the damages" applies. However, the factual uncertainty of scenario A, that is, the likelihood for the scenario to occur is only of probability 0.1. By multiplying the two estimations, we get the probability 0.095, which is then the probability for the mail order company to be liable for manufacturing defects.

Notice, importantly, that the likelihood we derive by combining the legal uncertainty and the factual uncertainty is not equal to the likelihood of being met with a legitimate legal claim. For scenario A, for example, there is a probability of 0.095 that the mail order company is liable and that a customer is in the position in which he or she can raise a legitimate legal claim against the company. However, if some customers for one reason or another do not demand compensation, the probability for the company to actually having to cover the damages is lower.

When we take legal aspects into account, we identify unwanted incidents that may be caused by legal norms in order to identify legal risks. Ultimately, our goal is therefore to estimate the likelihood of unwanted incidents such as *The mail order company covers the damages of the ordered goods*. The identification of relevant legal norms and the estimations of legal and factual uncertainties then serve to support and facilitate the estimation of the likelihood of unwanted incidents, and thereby the estimation of legal risks.

> **Definition 17.2** A *legal risk* is a risk that may be caused by one or more legal norms; the impact or significance of a legal norm is determined by estimating both the factual uncertainty of the circumstances that serve as the antecedent of the legal norm and the legal uncertainty of whether the norm applies to the circumstances.

17.3 Modelling Legal Aspects Using the CORAS Language

In the following, we introduce the legal CORAS language, which is an extension of the CORAS risk modelling language that gives support for the modelling and analysis of legal aspects. The objectives of this extended language are similar to the objectives with the standard CORAS language. The language should support the communication and common understanding between personnel of various backgrounds, the language should facilitate the risk analysis process where also legal aspects are taken into account, and the language should support the documentation of the results.

Needless to say, the analysis of legal aspects requires the participation of personnel with legal expertise, for example, lawyers. Additionally, the usual domain experts and professionals must be brought in, such as personnel responsible for system development and maintenance, decision makers, system users, etc. The legal CORAS language therefore aims to support the communication between and the combined efforts of such a heterogeneous group of people.

17.3.1 Legal CORAS Diagrams

Basically, we extend the CORAS language by introducing a construct for specifying legal norms. As we shall see, by the combination of this construct with the standard CORAS language we may capture legal norms, the source of legal norms, the antecedent and the consequent of legal norms, the legal and factual uncertainties, as well as the risks that are caused by the legal norms.

The diagram in Fig. 17.5 shows the new language construct and exemplifies its use. The example is a case in which a company needs to take into account data protection laws and regulations since the company is storing and processing personal data.

17.3 Modelling Legal Aspects Using the CORAS Language

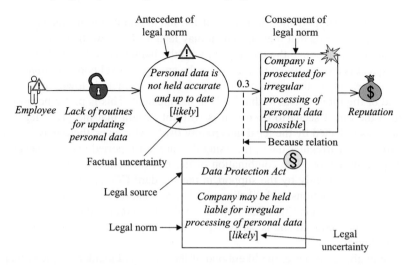

Fig. 17.5 Modelling legal aspects

The new construct for specifying legal norms is the rectangle with two compartments. We write the legal norm in the lower compartment and the source of the legal norm in the upper compartment.

The antecedent and the consequent of a legal norm are specified using the standard CORAS language. Both the antecedent and the consequent can be a threat scenario or an unwanted incident, and the antecedent is related to the consequent by the leads-to relation. The legal norm serves as an explanation of how the antecedent can lead to the consequent and is therefore specified as an annotation on the leads-to relation. We refer to the dashed line that relates a legal norm to the annotated leads-to relation as the because relation. The legal norm in Fig. 17.5 with its antecedent and consequent can therefore be read as follows:

> Personal data not held accurate and up to date leads to company being prosecuted for irregular processing of personal data because the company may be held liable for such irregular processing as stated in the Data Protection Act.

Having identified a scenario that may be the antecedent of a legal norm, we must estimate the factual uncertainty. In other words, we must answer the question of what is the likelihood for the scenario to occur. The antecedent is specified as a threat scenario or an unwanted incident for which we already have support in the CORAS language for likelihood specification. We therefore specify the factual uncertainty by annotating the threat scenario with a frequency or a probability. In the example of Fig. 17.5, the factual uncertainty is specified as *likely*.

The legal uncertainty is the uncertainty of whether the identified threat scenario can be judged to fulfil the antecedent of the legal norm. We specify the legal uncertainty by annotating the legal norm with a likelihood, either quantitative or qualitative. In the diagram of Fig. 17.5, the legal uncertainty is specified as *likely*.

As explained above, in order to understand the impact of a legal norm on the general risk level, we use the combination of the legal uncertainty and the factual uncertainty to support the estimation of the likelihood of the unwanted incidents and threat scenarios that the antecedent may lead to. Once we have conducted this estimation, we annotate the consequent with the likelihood value. In Fig. 17.5, the consequent is the unwanted incident with the annotated likelihood *possible*.

In this example, the likelihood of the unwanted incident is lower than the likelihood of the threat scenario that leads to the incident. This is partly because even though the threat scenario occurs, it is still not certain that the given legal norm applies, as reflected by the legal uncertainty *likely*. Moreover, there may be several occurrences of irregular data processing that are not reported and that never leads to legal prosecution. The combination of these two likelihood considerations can be specified explicitly by the support in the standard CORAS language for annotating leads-to relations with likelihoods. The annotation of the probability 0.3 on the leads-to relation in Fig. 17.5 shows the estimated likelihood of the antecedent to lead to the consequent, and is partly explained by the estimated legal uncertainty of the legal norm.

Although, in this case, the likelihood of the unwanted incident is lower than the likelihood of the preceding threat scenario, it may in other cases be higher. In the diagram of Fig. 17.5, for example, this could be the case if there were other threat scenarios and legal norms that are not documented in the diagram, and that may lead to the unwanted incident.

Since the legal CORAS language is an extension of the CORAS language, the leads-to relations of any syntactically correct CORAS diagram can be annotated with legal norms to yield a syntactically correct legal CORAS diagram. Conversely, the removal of the legal norms from a legal CORAS diagram yields a standard CORAS diagram. The result of removing the legal norm *Company may be held liable for irregular processing of personal data* from the diagram in Fig. 17.5 exemplifies the modelling of legal risks in standard CORAS. By the extensions of the legal CORAS language, we can make the legal dimension explicit and thereby facilitate the analysis and reasoning about the legal aspects and the legal risks.

17.3.2 How to Schematically Translate Legal CORAS Diagrams into English Prose

In Appendix B, we define the semantics of the CORAS language as a schematic translation of any ordinary CORAS diagram, or a fragment thereof, into a paragraph in English. In the following, we describe the semantics of legal CORAS in the same way as the presentation of the semantics of CORAS diagrams in Chap. 4. The reader is referred to Appendix B for a more formal and thorough description.

Since we already explained in Chap. 4 how to translate CORAS diagrams into English, and since legal diagrams are defined as an extension of these, we only explain how to translate the language constructs that are specific for legal CORAS in this section.

The only new element is the legal norm. The specification of legal norms includes the legal source of the norm, and may also be annotated with a legal uncertainty

17.3 Modelling Legal Aspects Using the CORAS Language

value. The legal uncertainty refers to the threat scenario or unwanted incident that make the antecedent of the legal norm. When translating the element of a legal norm into English, we omit the legal uncertainty value since this annotation is not relevant when representing the legal norm independent of its possible antecedents. The translation of legal norms therefore yields sentences of the following form:

> Legal norm ... is stated in the legal source ....

The description of the legal norm is inserted into the former open field, whereas the description of its legal source is inserted into the latter open field. The legal norm of the diagram in Fig. 17.5 is hence translated into the following sentence:

> Legal norm *Company may be held liable for irregular processing of personal data* is stated in the legal source *Data Protection Act*.

Legal norms are specified as annotations on leads-to relations. We therefore redefine the rule from Chap. 4 for how to translate the leads-to relation into English.

With the introduction of legal norms, the leads-to relation is still basically translated into a sentence of the following form:

> ... leads to ....

The description of the source of the relation is inserted into the former open field, and the description of the target is inserted into the latter. Both the source and the target can be a threat scenario, an unwanted incident or a risk.

In the setting of legal CORAS, the translation of the leads-to relation takes into account that the relation may or may not be annotated with a conditional likelihood, it may or may not be annotated with one or more vulnerabilities, and it may or may not be annotated with one or more legal norms. The translation furthermore takes into account that the legal norm may or may not be annotated with a legal uncertainty that specifies the certainty of whether the legal norms applies to the antecedent, in other words to the source of the leads-to relation. The schematic translation distinguishes between these cases and defines straightforward rules for how to translate the leads-to relation in each of them.

A leads-to relation annotated with a vulnerability and a legal norm, and with the conditional likelihood and the legal uncertainty unspecified is translated into a sentence of the following form:

> ... leads to ... with undefined conditional likelihood, due to ... and because ... applies with undefined legal uncertainty.

The description of source of the leads-to relation, which is also the antecedent of the legal norm, is inserted into the first open field. The description of the target of the leads-to relation, which is also the consequent of the legal norm, is inserted into the second open field. The description of the vulnerability is inserted into the third open field, and the description of the legal norm is inserted into the fourth open field.

A leads-to relation annotated with a legal norm, and where both the conditional likelihood as well as the legal uncertainty are specified, is translated into a sentence of the following form:

> ... leads to ... with conditional likelihood ... because ... applies with legal uncertainty ....

The description of the source is inserted into the first open field, the description of the target is inserted into the second open field, the conditional likelihood is inserted into the third open field, the description of the legal norm is inserted into the fourth open field, and the legal uncertainty is inserted into the fifth open field. The leads-to relation of the diagram in Fig. 17.5 is therefore translated into the following sentence:

> *Personal data is not held accurate and up to date* leads to *Company is prosecuted for irregular processing of personal data* with conditional likelihood 0.3 because *Company may be held liable for irregular processing of personal data* applies with legal uncertainty *likely*.

The reader is referred to Appendix B for the treatment of all the various combinations of annotations in the setting of legal CORAS. Each of these cases are straightforward generalisations of the rules presented in Chap. 4 for how to translate the leads-to relation into English.

Apart from the introduction of the legal norm and its legal source, as well as the generalisation of the leads-to relation to allow for the annotation of legal norms, the legal diagrams are equivalent to basic CORAS diagrams. The translation rules from Chap. 4 therefore apply to the remaining language elements and relations. The translation of the full legal diagram depicted in Fig. 17.5 is hence as follows:

> *Employee* is an accidental human threat.
> *Lack of routines for updating personal data* is a vulnerability.
> Threat scenario *Personal data is not held accurate and up to date* occurs with likelihood *likely*.
> Legal norm *Company may be held liable for irregular processing of personal data* is stated in the legal source *Data Protection Act*.
> Unwanted incident *Company is prosecuted for irregular processing of personal data* occurs with likelihood *possible*.
> *Reputation* is a direct asset.
> *Employee* exploits vulnerability *Lack of routines for updating personal data* to initiate *Personal data is not held accurate and up to date*.
> *Personal data is not held accurate and up to date* leads to *Company is prosecuted for irregular processing of personal data* with conditional likelihood 0.3 because *Company may be held liable for irregular processing of personal data* applies with legal uncertainty *likely*.
> *Company is prosecuted for irregular processing of personal data* impacts *Reputation*.

17.4 Analysing Legal Aspects through the Eight Steps of CORAS

In this section, we go through each of the eight steps of the CORAS method and explain how we can use the method to take into account and analyse legal aspects. To a large extent, we can use CORAS as it is, and some of the analysis steps therefore remain more or less unaltered.

In order to exemplify some of the issues that are relevant with respect to legal aspects, we revisit the running example that we used in Chaps. 5 through 12. In these examples, legal aspects are taken into account in the sense that compliance with data protection laws and regulations is one of the assets. In the following, we use this example case to show how legal norms that stem from a data protection act

17.4 Analysing Legal Aspects through the Eight Steps of CORAS

may cause risks to arise, and we show how to explicitly include such legal norms in the risk analysis.

We conduct Step 1, namely preparations for the analysis, in the same way as for a standard CORAS risk analysis. However, we need to clarify whether we should prepare for legal aspects to be taken into account during the analysis. Typical issues to consider are whether there are any laws and regulations, legal contracts, legally binding agreements, and so forth that may be of particular relevance for the target and for the parties of analysis. The following two examples demonstrate cases in which legal aspects are particularly relevant:

1. An enterprise A has entered a joint venture with another enterprise B, and their partnership and agreements are settled in a legal contract. The joint venture presupposes the exchange of trade secrets, and both parties can be held liable for compromising or misusing sensitive information about the other partner. Laws on intellectual property rights may furthermore be relevant in case some information is compromised or misused.

   Enterprise A decides to conduct a risk analysis of the joint venture. The objective of the analysis is twofold. First, the enterprise wants to understand the risks for enterprise B to misuse trade secrets and other information assets that belong to enterprise A. In particular, they are worried that the contract does not give sufficient legal protection. Second, the enterprise wants to understand the risks of themselves being held liable for unintentionally being in breach of the contract or for failing to respect the intellectual property rights of enterprise B.

2. A company is storing and processing personal information as part of its business process. Some of this information may according to the law be held as personal data, and some of it may also be personally identifiable information.

   The company decides to conduct a risk analysis in order both to understand to what extent their activities are subject to data protection laws and regulations, and to understand the risks that may be involved.

In case legal aspects are relevant, we address this in more detail during Step 2, which is the customer presentation of target. In particular, the presentation of the target should be specific about the legal aspects that must be included in the risk analysis. Such aspects may, for example, be relevant contracts, laws and regulations. We should also clarify during Step 2 whether lawyers or other experts on legal issues should be brought in. It is crucial for a risk analysis that we involve the necessary expertise with respect to all parts of the target, and this is obviously also the case when legal aspects are an issue.

Step 3, refining the target description, is where the analysts present the target as they have understood it. To the extent that legal aspects are relevant, we must at this point make sure that we have understood which part of the target is relevant for legal aspects, and also which contracts, legal agreements and so forth that must be considered during the analysis. These contracts, agreements and the like must be explicitly included in the target description.

The asset identification and the high-level analysis are also conducted during Step 3. These tasks are conducted in the same way as in a standard CORAS risk

Table 17.1 Legal uncertainty scale

Likelihood value	Description
Certain	[0.7, 1]
Likely	[0.3, 0.7)
Possible	[0.1, 0.3)
Unlikely	[0.01, 0.1)
Rare	[0, 0.01)

analysis, but we must obviously make sure to include assets that may be particularly relevant for the legal aspects in question.

Step 4, namely approval of target description, is entirely conducted as for a standard CORAS risk analysis apart from one thing. It is during this step that we set the likelihood scale that we use in the estimation of likelihoods of unwanted incidents and threat scenarios. When legal aspects are taken into account, we must also estimate the legal uncertainty of legal norms. During Step 4, we therefore need to establish a scale for legal uncertainty.

Legal uncertainties are specified as likelihoods, and the task of establishing the likelihood scale is as described in Chap. 8. The likelihood values can be given as probabilities or frequencies, and they may be quantitative or qualitative, depending on what is most suitable. It may often be that we can use the likelihood scale that we define for threat scenarios and unwanted incident for estimating legal uncertainty also. However, if we choose frequency values for the former and rather want to use probability values for the latter, we need to define a separate scale for legal uncertainties. If we work with qualitative values, we should for convenience use the same names for the values in the definition of the probability scale.

Example 17.1 After having established the likelihood scale to be used for assigning likelihoods to threat scenarios and unwanted incidents, the analysis leader of the AutoParts risk analysis explains that they next need to establish a scale for assigning legal uncertainty estimates to the relevant legal norms.

The likelihood scale for the threat scenarios and unwanted incidents is qualitative with values *rare*, *unlikely*, *possible*, *likely* and *certain*. Each of the values is furthermore described in terms of frequencies. The participants decide to use the same qualitative values for legal uncertainty, but they choose to describe them in terms of probabilities instead of frequencies. The resulting scale for legal uncertainties is shown in Table 17.1.

Although they choose to use qualitative values for legal uncertainties they may still use exact values in the estimations in case the exact values are known. If there is no doubt what so ever that a certain legal norm applies to some given circumstances, for example, they may assign the value 1 to the norm rather than the less precise description *certain*.

The next step, namely Step 5, is risk identification. As before we conduct the subtasks of identifying threats, unwanted incidents, threat scenarios and vulnerabilities in order to identify risks and the causes of risks. We furthermore use the

17.4 Analysing Legal Aspects through the Eight Steps of CORAS

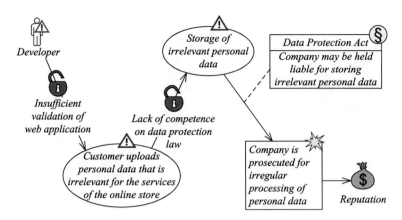

Fig. 17.6 Identification of legal norms

standard CORAS language to support the tasks and to document the results. In case legal aspects are relevant, we also identify relevant legal norms as well as scenarios to which these norms may apply. In support of this, we use the legal CORAS language.

Example 17.2 In order to use AutoParts' services that are offered through the online store, the customers of AutoParts first need to register as users. During user registration the customers fill in a registration form and thereby upload personal data. The lawyer in the target team explains that the data protection act prohibits the storage and processing of any personal data that are not necessary or relevant for the services they provide. She therefore wants the scenario *Storage of irrelevant personal data* to be considered in the analysis in order to evaluate the significance of the data protection act in this respect.

The threat diagram in Fig. 17.6 shows a possible way for this scenario to occur. Additionally, the diagram explicitly shows that the legal norm *Company may be held liable for storing irrelevant personal data* may be applied to this scenario.

One of the unwanted incidents to consider with regard to this is *Company is prosecuted for irregular processing of personal data*. Although legal prosecution may not necessarily lead to a sentence, it is still an unwanted incident since is may affect AutoParts' reputation, as shown in the diagram.

The identification, documentation and evaluation of this risk may of course be conducted without explicitly specifying the relevant legal norm and its source. By including the relevant legal norm, however, the sources of the risk is better documented and the basis for evaluating the risk is more solid.

After having completed the risk identification, including the identification of legal norms, risk estimation of Step 6 is next. To the extent that legal aspects are irrelevant for the identified risks, we conduct this step according to the standard CORAS method. The objective of the step is to estimate the likelihood of unwanted incidents and the consequence for the assets they harm. For each of the identified

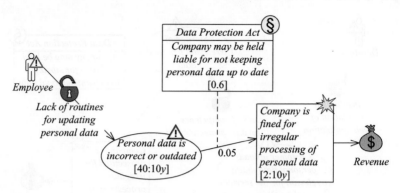

Fig. 17.7 Estimation of legal and factual uncertainty

legal norms, we additionally estimate the legal uncertainty, that is, the likelihood for the norm to apply to the threat scenario that serves as its antecedent. In turn, we use the estimation of the legal uncertainty to support the estimation of the likelihoods of the threat scenarios and unwanted incidents that may be caused by the legal norm.

In order to see how we estimate and reason about legal uncertainties, consider the threat diagram of Fig. 17.7 in which there are only exact likelihood values. The legal norm is that *Company may be held liable for not keeping personal data up to date*. On the one hand, we need to determine whether this norm applies to the antecedent *Personal data is incorrect or outdated*. This is a question of legal uncertainty. On the other hand, we need to determine the likelihood of the threat scenario that serves as the antecedent to occur. This is a question of factual uncertainty.

In the diagram, the legal norm is assigned the legal uncertainty 0.6, that is, the estimated probability for a legal decision maker to judge the norm to apply to the scenario is 0.6. To determine the chance for the company to be liable for the threat scenario, we must also consider the factual uncertainty. In the diagram the factual uncertainty is estimated to $40:10y$, that is, the frequency of 40 occurrences per ten years. Since the estimated probability for the norm to apply is 0.6, we have that there are $0.6 \cdot 40 = 24$ occurrences per ten years of personal data that is incorrect and outdated, and for which the company is liable at law.

Knowing the likelihood $24:10y$ for the company being liable for the scenario in question, we use this to support the estimation of the unwanted incident *Company is fined for irregular processing of personal data*. If we assume that the company is never fined unless liable by law, the likelihood of the unwanted incident is at most $24:10y$. It may, however, be that there are occurrences of the threat scenario that go by undetected or that are not reported. We must therefore take into account the probability of violating the law without being prosecuted or met by other legal claims.

For the diagram of Fig. 17.7, it is estimated that less that one out of ten occurrences of the threat scenario for which the company is liable results in penalty. Therefore, of the 24 occurrences each ten years, only two results in the company being fined, as shown by the frequency $2:10y$ assigned to the unwanted incident. From these estimations, it can furthermore be concluded that the probability for the

17.4 Analysing Legal Aspects through the Eight Steps of CORAS

threat scenario to lead to the unwanted incident is 0.05, which explains the annotation on the leads-to relation in the diagram.

Importantly, the conclusion that the frequency of the unwanted incident is $2 : 10y$ is based on the assumption that the diagram is complete in the sense that is shows all possible causes of the incident to occur. In case there are other threat scenarios that may lead to the incident, the frequency may be higher.

Example 17.3 The threat diagram of Fig. 17.8 shows some of the further unwanted incidents, legal norms, threats, and so forth that were identified in the analysis for AutoParts.

Notice, incidentally, that the unwanted incident *Online store down due to SW flaw* is not related to legal aspects. One and the same analysis, as well as one and the same threat diagram, may well address both standard risks and legal risks simultaneously.

Two of the unwanted incidents of the diagram represent legal risks since they are caused by legal norms, namely *Company is prosecuted for irregular processing of personal data* and *Company is fined for irregular processing of personal data*. The legal prosecution is an unwanted incident since it harms reputation. This unwanted incident may lead to the unwanted incident of being fined, which harms both reputation and revenue.

There are two paths in the diagram leading to the unwanted incidents. Since the scenarios of the two paths are independent, the likelihood of the unwanted incident *Company is prosecuted for irregular processing of personal data* can be estimated by adding the contribution from each of the paths.

As to the threat scenario *Personal data is incorrect and outdated*, the likelihood is estimated to *likely* which is the frequency interval $[20, 49] : 10y$ as defined in Table 8.10. The legal uncertainty, that is, the probability for the legal norm *Company may be held liable for not keeping personal data up to date* to apply to the threat scenario as antecedent, is estimated to *likely*. By Table 17.1, this value is defined by the probability interval $[0.3, 0.7\rangle$. The likelihood of AutoParts being liable for the threat scenario is then derived by combining the factual uncertainty with the legal uncertainty. This yields a frequency in the interval of at least $20 \cdot 0.3 = 6$ occurrences per ten years and at most $49 \cdot 0.7 = 34.3$ occurrences per ten years, that is, $[6, 34] : 10y$ as an approximation.

The lawyer explains that most minor data protection incidents go undetected or unreported in most companies. As a conservative estimate, the participants make the assumption that one out of ten liability cases for the threat scenario in question leads to prosecution. The contribution from the threat scenario and the legal norm of this path of the diagram to the unwanted incident *Company is prosecuted for irregular processing of personal data* is therefore $[0.6, 3.4]$.

By the same lines of reasoning, the participants next estimate the contribution from the threat scenario *Storage of irrelevant personal data* and the legal norm *Company may be held liable for storing irrelevant personal data*. As a result, they get the frequency interval $[0, 0.07]$, which practically is no contribution at all.

Taken together, they find that the result of the estimations lies in the intersection between the likelihoods *rare* and *unlikely*. Since the estimated interval gravitates

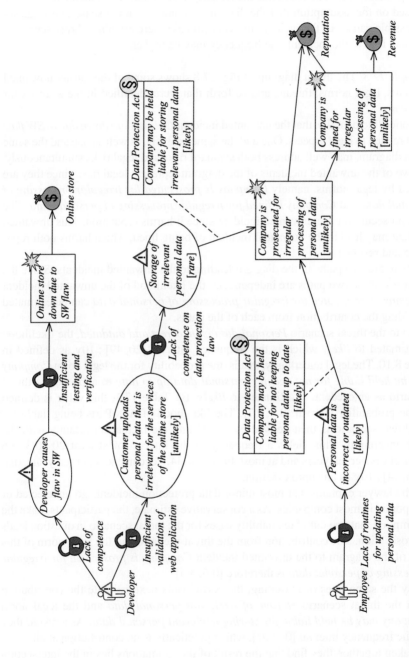

Fig. 17.8 Identification and estimation of risks and legal norms

towards *unlikely* they choose to assign this value to the unwanted incident, as shown in Fig. 17.8. Thereby, they also take into account that there may be other scenarios that could lead to prosecution that are not documented in the threat diagram.

After having completed the likelihood and consequence estimation of Step 6, including the estimation of legal and factual uncertainties for each of the identified legal norms, Step 7 follows next. This is risk evaluation which is conducted according to the standard CORAS method, irrespective of whether legal aspects are taken into account. Also the final step, namely risk treatment of Step 8 is conducted as for a standard risk analysis.

17.5 Summary

For some risk analyses, we find that it may be relevant and necessary to take legal aspects explicitly into account in order to understand, evaluate and document the general risk picture. A legal risk is a risk that may be caused by one or more legal norms. When we are taking legal aspects into account, we must therefore identify relevant legal norms and evaluate their significance.

A legal norm consists of an antecedent and a consequent, and the legal norm is triggered if some circumstances occur that fulfil the antecedent. There is, however, an uncertainty as to whether a given legal norm applies for a given target of analysis. This uncertainty can be decomposed into the legal uncertainty and the factual uncertainty.

In this chapter, we have demonstrated the support in the CORAS risk analysis method and language to document and analyse legal aspects. By introducing a new construct to the CORAS language for specifying legal norms, we gain the expressiveness to specify not only the relevant legal norms, but also the source of legal norms, the antecedent and the consequent of legal norms, as well as the risks that may be caused by legal norms.

The impact of a legal norm is estimated by estimating the legal uncertainty on the one hand and the factual uncertainty on the other hand, and subsequently combining these uncertainties. We have in this chapter shown how the support for specifying legal and factual uncertainties, as well as the support in the CORAS language to reason about likelihoods, facilitate the evaluation of the impact of legal norms.

To a large extent, the standard CORAS risk analysis method can be used to take legal aspects into account. It is only during Step 5 and Step 6, that is, during risk identification and risk estimation, respectively, that we need to take additional and particular measures to adequately address the legal aspects. During Step 5, this amounts to identify and document relevant legal norms, and during Step 6 it amounts to estimate legal uncertainties and to combine these uncertainties with factual uncertainties.

Chapter 18
The CORAS Tool

A unique feature of the CORAS method is the utilisation of the CORAS language to support the risk analysis process. In particular, we use CORAS diagrams intensively in the workshops where information is gathered through structured brainstorming. Moreover, we also use the CORAS language to document the risk analysis and present the risk analysis results. Appropriate tool support is advantageous in particular during the structured brainstorming sessions, where CORAS diagram are made on-the-fly to support, stimulate and document the discussions and the findings.

During such sessions the target team, under the guidance of the analysis leader, cooperates in developing the diagrams. At the same time, the analysis secretary is responsible for doing the actual modelling by drawing the diagrams as the discussion progresses. To avoid interrupting the flow of discussion, it is important that the diagrams are drawn efficiently, and that the diagrams are clearly presented. Good tool support is highly beneficial for achieving this.

The CORAS tool is designed to support on-the-fly modelling using all five kinds of basic CORAS diagrams, as well as high-level CORAS, dependent CORAS and legal CORAS. The CORAS tool hence facilitates the entire CORAS risk analysis process.

In Sect. 18.1, we give a description of the main functionality of the tool. Then, in Sect. 18.2, we describe how to use the CORAS tool during risk analysis with particular focus on the risk identification of Step 5 of the CORAS method. In Sect. 18.3, we discuss integration with other tools before we conclude in Sect. 18.4 by summarising.

18.1 Main Functionality of the CORAS Tool

The CORAS tool has six main parts, as can be seen in the screenshot presented in Fig. 18.1:

This chapter was coauthored by Fredrik Seehusen, SINTEF ICT.
Email: Fredrik.Seehusen@sintef.no

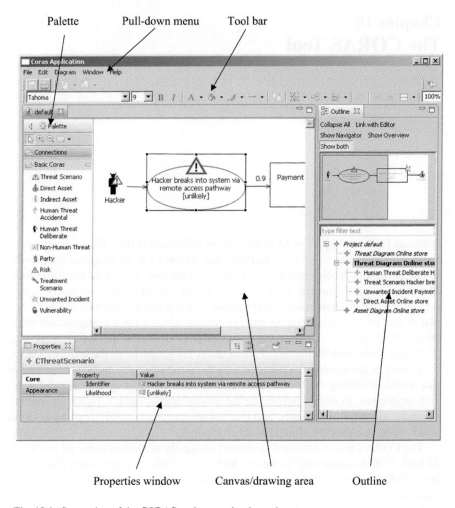

Fig. 18.1 Screenshot of the CORAS tool presenting its main parts

- *Pull-down menu*: Offers standard functions such as open, save, copy, cut, paste, undo, and print.
- *Tool bar*: Offers easy access to the standard functions of the pull-down menu.
- *Palette*: Contains all the model element and relations for drawing CORAS diagrams.
- *Drawing area*: The area or canvas for drawing the CORAS diagrams.
- *Properties window*: Lists the properties of selected elements. Can be used to edit the values of the properties.
- *Outline*: Presents a project and its diagrams as a tree.

Except for the pull-down menu and the tool bar, all parts of the tool can be closed or hidden. In the examples presented below, we hide the palette, the properties window and the outline, and show only the drawing area, in order to get more space for the diagrams.

In the CORAS tool, a *project* is a collection of diagrams, and each diagram must belong to a project. In order to start modelling, we must therefore first create a project.

The outline contains a tree representation of the project. The diagrams of the project are listed at the first level, and under each diagram all the elements of the diagram are listed. When a new element is created in the drawing area, it is automatically added to the tree under the correct diagram. A diagram element of the outline tree may be copied into the drawing area by simple drag-and-drop functionality.

The drawing area is the part of the CORAS tool where the diagrams are made. Basically, there are three things we can do in the drawing area, namely create new elements, edit elements, and delete elements. Elements of a diagram can be repositioned, resized, or renamed. For some elements and relations, we can also specify likelihoods or consequences.

18.2 How to Use the CORAS Tool During Risk Analysis

The CORAS tool is intended to be used whenever CORAS diagrams are drawn throughout the eight steps of the CORAS method. In general, there are three kinds of usage scenarios:

- *Before meeting*: Initial CORAS diagrams are drawn by the risk analysis team in preparation for or as input to a meeting or workshop.
- *During meeting*: CORAS diagrams are created on-the-fly, visible to everyone participating at the meeting or workshop.
- *After meeting*: CORAS diagrams are revised, polished, and analysed by the analysis team after a meeting or workshop has taken place.

In the following, we describe each of these scenarios in more detail.

18.2.1 Initial Modelling Before a Meeting

When preparing for a meeting, we often make initial CORAS diagrams that are used as starting point for discussion at the meeting. These diagrams should be drawn and stored in the CORAS tool. Later, during the meeting, we use the CORAS tool to expand, complete and revise the diagrams according to suggestions from the target team.

Example 18.1 For the sake of example, we assume that the analysis team as preparations for the risk identification of Step 5 sketches some initial threat diagrams showing the already identified assets together with relevant threats as identified during the high-level analysis of Step 3. Such a diagram is presented in Fig. 18.2 where

Fig. 18.2 Initial threat diagram made in preparation for a risk identification workshop

two of the assets are placed to the right and three relevant threats are placed to the left. Expanding the diagram to explain and document how the asset and threats are connected with threat scenarios and incidents is intended to be conducted on-the-fly during the risk identification workshop.

18.2.2 On-the-fly Modelling During a Meeting

Usually we use a beamer of some kind to make the tool available for all participants to see during a meeting. In some cases, we also use a second beamer to display the models of the target description as an aid during discussions. Under the guidance of the analysis leader, the analysis secretary uses the CORAS tool to add new elements and edit the CORAS diagrams as the discussion proceeds.

When doing on-the-fly modelling, it is important that diagrams are efficiently drawn and clearly presented. Otherwise the flow of discussion may be interrupted by

18.2 How to Use the CORAS Tool During Risk Analysis

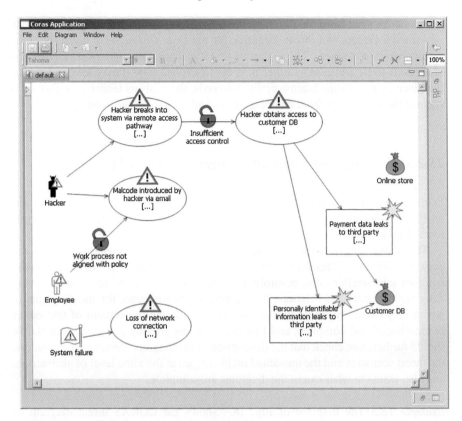

Fig. 18.3 Threat diagram drawn on-the-fly during risk identification

unnecessary delays and misunderstandings. The CORAS tool facilitates risk modelling with CORAS diagrams because it supports exactly the constructs that are needed. The CORAS tool also ensures that all the CORAS diagram elements are uniformly and clearly presented.

Example 18.2 The purpose of the risk identification of Step 5 is to identify threats, threat scenarios, vulnerabilities, and unwanted incidents. Suppose the diagram in Fig. 18.2 of the previous example is used as the starting point risk identification. The analysis secretary loads this diagram into the CORAS tool and displays it to all the participants. As the target team makes suggestions for threat, threat scenarios, and so fourth, the analysis secretary documents these by adding them to the diagram and by drawing the appropriate relations.

Figure 18.3 presents the initial diagram of Fig. 18.2 as it may look at some point during a brainstorming. The target team has suggested four threat scenarios and two unwanted incidents. These have been plotted into the diagram by the analysis secretary and furthermore related to the relevant threats and assets of the initial diagram. The target team has furthermore suggested two relevant vulnerabilities.

As can be seen from the diagram, the target team members have yet to identify unwanted incidents and threat scenarios for the asset *Online store* given the three threats presented to them. The suggested threat scenario *Loss of network connection* also needs to be addressed further in order to explain how it may lead to harm to the given assets. As the brainstorming proceeds, the analysis leader may choose to remove these elements from this diagram and address them separately.

18.2.3 Revising and Analysing Diagrams After a Meeting

During a meeting, we usually do not have the time to properly revise, polish and analyse the diagrams in detail. This is therefore most often done by the analysis team after the meeting.

The analysis team should clean up the diagrams and make them presentable and easy to understand. This may include splitting up diagrams that are too large and complex into smaller ones, possibly using high-level CORAS, and also removing dangling elements and relations that turned out as irrelevant for the diagrams in which they occur. We must furthermore make sure that the content of text boxes is meaningful and complete, and that the diagrams are tidy and easy to read. We should furthermore check that the description of the diagram elements, for example, the threat scenarios and the unwanted incidents, are at the same level of abstraction, and also assess to what extent the diagrams are complete.

Example 18.3 The diagram of Fig. 18.4 shows the CORAS threat diagram of Fig. 18.3 as it may look at the end of the brainstorming session. The diagram is rather messy as the analysis team did not have time to clean up, polish and restructure it during the meeting.

The revised diagram is shown in Fig. 18.5. The elements and relations that are not relevant for the unwanted incidents occurring in the diagram have been removed. The remaining elements and relations have been restructured in order to make the diagram easier to read and understand. The analysis team can then use this diagram as input to the subsequent analysis steps.

18.3 Integration with Other Tools

The CORAS approach may be used within many different problem domains, each of which may require more specific tool support. The CORAS tool has therefore been designed as a plug-in to Eclipse such that it is extensible and flexible. It can easily be integrated with other tools or extended with new features.

To ensure that the models created in the CORAS tool are stored in a standard format, the tool is based on Eclipse Modeling Framework (EMF). EMF provides

18.4 Summary

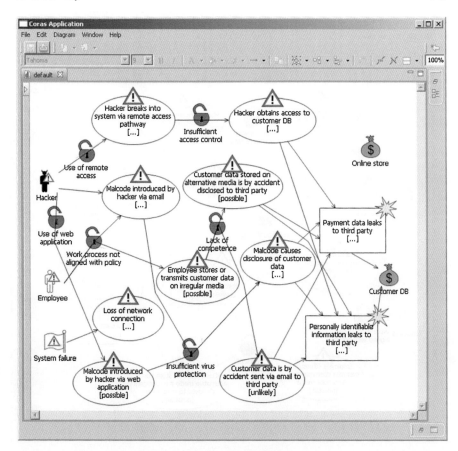

Fig. 18.4 Diagram showing the outcome of a risk identification workshop

support for processing structured data models based on eXtensible Markup Language (XML).

To ensure extensibility, the CORAS tool has been made available under an open source license. This allows third parties to contribute to the development of the tool, or to further develop the tool for their own purposes.

18.4 Summary

The CORAS language plays an essential role in the CORAS method, both as support during structured brainstorming and for documenting the outcome of risk analyses. It is therefore important to have a tool that supports the drawing and analysis of CORAS diagrams, particularly on-the-fly during meetings.

The CORAS tool supports the modelling of all five kinds of basic CORAS diagrams, as well as diagrams in high-level CORAS, dependent CORAS and legal

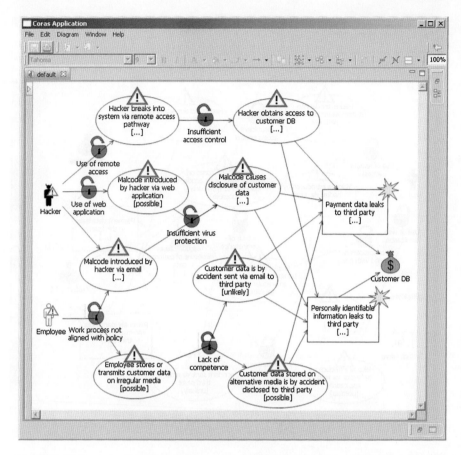

Fig. 18.5 Diagram showing the revised outcome of a risk identification workshop

CORAS. The CORAS tool has been specifically designed to support on-the-fly modelling during meetings, as well as model revising and analysis between meetings.

The CORAS tool is open source and developed as a plug-in to Eclipse, and it supports a standard data format. The CORAS tool can therefore be easily integrated with other tools or further developed by the open source community.

Chapter 19
Relating CORAS to the State of the Art

In Chap. 2, we already presented some important approaches to risk analysis and also explained how the CORAS approach relates to them. At that point, we could not do a detailed comparison since this requires an understanding of CORAS that the reader will obtain only after having read a major bulk of the book. In this chapter, we therefore revisit these other approaches and also give a more detailed account of the relation between CORAS and state of the art. In particular, we structure the presentation of this chapter according to the various features of CORAS as presented in Part II and Part III of this book.

Section 19.1 relates the CORAS language to other risk modelling languages. Section 19.2 presents existing risk analysis methods and relates these to the CORAS method as presented in Chaps. 5 through 12. Approaches to support the analysis and reasoning about likelihood are presented, and related to the CORAS approach of Chap. 13, in Sect. 19.3. The relationship to state of the art for Chap. 14 on modelling languages that offer support for high-level and hierarchical specifications is presented Sect. 19.4. In Sect. 19.5, we do the same for Chap. 15 by presenting existing approaches to managing change in relation to risks. Section 19.6 presents approaches to analyse dependencies and relates these to the CORAS approach as presented in Chap. 16. Finally, the relationship of Chap. 17 to the state of the art for legal risk management is presented in Sect. 19.7.

CORAS is a comprehensive approach to risk analysis that combines several analysis methods, techniques and modelling features, and has been developed to provide support and practical guidance for the complete risk analysis process. The structuring of this chapter according to the various facilities and features of CORAS demonstrates the uniqueness of CORAS in being a very much self contained approach to risk analysis.

19.1 Risk Modelling

The CORAS language draws upon a long history of modelling languages and has many sources of inspiration. First and foremost, there are several general modelling languages, among which the UML [59] is best known and most widely used.

UML has mechanisms, called profiles, that allow its users to make extensions to the language itself. There exist several profiles that extend UML with the possibility of modelling security properties or risks related information. Examples of the former are SecureUML [45] and UMLSec [41]. Examples of the latter are abuse-cases by McDermott and Fox [52], who first proposed the idea of applying specialised use-cases for the purpose of threat identification, and misuse-cases [72] that extend UML use-cases with means to elicit security requirements. The CORAS language for risk modelling in its early form was also defined as a UML profile [47], and standardised as a part of the UML Profile for Modeling Quality of Service and Fault Tolerance Characteristics and Mechanisms (UML QoS&FT) [58].

In difference from general modelling languages such as these, the CORAS language is a domain specific language. This means that the CORAS language is tailored for its specific task, namely risk modelling. In fact, misuse-cases was an important early source of inspiration for the CORAS language, but the CORAS language is richer because it is designed to support a full scale asset-driven defensive risk analysis process.

The fact that the CORAS language has been designed specifically for risk modelling has several benefits. First and foremost, it supports efficient modelling because it contains just the constructs needed to support the risk modelling. This also makes the language easy to learn and the models easy to comprehend. In addition, because the language is restricted to risk modelling, the diagrams become more homogeneous than what would be the case if the users were given the freedom of general modelling languages.

For traditional risk and dependability analysis, there exist various tree-based languages and other notations to support different parts of the analysis and documentation process. These notations typically use a small set of constructs to describe, for example, how a set of circumstances can lead to an unwanted incident, or what may be the consequences of an incident.

Most prominent of the tree notations are fault trees and event trees. The fault tree [30] notation is used in fault tree analysis (FTA) to describe the causes of an event. Fault trees are well known and widely used within risk analysis, and are becoming more common in security analysis, typically for systems for which security standards are high. The notation enables structuring of the order of events, where the top node represents an unwanted incident, or failure. The child nodes represent different events that can lead to the top event and are related to the top node with logical AND and OR gates. Fault trees can be used purely qualitatively as a means for decomposing an unwanted incident and identifying its contributing subevents, and they can be used quantitatively to analyse the probability of the unwanted incident. Quantitative analysis with fault trees is discussed further in Sect. 19.3.

Event trees [31] of event tree analysis (ETA) works in the opposite direction as fault trees; from an initial unwanted incident the consequences of the incident are explored. This exploration takes the form of a binary tree where each branching point is a split between the presence/occurrence or non-presence/non-occurrence of a vulnerability, barrier or event. As with fault trees, event trees can be used purely qualitatively to explore consequences, but can also be sub-

19.1 Risk Modelling

jected to quantitative analysis to estimate the likelihood of consequences, see Sect. 19.3.

A variant of fault trees specific to the security domain is attack trees [69, 70]. An attack tree has an attack goal as it top node and the branches of the tree explores different ways that this goal may be reached by an attacker. The focus is more on human behaviour than system behaviour since it represents different ways of attacking a system. A similar variant is the threat trees found in [79].

Event trees and fault trees can be viewed as complementary in the sense that event trees focus on the potential outcomes or consequences of an event, while fault trees focus on the causes of the event. Cause-consequence diagrams [57] of CCA can be seen as a combination of fault trees and event trees. Starting from an unwanted incident the causes are identified in the fashion of a fault tree and the consequences are explored similar to an event tree.

Microsoft has developed an approach called Threat Modeling [54, 79] which proposes Data flow diagrams (DFDs) to graphically represent the target of analysis. Based on such diagrams, the target is decomposed into components that are analysed with respect to identified threats. The method uses threat trees for the modelling and analysis of attack paths. The approach has been criticised for relying heavily on checklists [53], as it is based on the STRIDE model [26]. STRIDE is a checklist approach to the systematic walk-through of the target model to identify risks, and the name is an acronym for Spoofing, Tampering, Repudiation, Information disclosure, Denial of Service, and Elevation of privilege.

The OCTAVE method for security analysis [1] has its own tree notation which has much in common with event trees, but also fault trees. The notation supports the modelling of both the causes for and the outcomes of incidents. The OCTAVE threat tree can be seen as taking the tree notation one step closer to security risk analysis, in particular information security.

The CORAS language can in two ways be said to be more general than these tree-based notations. Most important, the CORAS language is tightly integrated in the risk analysis method and is designed to support all phases of the analysis process, while the tree-based notations focus on more specific and limited parts of the analysis. Second, the CORAS language is flexible with respect to the level of abstraction in the analysis and the CORAS diagrams have the form of directed acyclic graphs. The CORAS language can therefore be used to simulate all the tree-based notations, including cause-consequence diagrams, by making restrictions on the layout of the CORAS diagrams and the text of its elements.

Markov models and Bayesian networks are other notations that have been applied to support risk analysis. A Markov model [27, 32, 43] describes a system by a set of operation and failure states and the transitions between these states. The modelling may also include describing the different barriers that should prevent an attack or reduce the consequences of an attack. The transitions are annotated with probabilities, after which the models are subjected to a Markov analysis. This is discussed further in Sect. 19.3.

A Bayesian network [7, 18] is a directed, acyclic graph where the nodes represent causes or contributing factors with respect to related nodes. A Bayesian network is a graphical, probabilistic model that can be used both qualitatively to represent relations between causes and effects, and quantitatively to compute probabilities. The quantitative analysis of Bayesian networks is discussed in Sect. 19.3.

Markov models do in a sense belong to a different paradigm than the CORAS language. While the CORAS language describes the causal relationship between scenarios and incidents, Markov models characterise the possible transitions between different states of a system. They are therefore not directly comparable. It is, however, clear that the CORAS language is more general than Markov models. We could potentially construct a Markov model using a CORAS diagram if the latter described events at the appropriate abstraction level, but there is no obvious way to do the opposite.

Bayesian networks, on the other hand, are quite similar to CORAS diagrams. Both approaches describe causal relationships in terms of directed acyclic graphs. An important difference, however, is that Bayesian networks are first and foremost designed to do probabilistic reasoning, while the CORAS language also has a strong emphasis on communicating scenarios, incidents and risks. We therefore believe that the CORAS language pragmatically is easier in use in settings like risk identification workshops. We do, however, clearly see the possibility of interpreting a CORAS diagram as a Bayesian network for the purpose of doing Bayesian reasoning.

19.2 Risk Analysis Methods

Most risk analysis methods follow a process similar to the one described in Fig. 2.1, and several methods moreover include pragmatics for how to conduct one or more of the phases of the risk management process. Such pragmatics may, for example, be to instantiate the risk analysis process with techniques for risk identification and risk estimation, or to provide guidelines for how to carry out each of the activities of the risk analysis process in practice.

As already described in Chap. 2, OCTAVE and CRAMM [5] are full-fledged risk analysis methods. The latter has more recently been redeveloped by Siemens Enterprise Communications Limited [13].

The Facilitated Risk Analysis and Assessment Process (FRAAP) [61, 62] is a method for analysis of information security related issues. A FRAAP assessment team consists of representatives with competence on technical aspects, as well as business and management aspects. The process focuses on threats, vulnerabilities and consequences towards data integrity, confidentiality and availability. The outcome is an overview of risks and how they should be mitigated by existing or new controls (treatments). The final report contains a complete documentation of the process, including an action plan for the recommended treatments.

The Microsoft security risk management process [55] consists of four phases. The first of these phases is concerned with gathering and analysing data about assets, threats, vulnerabilities, existing security controls and possible treatments. The remaining three phases are concerned with selecting, implementing, and monitoring treatments, respectively.

The Microsoft Security Development Lifecycle [54] is a software security assurance process that integrates security and privacy into all phases of the software development process from requirements, through design, implementation and verification, to release. A risk analysis is performed during the design phase to identify software threats and vulnerabilities, where the focus is on security issues and privacy risks. The risk analysis makes use of the Microsoft Threat Modeling, including STRIDE, for the systematic identification assets, threats and vulnerabilities, as well as means for threat mitigation.

The U.S. National Institute of Standards and Technology (NIST) publishes standards and best practice guidelines for a wide range of IT security related topics. The NIST SP800-30 Risk Management Guide for Information Technology Systems [77] provides a foundation for the development of an effective risk management program, containing both the definitions and the practical guidance necessary for assessing and mitigating risks identified within IT systems. The publication is therefore more like a guideline for risk management that can be complemented with a risk analysis process than a full fledged risk analysis method.

Risk analysis processes and methods such as those described above are more or less self-contained in the sense of providing concrete and practical guidelines for how to carry out the various analysis phases and steps in practice and for how to document the results. To various extents, the processes may also be supported by analysis techniques, for example, to determine likelihoods, do consistency checking and evaluate risks.

CORAS is based on the ISO 31000 standard [36] which is preceded by AS/NZS 4360 [75]. Other risk analysis methods are usually based on these or similar standards, and for this reason the overall process of many risk analysis methods will be similar to the overall process of the CORAS method. What truly distinguishes CORAS from other risk analysis methods is the strong focus on modelling, both of the target and of the risks. The CORAS language is tightly integrated with the methodology and guidelines as a means for communication and documentation, and for supporting the risk analysis process. In contrast, other risk analysis methods usually depend on predefined table formats, and to some degree on notations such as fault trees and event trees, for these purposes.

19.3 Likelihood Analysis

Likelihood estimation is crucial in risk analyses since it is a prerequisite for assessing and evaluating risks. At the very least a risk analysis method must be supported by techniques or practical guidelines for how to determine the likelihood of unwanted incidents, but there should also be support for determining the likelihood of the events and scenarios that may lead to unwanted incidents. Such support will not only facilitate the estimation of the likelihood of the unwanted incidents, but also provide valuable information and documentation about the complete risk picture and the most important sources of risks.

In FTA, the fault trees serve as a basis for estimating the likelihood of the unique top node. Each branch of a fault tree can be of any depth, and the likelihood of a given node is calculated by considering the preceding nodes and whether the preceding nodes are combined by AND gates or OR gates. FTA requires the assignment of numerical statistical data to each leaf node and the leaf nodes must furthermore be independent of each other since the representation is based on Boolean combinations of events. In case there are such dependencies, other methods for computing the likelihoods must be applied [50]. Reasoning about likelihoods using attack trees is similar to that of FTA.

In ETA, the event trees represent the consequences of unwanted incidents, where the various possible consequences depend on the success or failure of system components. Event trees therefore have binary branching where the two branches are mutually exclusive. Because of this, the construction of an event tree always faces the risk of size explosion. The probability of each consequence is calculated by aggregating the probabilities of the component successes or failures that lead to this consequence.

As explained in Sect. 19.1, we can use CORAS diagrams to simulate different kinds of tree-based notations. The rules for doing reasoning on the likelihoods in CORAS diagrams are based on the same probability theory as rules for calculating probabilities in fault trees and event trees. We can therefore apply CORAS diagrams to do the same probability calculations as in the tree-based notations if we carefully place some restrictions on the possible interpretations of CORAS diagrams. On the other hand the CORAS language, with its rules for reasoning about likelihood, is more general than fault trees and event trees. There are therefore situations, especially cases where we do not have complete models or precise likelihood estimates, in which CORAS diagrams can be used for likelihood calculations that we cannot do with the tree-based notations.

The CORAS method and Markov analysis are not directly comparable, but as we mention in Sect. 19.1, we can imagine Markov models being created on the basis of CORAS diagrams. In that case, we could apply Markov analysis to calculate the likelihood of certain kinds of unwanted incidents and feed these back into the CORAS diagrams. However, because Markov analysis is tailored more specifically towards reliability, we suspect that the cases where Markov analysis is applicable are quite restricted compared to the variety of targets and concerns that we are likely to encounter in risk analysis in general.

Reasoning about likelihood using Bayesian networks may in some sense be more powerful than using the CORAS approach for this purpose, but do at the same time require more complete and specific input. In other words, we can do more calculations using Bayesian networks, but this requires the specification of input values in more detail than what is needed for doing likelihood reasoning in using CORAS. As discussed in Sect. 19.1, CORAS diagrams and Bayesian networks are quite similar with respect to the modelling. Because of these similarities, we have the possibility of interpreting CORAS diagrams as Bayesian networks. In this case, we could apply Bayesian reasoning to CORAS diagrams, but we would then also need to provide the input data required by the former.

19.4 High-level Risk Modelling

The usefulness of high-level modelling and hierarchical diagrams is apparent in several domains. Many of the existing modelling languages, for example within system development, support high-level modelling in order to allow specification at any level of detail while maintaining high-level overviews even of large and complex systems.

Several kinds of UML diagrams allow hierarchical modelling for arbitrary levels of detail. UML internal structure diagrams show the decomposition of a class into a collection of parts and their connections, and each part can be further decomposed in a hierarchical manner. UML collaboration diagrams show the contextual relationship among objects that interact in order to implement a certain system functionality and can be nested such that larger collaborations are built from smaller ones. UML component diagrams represent system components as modular pieces the visible behaviour of which is represented as a set of interfaces. Each component can be decomposed into subcomponents the internal structure of which is described in a separate component diagram, which in turn can be decomposed further to form a hierarchy. UML state machine diagrams describe the sequences of the states of an object, where state transitions are triggered by events. Also state machine diagrams support hierarchical modelling by the use of so-called submachines, which are state machine diagrams that can be referred to in a more high-level diagram in which the details of the submachines are hidden. UML sequence diagrams allow hierarchical modelling by so-called interaction uses. An interaction use is a reference to another sequence diagram which in turn can contain other interaction uses. The interaction uses can be nested to any depth. Sequence diagrams can furthermore be described at higher levels by the use of UML interaction overview diagrams which incorporate sequence diagrams and/or interaction uses.

The Statechart notation [22] precedes UML state machine diagrams and has several facilities for making large specifications of complex systems manageable and comprehensible. One of these features is the support for hierarchical specifications where states can be clustered to form a superstate. A superstate can be understood as an abstraction of the substates it contains, and the specifications can thereby be viewed at different levels of detail.

Message Sequence Charts (MSC) [38] is a notation most of which has been adopted and extended by UML sequence diagrams. Similar to interaction uses in UML, MSCs can be referred to by other MSCs, and similar to UML interaction overview diagrams high-level MSCs (HMSCs) give an overview of how MSCs can be combined. HMSCs can be hierarchical by combining other HMSCs.

The Specification and Description Language (SDL) [39] is aimed particularly at the specification of telecommunication systems. SDL is based on finite state machines and includes concepts for behaviour and data description. The concept of block refers to an agent that contains one or more blocks or processes, and blocks can be decomposed recursively over any number of levels thereby supporting hierarchical specifications.

DFDs are used in threat modelling in [79] for the purpose of increasing the understanding of the operation of the system under analysis. One of the advantages of

DFDs that is emphasised is their hierarchical nature which allows more important or more interesting parts of the system to be described in more detail. Essentially, a processing node of a diagram can be expanded into lower-levels in separate diagrams. A processing node represents a task in the system, and any node can be expanded in arbitrary depth of hierarchies by iterative nesting.

As seen from the overview of this section, high-level and hierarchical modelling is supported by a range of languages. High-level CORAS obviously draws inspiration from several of these. The motivation for high-level CORAS is, as for the other languages mentioned above, to make specifications scalable. In particular, high-level CORAS has been designed to make the analysis and risk modelling of large systems manageable and more easily comprehensible, and to allow the level of details to be increased or decreased when suitable. The distinguishing feature of high-level CORAS in relation to the state of the art is of course the support for the hierarchical modelling of risk related information. None of the risk modelling languages presented in Sect. 19.1 provide specific support for hierarchical modelling.

19.5 Change Management

For targets that undergo change or that are inherently evolving, also the risks are changing and evolving and should be understood, analysed and documented as such. For most of the established risk analysis methods, such as those presented in Sect. 19.2, there is little or no support for analysing risks as changing and evolving over time; the analyses rather focus on a particular state or configuration of the target at a particular point in time.

The extent to which existing risk analysis methods supports the analysis and representation of changing risks depends of course on the relevant perspective of change. Recall from Chap. 15 the classification of changes into the maintenance perspective, the before-after perspective and the continuous evolution perspective. In principle, any risk analysis method can be used to analyse changing risks from the maintenance and the before-after perspective; in the former case the aim is basically to update a previous risk analysis, and in the latter case the changes are planned so that separate analyses can be conducted for the current situation (before) and the situation after the planned changes have been implemented (after). For the management of changing risks to be efficient, however, there should be methodical support for dealing with change in a systematic and focused manner.

Some approaches have support for associating elements of risk models to parts of the target description, which may facilitate the identification and documentation of risk changes due to target changes. UML-based approaches such as misuse cases may utilise built-in mechanisms in the UML for relating elements from different UML diagrams. ProSecO [29] relates risks to elements of a functional model of the target. The model elements are furthermore related to security objectives and security requirements, and risks are related to threats and security controls.

With respect to the before-after perspective, ProSecO provides some support for modelling the various phases or states of a change process since all the elements of the security model have a state. When the models change, elements of the model may be transferred to states that indicate the need for additional risk analysis. ProSecO furthermore provides some support for the modelling of the change process by means of state machines. Other academic studies have focused on either maintenance [46, 71] or variants of reassessment [21, 44].

In difference from the other approaches, CORAS provides a systematic approach to change management in the context of risk analysis. This includes methodological guidelines on how to maintain and update risk analysis documentation to reflect both past and future changes to the target without starting from scratch and doing a full reassessment, something that to a large extent is missing in other risk analysis methodologies.

The continuous evolution perspective is the most general of the three perspectives on change as it incorporates time into the picture and thereby refers to the instance of the evolving risk picture at any given point in time. From this perspective, the risks are continuously evolving and need to be represented and analysed as such. Of the existing approaches to risk analysis and risk modelling, the support for the continuous evolution perspective is virtually non-existent. Some modelling languages do, however, provide support for updating the values that are annotated to the diagrams in the sense that by changing the input values, the derived output values can be automatically updated. These includes fault trees, Markov models, and Bayesian networks. In [16], influence diagrams [28], originally a graphical language designed to support decision making by specifying the factors influencing a decision, are connected to the leaf nodes in fault trees. The influencing factors contribute to the probabilities of the leaf nodes and the influence thus propagates to the unwanted incidents specified at the root of the tree. Similar, but simpler, are the risk influence diagrams in [3], where influencing factors are connected to the nodes in event trees.

The support in CORAS for managing the evolving risks of changing and evolving targets is a clear contribution to the state of the art. The problem of dealing with change and the need for methods that specifically target risks that continuously evolve over time is increasingly recognised, but still virtually not addressed by existing approaches. The approach in this book is inspired by an application of CORAS in risk monitoring [67] and has some similarities with the use of influence diagrams mentioned above. However, CORAS is the sole approach to provide means for predicting the evolution of risks related to an evolving target.

19.6 Dependency Analysis

Several approaches to component-based hazard analysis describe system failure propagation by matching ingoing and outgoing failures of individual components. Giese et al. [19, 20] have defined a method for compositional hazard analysis of restricted UML component diagrams and deployment diagrams. They employ FTA to describe hazards and the combination of component failures that can cause them. For each component, they describe a set of incoming failures, outgoing failures, local failures and the dependencies between failures. Failure information of components can be composed by combining their failure dependencies.

The approach of Giese et al. is similar to dependent CORAS in the sense that it is model-based, as they do hazard analysis on UML diagrams. Their approach has an assumption-guarantee [40, 56] flavor, as incoming failures can be seen as a kind of assumptions. There are, however, some important differences. The approach of Giese et al. is limited to analysis targeting hazards caused by software or hardware failures. The CORAS method has a much broader scope. The CORAS threat diagrams document not only system failures, but also the threats that may cause them, such as for example human errors, and the consequences they may have. Furthermore, the hazard analysis of Giese et al. is linked directly to the system components. CORAS diagrams are not linked directly to system components, as the target of an analysis may be restricted to an aspect or particular feature of a system. The modularity of dependent CORAS diagrams is achieved by the assumption-guarantee structure of the diagrams, not by the underlying component structure. Composition is furthermore performed on risk analysis documentation, not on components. Finally, CORAS does not require any specific kind of system specification diagram as input for the risk analysis, the way the approach of Giese et al. does.

Papadoupoulos et al. [60] apply a version of FMEA/FMECA [9] that focuses on component interfaces to describe the causes of output failures as logical combinations of internal component malfunctions or deviations of the component inputs. They describe propagation of faults in a system by synthesising fault trees from the individual component results. Kaiser et al. [42] propose a method for compositional FTA. Component failures are described by specialised component fault trees that can be combined into system fault trees via input and output ports. Fenton et al. [17, 18] address the problem of predicting risks related to introducing a new component into a system by applying Bayesian networks to analyse failure probabilities of components. They combine quantitative and qualitative evidence concerning the reliability of a component and use Bayesian networks to calculate the overall failure probability. The approach is, however, not compositional; Bayesian networks are applied to predict the number of failures caused by a component, but there is no attempt to combine such predictions for several components.

A clear advantage of dependent CORAS as compared to other approaches is that support for the modelling, analysis and reasoning about dependencies is directly integrated into the overall CORAS process. Dependent CORAS is therefore not only a means to model and reason about dependencies, but offers a comprehensive approach to a full scale risk analysis in which assumptions are documented and specifically taken into account during the analysis.

19.7 Legal Risk Management

As opposed to traditional risk management, legal risk management is still very preliminary and immature, and there are no well-defined, established or standardised methodologies or processes for legal risk management [49]. The very notion of legal

19.7 Legal Risk Management

risk is also used in many different meanings and in different contexts [48]. Various frameworks and methods for legal risk management have, however, been proposed. Many of these proposals focus on risk management and risk analysis as activities that should be led and conducted by lawyers. This can be understood as a response to a prediction made by Susskind [78] of a shift in the services offered by lawyers from reactive legal problem solving towards proactive legal risk management.

Wahlgren [83] shows that many of the services offered by practicing lawyers somehow involve or are related to handling risks, and that this is so both for general legal problems, for example in analysing the law or legal decisions in general, and for specific legal problems such as in analysing a contract. Due to this inherency of risks in legal issues and legal problems, Wahlgren proposes the use of selected risk analysis techniques in legal problem solving. These techniques, such as FTA, risk matrices, checklists, and so forth, can then complement established legal methods. Several potential risk analysis techniques are reviewed for this purpose, but Wahlgren does not propose any general legal risk analysis process or a comprehensive analysis method that comprise these techniques.

Some of the approaches to legal risk management are aligned with established risk analysis processes, where the basic idea is that legal risks should be understood and addressed by following similar principles. Reid's framework for legal risk management [68] is based on the AS/NZS 4360 standard for risk management and refines the standard's subprocesses in the context of legal risks. The focus of this work is on how to identify and manage legal risks, and not on how the identified risks could be managed by legal means. The work furthermore primarily addresses legal risks in relation to eCommerce.

More recently, Standards Australia and Standards New Zealand published a handbook on legal risk management [76] that explains how AS/NZS 4360 can be applied by lawyers in their provisioning of services to their clients. The handbook reflects the above mentioned prediction of Susskind by its aim of assisting lawyers in the transition of legal services from its "largely reactive, dispute resolution focus to a client aligned, preventative law approach".

The identification and management of legal risk has also been addressed within the banking sector, see, for example, [6]. Related to this, McCormick proposes an approach to legal risk management within financial law [51]. The suggested approach is presented as a stepwise process with the phases of legal risk identification, legal risk assessment, and legal risk monitoring and mitigation. The process is, however, not described in much detail, and it is also not supported by a single, comprehensive methodology.

The more recent work by Mahler [49] addresses legal risk management from a broad and general perspective, although the primary focus is on legal contracts. The proposed process for legal risk is compliant with the ISO 31000 risk management standard and consists of five steps. (1) Establish the legal and factual context. (2) Identify risks, in particular legal risks. (3) Analyse risks, including legal and factual uncertainty analysis. (4) Evaluate risks. (5) Identify risk treatment options, including legal and factual controls. The process is intended to support lawyers in proactively managing the client's risks regarding a selected contractual relationship, and is also intended to be conducted by an interdisciplinary group of experts, including lawyers and non-lawyers. The work by Mahler has partly been inspired by the CORAS risk analysis process, and the legal risk management process is furthermore supported by a graphical language for the modelling and documentation of the identified legal risks inspired by the CORAS risk modelling language. This language in turn served the main source of inspiration for the legal CORAS language presented in this book. Some of the early synergies between the development of legal CORAS and the work of Mahler are presented in a joint paper [82].

The explicit support in CORAS for modelling and analysing legal aspects as presented in Chap. 17 is unique in the setting of methods for traditional risk analysis, such as those presented in Sect. 19.2. The approaches to legal risk management presented in this section differ from CORAS in that they are intended for assisting lawyers in their provisioning of legal services to their clients, while the handling of legal issues in CORAS is intended to assist risk analysts in taking relevant legal aspects into account in a more conventional risk analysis. A lawyer practicing proactive or preventive law may therefore need to address legal subtleties for which there is no specific support in CORAS. CORAS is, however, not intended for reasoning about legal subtleties, but rather for identifying and evaluating risks in general. With legal CORAS, there is explicit support for taking relevant legal norms into account and for estimating the impact of legal uncertainty on the risk level. The handling of legal issues is integrated into the CORAS risk analysis process, something that differentiates CORAS from both other traditional risk analysis methodologies and other approaches to legal risk management.

Appendix A
The CORAS Language Grammar

In this appendix, we provide the grammar, or syntax, of the CORAS risk modelling language, adapted from [14]. We start by defining the grammar for basic CORAS diagrams in Sect. A.1. In Sect. A.2, extend the grammar to high-level CORAS diagrams, in Sect. A.3 we extend the grammar to dependent CORAS diagrams, and in Sect. A.4 we extend the grammar to legal CORAS diagrams.

In each of these sections, the grammar is presented in two different ways. First, we define a meta-model of the language in Object Management Group (OMG) style, and second, we define a textual grammar using the Extended Backus-Naur From (EBNF) notation.

A.1 Basic CORAS

This section defines the syntax of basic CORAS diagrams, first by means of an OMG style meta-model in Sect. A.1.1, and then by means of an EBNF grammar in Sect. A.1.2. In Sect. A.1.3, we provide examples of the use of the ENBF grammar.

A.1.1 Meta-model

The meta-model of basic CORAS diagrams is divided into the three packages: Elements, Relations, and Diagrams, as shown in Fig. A.1. As the name indicates, these packages define the language elements, the relations and the diagrams of the CORAS language, respectively. In the following sections, we go through each of these packages.

A.1.1.1 Elements

The class diagram of Fig. A.2 defines the meta-model for the language elements of the CORAS language. The boxes represent classes and the arrows with the white

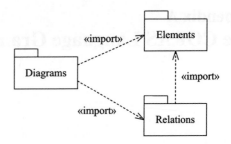

Fig. A.1 Packages of the CORAS meta-model

arrowhead specifies generalisation in such a way that *Element* is a generalisation of *Treatable element* and *Treatable element* is a specialisation of *Element*. The classes with names written in italic are abstract classes. The remaining classes are concrete classes and represent the actual elements of the language. Some of the classes have attributes, and these attributes are inherited down the specialisation tree.

The specialisation tree provides a classification of the elements. All elements are specialisations of the class *Element* and have an identifier. Threats, threat scenarios, unwanted incidents, risks, vulnerabilities, and assets are classified as "treatable elements", which means that they are elements to which we can specify treatments. Threat scenarios, unwanted incidents and risks are classified as "core elements" (a technicality used in the definitions of relations), and threat scenarios and unwanted incidents are classified as "elements with likelihood" (and therefore inherit the likelihood attribute). Threats are specialised into deliberate human threats, accidental human threats and non-human threats, while assets are specialised into direct assets and indirect assets. In addition, treatment scenarios are classified as "treating elements" (another technicality for the definition of relations).

A.1.1.2 Relations

Figure A.3 provides the meta-model of the relations of the CORAS language. Also this model contains a specialisation tree, where the relations are defined as specialisations of the abstract class *Relation*. In this tree, the initiates and leads-to relations are classified as "relations with likelihood" and inherit a likelihood attribute, while the impacts and harm relations are classified as "relations with consequence" and inherit a consequence attribute. The treats relation is a specialisation directly under the *Relation* class and has treatment category as attribute.

In addition, the meta-model of Fig. A.3 shows a number of classes imported from the Elements package, and defines the relationship between the relations and the imported elements. Each relation has exactly one source element and exactly one target element:

- The harm relation has asset both as source and target.
- The initiates relation has a threat as source and a core element as target.
- The leads-to relation has core element both as source and target.

A.1 Basic CORAS

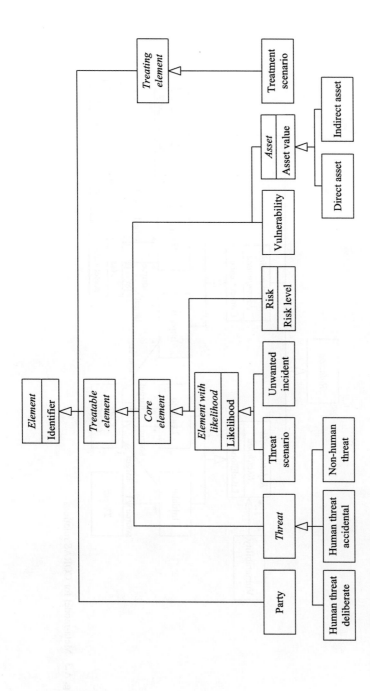

Fig. A.2 Meta-model for CORAS elements

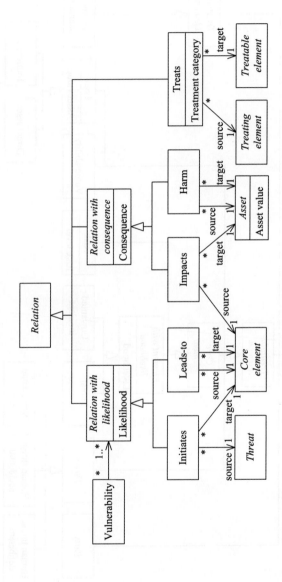

Fig. A.3 Meta-model for CORAS relations

A.1 Basic CORAS

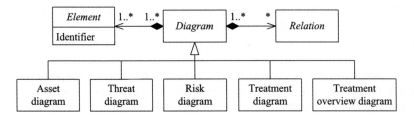

Fig. A.4 Meta-model for CORAS diagrams

- The impacts relation has a core element as source and an asset as target.
- The treats relation has a treating element as source and a treatable element as target.

The association from the *Relation with likelihood* class to the Vulnerability class shows that an initiates or leads-to relation (through inheritance) has zero or more vulnerabilities. The cardinalities of the association also specifies that a vulnerability belongs to at least one relation, which means that we cannot have free-standing vulnerabilities that are not assigned to a relation in the CORAS diagrams.

The relations are defined in their most general form. There are, however, some restrictions that apply to the relations:

- The source and target of a relation cannot be the same element.
- Initiates relations from threats to risks cannot be annotated with likelihood and vulnerabilities.
- Leads-to relations from a risk must go to another risk and cannot be annotated with likelihood or vulnerabilities.
- Leads-to relations to risks cannot be annotated with likelihood.
- Impacts relations from threat scenarios to assets cannot be annotated with consequence.
- Impacts relations from risks to assets cannot be annotated with consequence.

A.1.1.3 Diagrams

Figure A.4 shows the meta-model for CORAS diagrams. From the model, we see that a diagram is composed of one of more elements and zero or more relations. Diagrams are defined by means of an abstract class, which is specialised into the five concrete diagram types of the basic CORAS language.

The abstract diagrams defined by the *Diagram* class can contain any kinds of elements and relations. However, this is not the case for the concrete diagrams of the CORAS language. The concrete diagrams are defined by means of a number of restrictions on the diagrams:

- Asset diagrams can contain only parties, assets, and harm relations, and must contain exactly one party and at least one asset.

- Threat diagrams can contain any elements or relations except for parties, risks, treatment scenarios, and treats relations.
- Risk diagrams can contain only threats, risks, assets, initiates relations, leads-to relations and impacts relations.
- Treatment diagrams can contain any elements or relations except parties.
- Treatment overview diagrams can contain only threats, risks, assets, treatments, initiates relations, leads-to relations, impacts relations, and treats relations.

A.1.2 EBNF Grammar

In this section, we define a textual syntax for the CORAS language using the Extended Backus-Naur Form (EBNF). The presentation follows the same structure as in the previous section.

In the definitions, we use two undefined terms *identifier* and *value*. The term *identifier* is assumed to represent any alphanumerical string. The term *value* is a generic term representing any value that may be defined and used in a risk analysis, and is assumed to always contain the undefined value $\bot$.

We start by defining a number of different types of values used in the definition of the grammar. For each of these definitions, we use a terminal symbol (written in sans serif font) giving the type, and the term *value* to represent the value. The definitions are as follows:

$$asset\ value = \mathsf{av}(value);$$

$$likelihood = \mathsf{l}(value);$$

$$conditional\ likelihood = \mathsf{cl}(value);$$

$$consequence = \mathsf{c}(value);$$

$$risk\ level = \mathsf{rl}(value);$$

A.1.2.1 Elements

In the following, we give the definitions of the syntax of all the concrete elements of the CORAS language. As with the definitions of the value types, we use a terminal symbol (in sans serif) to give the type of the element. As can be seen, all the elements have an identifier, and some of them a specific kind of value.

$$party = \mathsf{p}(identifier);$$

$$direct\ asset = \mathsf{da}(identifier,\ asset\ value);$$

$$indirect\ asset = \mathsf{ia}(identifier,\ asset\ value);$$

$$human\ threat\ deliberate = \mathsf{htd}(identifier);$$

A.1 Basic CORAS

$$human\ threat\ accidental = \mathsf{hta}(identifier);$$
$$non\text{-}human\ threat = \mathsf{nht}(identifier);$$
$$vulnerability = \mathsf{v}(identifier);$$
$$threat\ scenario = \mathsf{ts}(identifier,\ likelihood);$$
$$unwanted\ incident = \mathsf{ui}(identifier,\ likelihood);$$
$$risk = \mathsf{r}(identifier,\ risk\ level);$$
$$treatment\ scenario = \mathsf{tms}(identifier);$$

In addition to defining the syntax of the concrete elements, we define a number of terms grouping or classifying different elements. These are used when we define the relations below and more of less match the specialisation hierarchy of the class diagram in Fig. A.2 (the exception being that we do not have a term for *Element with likelihood*; this is not needed as, in the EBNF grammar, the likelihood attributes are defined as part of the concrete elements and not as a part of an abstract element).

$$asset = direct\ asset\ |\ indirect\ asset;$$
$$threat = human\ threat\ deliberate | human\ threat\ accidental\ |$$
$$non\text{-}human\ threat;$$
$$core\ element = threat\ scenario\ |\ unwanted\ incident\ |\ risk;$$
$$treatable\ element = asset\ |\ threat\ |\ core\ element\ |\ vulnerability;$$
$$treating\ element = treatment\ scenario;$$
$$element = party\ |\ treatable\ element\ |\ treating\ element;$$

A.1.2.2 Relations

The relations of CORAS diagrams can have different annotations. Some of them, specifically *likelihood*, *conditional likelihood* and *consequence*, have already been defined. In addition, as a preliminary to defining the relations, we need two more terms for the annotations: *vulnerabilities*, that represents sets of vulnerabilities, and *treatment category*, that is a set of constants representing the available treatment categories.

$$vulnerabilities = \{vulnerability\};$$
$$treatment\ category = \mathsf{none}\ |\ \mathsf{avoid}\ |\ \mathsf{reduce\ consequence}\ |\ \mathsf{reduce\ likelihood}\ |$$
$$\mathsf{transfer}\ |\ \mathsf{retain};$$

With the classification of elements and the annotations in place, we can define the syntax of relations. The arrows in these definitions are terminal symbols representing the relations. Compared to the class diagram in Fig. A.3, the term on the left-hand side of the arrow is the source element and the term on the right-hand side is the target element. As we can see, the annotation are written on top of the arrows.

$$harm = asset \xrightarrow{consequence} asset;$$

$$initiates = threat \xrightarrow{vulnerabilities,\ likelihood} core\ element;$$

$$leads\text{-}to = core\ element \xrightarrow{vulnerabilities,\ conditional\ likelihood} core\ element;$$

$$impacts = core\ element \xrightarrow{consequence} asset;$$

$$treats = treating\ element \xrightarrow{treatment\ category} treatable\ element;$$

It should be noted that, as in Sect. A.1.2.2, these definitions capture the relations in their most general form, and that the restrictions given in Sect. A.1.1.2 apply also here.

A.1.2.3 Diagrams

As in Sect. A.1.1.3, we gave the definition of a diagram in general instead of providing definitions of each kind of basic diagram. A diagram is also here defined in the general form, as a pair of a non-empty set of elements and a set of relations.

$$diagram = \big(\{element\}^{-}, \{relation\}\big);$$

However, the same restrictions that where given in Sect. A.1.1.3 in order to define the different kinds of diagram also apply to this definition.

A.1.3 Examples

In the following, we provide examples of each of the diagram types, expressed with the EBNF grammar defined above. Throughout the example, we use arrows with no annotations $e \to e'$ as shorthand for relations annotated with the empty set of vulnerabilities and/or the undefined value, that is, as shorthand for $e \xrightarrow{c(\perp)} e'$, $e \xrightarrow{\emptyset,\ cl(\perp)} e'$, $e \xrightarrow{none} e'$, and so forth.

A.1.3.1 Asset Diagram

In this section, we give the textual syntax of the asset diagram found in Fig. 4.8 on p. 55. We use the following shorthand notation:

$$c = \mathsf{p}(Company)$$

A.1 Basic CORAS

$$is = \mathsf{da}(\textit{Integrity of server}, \mathsf{av}(\textit{high}))$$
$$ci = \mathsf{da}(\textit{Confidentiality of information}, \mathsf{av}(\textit{critical}))$$
$$as = \mathsf{da}(\textit{Availability of server}, \mathsf{av}(\textit{critical}))$$
$$cr = \mathsf{ia}(\textit{Company's reputation}, \mathsf{av}(\bot))$$

In the textual syntax, the diagram is represented as follows:

$$(\{c, is, ci, as, cr\}, \{is \to cr, is \to ci, ci \to cr, as \to cr\})$$

A.1.3.2 Threat Diagram

In this section, we give the textual syntax of the threat diagram found in Fig. 4.9 on p. 57. We use the following shorthand notation:

$$h = \mathsf{htd}(\textit{Hacker})$$
$$cv = \mathsf{nht}(\textit{Computer virus})$$
$$vn = \mathsf{v}(\textit{Virus protection not up to date})$$
$$unl = \mathsf{l}(\textit{unlikely})$$
$$pos = \mathsf{l}(\textit{possible})$$
$$lo = \mathsf{l}(\textit{low})$$
$$hi = \mathsf{c}(\textit{high})$$
$$si = \mathsf{ts}(\textit{Server is infected by computer virus}, pos)$$
$$ha = \mathsf{ui}(\textit{Hacker gets access to server}, unl)$$
$$vb = \mathsf{ui}(\textit{Virus creates back door to server}, pos)$$
$$sd = \mathsf{ui}(\textit{Server goes down}, unl)$$
$$is = \mathsf{da}(\textit{Integrity of server}, \mathsf{av}(\bot))$$
$$ci = \mathsf{da}(\textit{Confidentiality of information}, \mathsf{av}(\bot))$$
$$as = \mathsf{da}(\textit{Availability of server}, \mathsf{av}(\bot))$$

In the textual syntax, the diagram is represented as follows:

$$(\{h, cv, vn, si, ha, vb, sd, is, ci, as\},$$
$$\{h \xrightarrow{l(0.1)} ha, cv \xrightarrow{\{vn\}} si, si \to vb, si \xrightarrow{cl(0.2)} sd, vb \to ha,$$
$$ha \xrightarrow{hi} ci, vb \xrightarrow{hi} is, sd \xrightarrow{lo} is, sd \xrightarrow{hi} as\})$$

A.1.3.3 Risk Diagram

In this section, we give the textual syntax of the risk diagram found in Fig. 4.10 on p. 58. We use the following shorthand notation:

$$h = \mathsf{htd}(Hacker)$$
$$cv = \mathsf{nht}(Computer\ virus)$$
$$ac = \mathsf{rl}(acceptable)$$
$$uac = \mathsf{rl}(unacceptable)$$
$$HA = \mathsf{r}(HA\text{: }Hacker\ gets\ access\ to\ server,\ uac)$$
$$VB = \mathsf{r}(VB\text{: }Virus\ creates\ back\ door\ to\ server,\ uac)$$
$$SD1 = \mathsf{r}(SD1\text{: }Server\ goes\ down,\ uac)$$
$$SD2 = \mathsf{r}(SD2\text{: }Server\ goes\ down,\ ac)$$
$$is = \mathsf{da}(Integrity\ of\ server,\ \mathsf{av}(\bot))$$
$$ci = \mathsf{da}(Confidentiality\ of\ information,\ \mathsf{av}(\bot))$$
$$as = \mathsf{da}(Availability\ of\ server,\ \mathsf{av}(\bot))$$

In the textual syntax, the diagram is represented as follows:

$$(\{h, cv, HA, VB, SD1, SD2, is, ci, as\},$$
$$\{h \rightarrow HA, cv \rightarrow VB, cv \rightarrow SD1, cv \rightarrow SD2, VB \rightarrow HA,$$
$$HA \rightarrow ci, VB \rightarrow is, SD1 \rightarrow as, SD2 \rightarrow is\})$$

A.1.3.4 Treatment Diagram

In this section, we give the textual syntax of the treatment diagram found in Fig. 4.11 on p. 60. We use the following shorthand notation:

$$h = \mathsf{htd}(Hacker)$$
$$cv = \mathsf{nht}(Computer\ virus)$$
$$vn = \mathsf{v}(Virus\ protection\ not\ up\ to\ date)$$
$$pos = \mathsf{l}(possible)$$
$$ac = \mathsf{rl}(acceptable)$$
$$uac = \mathsf{rl}(unacceptable)$$
$$si = \mathsf{ts}(Server\ is\ infected\ by\ computer\ virus,\ pos)$$
$$HA = \mathsf{r}(HA\text{: }Hacker\ gets\ access\ to\ server,\ uac)$$

A.1 Basic CORAS

$$VB = \mathsf{r}(VB\text{: }Virus\ creates\ back\ door\ to\ server,\ uac)$$
$$SD1 = \mathsf{r}(SD1\text{: }Server\ goes\ down,\ uac)$$
$$SD2 = \mathsf{r}(SD2\text{: }Server\ goes\ down,\ ac)$$
$$\quad is = \mathsf{da}(Integrity\ of\ server,\ \mathsf{av}(\bot))$$
$$\quad ci = \mathsf{da}(Confidentiality\ of\ information,\ \mathsf{av}(\bot))$$
$$\quad as = \mathsf{da}(Availability\ of\ server,\ \mathsf{av}(\bot))$$
$$\quad im = \mathsf{tms}(Install\ monitoring\ software)$$
$$\quad ir = \mathsf{tms}(Implement\ new\ routines\ for\ updating\ virus\ protection)$$

In the textual syntax, the diagram is represented as follows:

$$\big(\{h, cv, vn, si, HA, VB, SD1, SD2, is, ci, as, im, ir\},$$
$$\{h \to HA,\ cv \xrightarrow{\{vn\}} si,\ si \to VB,\ si \to SD1,\ si \to SD2,$$
$$VB \to HA,\ HA \to ci,\ VB \to is,\ SD1 \to as,\ SD2 \to is,$$
$$im \xrightarrow{\text{reduce consequence}} HA,\ ir \xrightarrow{\text{reduce likelihood}} vn\}\big)$$

A.1.3.5 Treatment Overview Diagram

In this section, we give the textual syntax of the treatment overview diagram found in Fig. 4.12 on p. 62. We use the following shorthand notation:

$$h = \mathsf{htd}(Hacker)$$
$$cv = \mathsf{nht}(Computer\ virus)$$
$$ac = \mathsf{rl}(acceptable)$$
$$uac = \mathsf{rl}(unacceptable)$$
$$HA = \mathsf{r}(HA\text{: }Hacker\ gets\ access\ to\ server,\ uac)$$
$$VB = \mathsf{r}(VB\text{: }Virus\ creates\ back\ door\ to\ server,\ uac)$$
$$SD1 = \mathsf{ui}(SD1\text{: }Server\ goes\ down,\ uac)$$
$$SD2 = \mathsf{ui}(SD2\text{: }Server\ goes\ down,\ ac)$$
$$\quad is = \mathsf{da}(Integrity\ of\ server,\ \mathsf{av}(\bot))$$
$$\quad ci = \mathsf{da}(Confidentiality\ of\ information,\ \mathsf{av}(\bot))$$
$$\quad as = \mathsf{da}(Availability\ of\ server,\ \mathsf{av}(\bot))$$
$$\quad im = \mathsf{tms}(Install\ monitoring\ software)$$
$$\quad ir = \mathsf{tms}(Implement\ new\ routines\ for\ updating\ virus\ protection)$$

In the textual syntax, the diagram is represented as follows

$$\bigl(\{h, cv, HA, VB, SD1, SD2, is, ci, as, im, ir\},$$
$$\{h \to HA, cv \to VB, cv \to SD1, cv \to SD2, VB \to HA,$$
$$HA \to ci, VB \to is, SD1 \to as, SD2 \to is, im \xrightarrow{\text{reduce consequence}} HA,$$
$$ir \xrightarrow{\text{reduce likelihood}} VB, ir \xrightarrow{\text{reduce likelihood}} SD1, ir \xrightarrow{\text{reduce likelihood}} SD2\}\bigr)$$

A.2 High-level CORAS

This section defines the syntax of high-level CORAS diagrams, first by means of an OMG style meta-model in Sect. A.2.1, and then by means of an EBNF grammar in Sect. A.2.2. In Sect. A.2.3, we provide examples of high-level diagrams using the EBNF grammar.

A.2.1 Meta-model

The meta-model of high-level CORAS is divided into two packages, Referring elements and Referenced diagrams, as shown in Fig. A.5. Referring elements imports from the basic CORAS package Elements, and Referenced diagrams imports from both Referring elements and the basic CORAS package Diagrams. No new relations are defined for high-level CORAS; the basic CORAS package Relations is imported by the high-level CORAS packages indirectly through the package Diagrams. In the following subsections, we present the meta-models contained in the packages Referring elements and Referenced diagrams.

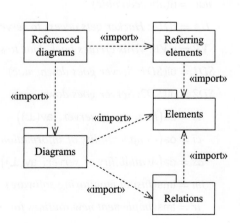

Fig. A.5 Packages of the CORAS meta-model

A.2.1.1 Referring Elements

The meta-model for referring elements is shown in Fig. A.6. In this meta-model, eight new concrete elements are defined: Referring threat scenarios, referring unwanted incidents, referring risks, referring treatment scenarios, in-gates, out-gates, treatment in-gates, and treatment out-gates. The elements are all placed as extensions of the specialisation tree from the Elements meta-model (see Fig. A.2).

In the model, in-gates and out-gates are classified as gates, which again are classified as "core element" from the Elements meta-model. Both of these types of gates furthermore have likelihood as attribute. Similarly, treatment in-gates and treatment out-gates are classified as treatment gates, which are specialisations of both "treatable elements" and "treating elements". By making the gates specialisations of "core elements", "treatable elements" and "treating elements" in this way, the definitions of the relations in the Relations meta-model (see Fig. A.3) carry over to the gates.

Referring threat scenarios, referring unwanted incidents and referring risks are classified as "referring core elements" that contain zero or more gates and zero or more treatment in-gates. In addition, referring risks have a risk level attribute. The referring core elements and referring treatment scenarios are specialisations of elements from the Elements package (see Fig. A.2). This gives them the identifier attribute and allows them to be parts of diagrams. In addition, referring treatment scenarios also have a non-empty set of treatment out-gates.

Because these referring elements are placed directly under *Element* in the specialisation tree, they cannot be the source or target of the relations. This means that relations to and from referring elements must use gates contained in the referring elements as source and target. However, the restrictions listed in Sect. A.1.1.2 carry over to the gates contained by the referring elements such that initiates relations from threats to gates contained in referring risks cannot be annotated with likelihood and vulnerabilities and so forth. With the new gate elements, we need some additional restrictions, however.

- If the source of a relation is a gate contained in a referring element, it must be an out-gate.
- If the target of relation is a gate contained in a referring element, it must be an in-gate.
- Gates that are not treatment gates cannot be source or target of treats relations.
- A referring risk can only have one out-gate.
- The identifiers of in-gates, out-gates, treatment in-gates and treatment out-gates in a referring element must be unique, i.e. two gates in the same referring element cannot have the same name.
- The in-gates and out-gates of referring risks cannot be annotated with likelihoods.

A.2.1.2 Referenced Diagrams

The meta-model for referenced diagrams is shown in Fig. A.7. The model defines four referenced diagrams: Referenced threat scenarios, referenced unwanted inci-

A The CORAS Language Grammar

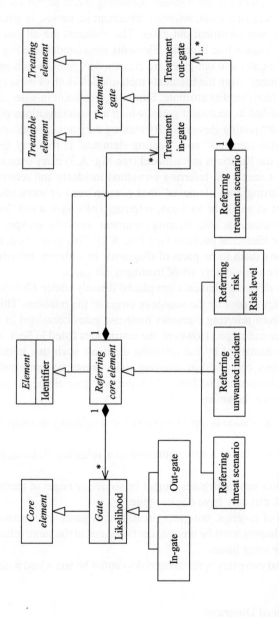

Fig. A.6 Meta-model for CORAS elements

A.2 High-level CORAS

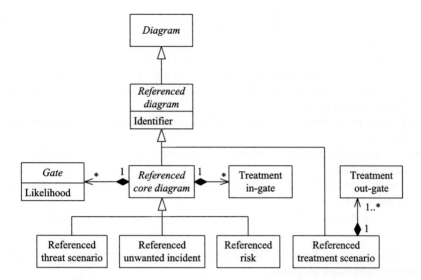

Fig. A.7 Meta-model for CORAS diagrams

dents, referenced risks, and referenced treatment scenarios. Referenced threat scenarios, referenced unwanted incidents and referenced risks are classified as "referenced core diagrams" that each contains a set of gates and a set of treatment in-gates. Further, referenced core diagrams and referenced treatment scenarios, which contain non-empty sets of treatment out-gates, are classified as "referenced diagrams", which gives them the identifier attribute. Finally, referenced diagrams are specialisations of diagrams from the meta-model of basic diagrams (see Fig. A.4). Because of this, the referenced diagrams contain elements and relations in the same way as the basic diagrams.

The restrictions placed on the different kinds of basic diagrams in Sect. A.1.1.3 carry over to referring elements, in such a way that for example threat diagrams, that may contain threat scenarios but not risks, can contain referring threat scenarios but not referring risks. In addition, we have similar restrictions on the referenced diagrams:

- Referenced threat scenarios can only contain threat scenarios, referring threat scenarios, unwanted incidents, referring unwanted incidents, threats and vulnerabilities.
- Referenced unwanted incidents can only contain unwanted incidents and referring unwanted incidents.
- Referenced risks can only contain risks and referring risks.
- Referenced treatment scenarios can only contain treatment scenarios and referring treatment scenarios.

The gates and treatment gates contained in referenced diagrams, that is, the gates at the border, may be the source and target of relations. (See the previous subsection.) However, we need to place some additional restrictions on relations in referenced diagrams.

- In referenced unwanted incidents, referenced risks and referenced treatment scenarios, relations may only go between gates (at the border) and basic elements or gates contained in referring elements, that is, not between elements contained in the referenced diagram.
- If a gate (at the border of a referenced diagram) is the source of a relation, it must be an in-gate.
- If a gate (at the border of a referenced diagram) is the target of a relation, it must be an out-gate.
- An out-gate (at the border of a referenced diagram) cannot be the target of an impacts relation.
- The in-gates and out-gates of referenced risks cannot be annotated with likelihoods.

We furthermore have the following restriction on the identifiers (names) of the gates of referenced diagrams:

- The identifiers of in-gates, out-gates, treatment in-gates and treatment out-gates in a referenced diagram must be unique, that is, two gates in the same referenced diagram cannot have the same name.

A.2.2 EBNF Grammar

In this section, we define a textual syntax for high-level CORAS in EBNF. In difference from the meta-model section above, this section also contains a separate subsection on relations.

A.2.2.1 Referring Elements

Before we define the syntax of the referring elements, we must define the syntax of the gates. Both regular gates and treatment gates are defined similar to other elements with a terminal symbol, giving the type of the element, and an identifier. In-gates and out-gates moreover have a likelihood. In addition, we define the terms *gate* and *treatment gate*.

$$in\text{-}gate = \mathsf{ig}(identifier, likelihood);$$

$$out\text{-}gate = \mathsf{og}(identifier, likelihood);$$

$$gate = in\text{-}gate \,|\, out\text{-}gate;$$

$$treatment\ in\text{-}gate = \mathsf{tig}(identifier);$$

$$treatment\ out\text{-}gate = \mathsf{tog}(identifier);$$

$$treatment\ gate = treatment\ in\text{-}gate \,|\, treatment\ out\text{-}gate;$$

A.2 High-level CORAS

When the gates have been defined, we can define the referring elements. These elements are also defined by means of terminal symbols giving the type of the elements. The referring threat scenarios, referring unwanted incidents and referring risks have an identifier, a set of gates and a set of treatment in-gates. Referring risks furthermore have a risk level. The referring treatment scenarios have an identifier and a non-empty set of treatment out-gates.

$$\textit{referring threat scenario} = \text{rts}(\textit{identifier}, \{\textit{gate}\}, \{\textit{treatment in-gate}\});$$

$$\textit{referring unwanted incident} = \text{rui}(\textit{identifier}, \{\textit{gate}\}, \{\textit{treatment in-gate}\});$$

$$\textit{referring risk} = \text{rr}(\textit{identifier}, \textit{risk level},$$
$$\{\textit{gate}\}, \{\textit{treatment in-gate}\});$$

$$\textit{referring treatment scenario} = \text{rtms}(\textit{identifier}, \{\textit{treatment out-gate}\}^{-});$$

With the new elements defined, we define the two terms *referring core element* and *referring element*, and redefine the term *element* from Sect. A.1.2.1 so that it contains the referring elements.

$$\textit{referring core element} = \textit{referring threat scenario} \mid \textit{referring unwanted incident} \mid$$
$$\textit{referring risk};$$

$$\textit{referring element} = \textit{referring core element} \mid \textit{referring treatment scenario};$$

$$\textit{element} = \textit{party} \mid \textit{treatable element} \mid \textit{treating element} \mid$$
$$\textit{referring element};$$

A.2.2.2 Relations

In the following, we redefine the initiates, leads-to, impacts and treats relations so that they also can be used with gates and referring elements, as well as inside referenced diagrams. We use dot-notation in order to refer to gates contained in referring elements when we specify relations. This means that, if e is an referring element and i and o are, respectively, an in-gate and an out-gate contained in e, we write $e.o \rightarrow \cdots$ or $\cdots \rightarrow e.i$ in order to specify that o and i is the source or target of relations. The redefined relations are:

$$\textit{initiates} = \textit{threat} \xrightarrow{\textit{vulnerabilities, likelihood}} \textit{core element} \mid \textit{out-gate} \mid$$
$$\textit{referring core element.in-gate};$$

$$\textit{leads-to} = \textit{core element} \mid \textit{in-gate} \mid \textit{referring core element.out-gate}$$
$$\xrightarrow{\textit{vulnerabilities, conditional likelihood}} \textit{core element} \mid \textit{out-gate} \mid$$
$$\textit{referring core element.in-gate};$$

$$impacts = core\ element\ |\ referring\ core\ element.out\text{-}gate$$
$$\xrightarrow{consequence} asset\ |\ out\text{-}gate;$$

$$treats = treating\ element\ |\ treatment\ in\text{-}gate\ |$$
$$referring\ treatment\ scenarios.treatment\ out\text{-}gate$$
$$\xrightarrow{treatment\ category} treatable\ element\ |\ treatment\ out\text{-}gate\ |$$
$$referring\ core\ element.treatment\ in\text{-}gate;$$

Note that most of the new restrictions on relations from Sect. A.2.1 are embedded into these definitions. All remaining restrictions carry over to the redefined relations.

A.2.2.3 Referenced Diagrams

Below, we define each of the referenced diagrams. A referenced threat scenario, referenced unwanted incident or referenced risk is a five-tuple of an identifier, a set of gates, a set of treatment in-gates, a non-empty set of elements and a set of relations. A referenced treatment scenario is a four-tuple of an identifier, a non-empty set of treatment out-gates, a non-empty set of elements and a set of relations. Note that in these definitions, the restrictions on which elements the diagrams are allowed to contain are part of the definitions.

$$referenced\ threat\ scenario = \big(identifier,\ \{gate\},\ \{treatment\ in\text{-}gate\},$$
$$\{threat\ |\ vulnerability\ |\ threat\ scenario$$
$$|\ referring\ threat\ scenario\ |\ unwanted\ incident\ |$$
$$referring\ unwanted\ incident\}^{-},\ \{relation\}\big);$$

$$referenced\ unwanted\ incident = \big(identifier,\ \{gate\},\ \{treatment\ in\text{-}gate\},$$
$$\{unwanted\ incident\ |$$
$$referring\ unwanted\ incident\}^{-},\ \{relation\}\big);$$

$$referenced\ risk = \big(identifier,\ \{gate\},\ \{treatment\ in\text{-}gate\},$$
$$\{risk\ |\ referring\ risk\}^{-},\ \{relation\}\big);$$

$$referenced\ treatment\ scenario = \big(identifier,\ \{treatment\ out\text{-}gate\}^{-},$$
$$\{treatment\ scenario\ |$$
$$referring\ treatment\ scenario\}^{-},\ \{relation\}\big);$$

A.2 High-level CORAS

Finally, we redefine the term *diagram* from Sect. A.1.2.3 to also contain the referenced diagrams, in addition to the general diagram.

$$diagram = (\{element\}^-, \{relation\}) \mid referenced\ threat\ scenario \mid$$
$$referenced\ unwanted\ incident \mid referenced\ risk \mid$$
$$referenced\ treatment\ scenario;$$

A.2.3 Examples

In the following, we provide examples of high-level threat, risk, treatment and treatment overview diagrams, all containing referring elements, and some of the referenced diagrams to which they refer. All the examples are expressed in the EBNF grammar defined above.

A.2.3.1 High-level Threat Diagram

In this section, we give the textual syntax of the high-level threat diagram found in Fig. 14.18 on p. 258 with its referenced diagrams found in Fig. 14.16 on p. 256, in Fig. 14.19 on p. 260, and in Fig. 14.22 on p. 263. We use the following shorthand notation:

$$h = \mathsf{htd}(Hacker)$$
$$cv = \mathsf{nht}(Computer\ virus)$$
$$vn = \mathsf{v}(Virus\ protection\ not\ up\ to\ date)$$
$$la = \mathsf{v}(Low\ awareness\ of\ email\ worms)$$
$$unl = \mathsf{l}(unlikely)$$
$$lo = \mathsf{c}(low)$$
$$me = \mathsf{c}(medium)$$
$$hi = \mathsf{c}(high)$$
$$ig_1 = \mathsf{ig}(i_1, \mathsf{l}(0.01))$$
$$ig'_1 = \mathsf{ig}(i_1, \mathsf{l}(0.02))$$
$$ig_2 = \mathsf{ig}(i_2, \mathsf{l}(0.1))$$
$$ig'_2 = \mathsf{ig}(i_2, \mathsf{l}(\bot))$$
$$ig_3 = \mathsf{ig}(i_3, \mathsf{l}(0.05))$$
$$og_1 = \mathsf{og}(o_1, \mathsf{l}(0.05))$$

$og'_1 = \mathsf{og}(o_1,\ \mathsf{l}([0.02, 0.06]))$

$og_2 = \mathsf{og}(o_2,\ \mathsf{l}(0.1))$

$og'_2 = \mathsf{og}(o_2,\ \mathsf{l}(0.02))$

$og_3 = \mathsf{og}(o_3,\ \mathsf{l}(0.05))$

$og_4 = \mathsf{og}(o_4,\ \mathsf{l}(0.1))$

$og'_4 = \mathsf{og}(o_4,\ \mathsf{l}(\bot))$

$og_5 = \mathsf{og}(o_5,\ \mathsf{l}(0.01))$

$og_6 = \mathsf{og}(o_6,\ \mathsf{l}(0.05))$

$hs = \mathsf{rts}(Hacker\ attack\ on\ server,\ \{ig_1, og_1, og_2\},\ \emptyset)$

$si = \mathsf{rts}(Server\ is\ infected\ by\ computer\ virus,\ \{ig_2, og_3, og_4\},\ \emptyset)$

$sd = \mathsf{rui}(Server\ goes\ down,\ \{ig_3, og_5, og_6\},\ \emptyset)$

$st = \mathsf{ts}(Server\ is\ infected\ by\ Trojan\ horse,\ \mathsf{l}(\bot))$

$se = \mathsf{ts}(Server\ is\ infected\ by\ email\ worm,\ \mathsf{l}(0.1))$

$es = \mathsf{ts}(Email\ worm\ is\ spreading,\ \mathsf{l}(0.08))$

$ss = \mathsf{ts}(Server\ is\ infected\ by\ SQL\ worm,\ \mathsf{l}(0.1))$

$so = \mathsf{ts}(Server\ is\ overloaded,\ \mathsf{l}(\bot))$

$sc = \mathsf{ui}(Server\ crashes,\ \mathsf{l}(0.01))$

$sr = \mathsf{ui}(Server\ does\ not\ respond,\ \mathsf{l}(0.05))$

$ha = \mathsf{ui}(Hacker\ gets\ access\ to\ server,\ \mathsf{l}([0.02, 0.06]))$

$vb = \mathsf{ui}(Virus\ creates\ back\ door\ to\ server,\ \mathsf{l}(0.02))$

$is = \mathsf{da}(Integrity\ of\ server,\ \mathsf{av}(\bot))$

$ci = \mathsf{da}(Confidentiality\ of\ information,\ \mathsf{av}(\bot))$

$as = \mathsf{da}(Availability\ of\ server,\ \mathsf{av}(\bot))$

The syntax of the high-level diagram is:

$(\{cv, vn, si, hs, sd, is, ci, as\},$

$\{cv \xrightarrow{\{vn\},\ \mathsf{l}(0.1)} si.ig_2, si.og_3 \xrightarrow{\mathsf{cl}(0.2)} hs.ig_1, si.og_4 \xrightarrow{\mathsf{cl}(0.2)} sd.ig_3,$

$hs.og_1 \xrightarrow{hi} ci, hs.og_2 \xrightarrow{hi} is, sd.og_5 \xrightarrow{lo} is, sd.og_6 \xrightarrow{me} as\})$

A.2 High-level CORAS

The syntax of the referenced threat scenario *Hacker attack on server* is:

$(Hacker\ attack\ on\ server, \{ig'_1, og'_1, og'_2\}, \emptyset, \{h, ha, vb\},$

$\{h \xrightarrow{unl} ha, vb \xrightarrow{cl(0.5)} ha, ig'_1 \xrightarrow{cl(1.0)} vb, ha \xrightarrow{cl(1.0)} og'_1, vb \xrightarrow{cl(1.0)} og'_2\})$

The syntax of the referenced threat scenario *Server is infected by computer virus* is:

$(Server\ is\ infected\ by\ computer\ virus, \{ig'_2, og_3, og'_4\}, \emptyset, \{la, st, se, es, ss, so\},$

$\{ig'_2 \xrightarrow{cl(1.0)} st, ig'_2 \xrightarrow{cl(1.0)} se, ig'_2 \xrightarrow{cl(1.0)} ss,$

$se \xrightarrow{\{la\},\ cl(0.8)} es, ss \xrightarrow{cl(0.8)} so, es \xrightarrow{cl(1.0)} so, so \xrightarrow{cl(1.0)} og'_4, st \xrightarrow{cl(1.0)} og_3\})$

The syntax of the referenced unwanted incident *Server goes down* is:

$(Server\ goes\ down, \{ig_3, og_5, og_6\}, \emptyset, \{sc, sr\},$

$\{ig_3 \xrightarrow{cl(0.2)} sc, ig_3 \xrightarrow{cl(1.0)} sr, sc \xrightarrow{cl(1.0)} og_5, sc \xrightarrow{cl(1.0)} og_6, sr \xrightarrow{cl(1.0)} og_6\})$

A.2.3.2 High-level Risk Diagram

In this section, we give the textual syntax of the high-level risk diagram found in Fig. 14.23 on p. 265 with its referenced diagrams found in Fig. 14.8 on p. 250 and in Fig. 14.9 on p. 251. We use the following shorthand notation:

$h = \mathsf{htd}(Hacker)$

$cv = \mathsf{nht}(Computer\ virus)$

$ac = \mathsf{rl}(acceptable)$

$uac = \mathsf{rl}(unacceptable)$

$ig_3 = \mathsf{ig}(i_3,\ \mathsf{l}(\perp))$

$og_5 = \mathsf{og}(o_5,\ \mathsf{l}(\perp))$

$og_6 = \mathsf{og}(o_6,\ \mathsf{l}(\perp))$

$HA = \mathsf{r}(HA\text{: }Hacker\ gets\ access\ to\ server,\ uac)$

$VB = \mathsf{r}(VB\text{: }Virus\ creates\ back\ door\ to\ server,\ uac)$

$SD1 = \mathsf{rr}(SD1\text{: }Server\ goes\ down,\ uac,\ \{ig_3, o_6\},\ \emptyset)$

$SD2 = \mathsf{rr}(SD2\text{: }Server\ goes\ down,\ ac,\ \{ig_3, og_5\},\ \emptyset)$

$SC1 = \mathsf{r}(SC1\text{: }Server\ crashes,\ \mathsf{rl}(\perp))$

$SC2 = \mathsf{r}(SC2\text{: }Server\ crashes,\ \mathsf{rl}(\perp))$

$SN1 = \mathsf{r}(SN1\text{: }Server\ does\ not\ respond,\ \mathsf{rl}(\perp))$

$$is = \mathsf{da}(Integrity\ of\ server,\ \mathsf{av}(\bot))$$
$$ci = \mathsf{da}(Confidentiality\ of\ information,\ \mathsf{av}(\bot))$$
$$as = \mathsf{da}(Availability\ of\ server,\ \mathsf{av}(\bot))$$

The syntax of the high-level diagram is:

$$(\{h, cv, HA, VB, SD1, SD2, is, ci, as\},$$
$$\{h \to HA, cv \to VB, cv \to SD1.ig_3, cv \to SD2.i_3, VB \to HA,$$
$$HA \to ci, VB \to is, SD1.og_6 \to as, SD2.og_5 \to is\})$$

The syntax of the referenced risk $SD1$: Server goes down is:

$$(SD1: Server\ goes\ down,\ \{ig_3, og_6\},\ \emptyset,\ \{SC1, SN1\},$$
$$\{ig_3 \to SC1, ig_3 \to SN1, SC1 \to og_6, SN1 \to og_6\})$$

The syntax of the referenced risk $SD2$: Server goes down is:

$$(SD2: Server\ goes\ down,\ \{ig_3, og_5\},\ \emptyset,\ \{SC2\},$$
$$\{ig_3 \to SC2, SC2 \to og_5\})$$

A.2.3.3 High-level Treatment Diagram

In this section, we give the textual syntax of the high-level treatment diagram found in Fig. 14.10 on p. 252 with its referenced diagrams found in Fig. 14.11 on p. 253 and Fig. 14.12 on p. 253. We use the following shorthand notation:

$$cv = \mathsf{nht}(Computer\ virus)$$
$$vn = \mathsf{v}(Virus\ protection\ not\ up\ to\ date)$$
$$la = \mathsf{v}(Low\ awareness\ of\ email\ worms)$$
$$ig_1 = \mathsf{ig}(i_1,\ \mathsf{l}(\bot))$$
$$ig_2 = \mathsf{ig}(i_2,\ \mathsf{l}(\bot))$$
$$ig_3 = \mathsf{ig}(i_3,\ \mathsf{l}(\bot))$$
$$og_1 = \mathsf{og}(o_1,\ \mathsf{l}(\bot))$$
$$og_2 = \mathsf{og}(o_2,\ \mathsf{l}(\bot))$$
$$og_3 = \mathsf{og}(o_3,\ \mathsf{l}(\bot))$$
$$og_4 = \mathsf{og}(o_4,\ \mathsf{l}(\bot))$$
$$og_5 = \mathsf{og}(o_5,\ \mathsf{l}(\bot))$$
$$og_6 = \mathsf{og}(o_6,\ \mathsf{l}(\bot))$$

A.2 High-level CORAS

$tig_1 = \text{tig}(t_1)$

$tig_2 = \text{tig}(t_2)$

$tog_3 = \text{tog}(t_3)$

$tog_4 = \text{tog}(t_4)$

$\quad hs = \text{rts}(\textit{Hacker attack on server},\ \{ig_1, og_1, og_2\},\ \{tig_1\})$

$\quad si = \text{rts}(\textit{Server is infected by computer virus},\ \{ig_2, og_3, og_4\},\ \{tig_2\})$

$\quad st = \text{ts}(\textit{Server is infected by Trojan horse},\ \mathsf{l}(\bot))$

$\quad se = \text{ts}(\textit{Server is infected by email worm},\ \mathsf{l}(\bot))$

$\quad es = \text{ts}(\textit{Email worm is spreading},\ \mathsf{l}(\bot))$

$\quad ss = \text{ts}(\textit{Server is infected by SQL worm},\ \mathsf{l}(\bot))$

$\quad so = \text{ts}(\textit{Server is overloaded},\ \mathsf{l}(\bot))$

$\quad HA = \text{r}(\textit{HA: Hacker gets access to server},\ \mathsf{rl}(\bot))$

$\quad VB = \text{r}(\textit{VB: Virus creates back door to server},\ \mathsf{rl}(\bot))$

$SD1 = \text{rr}(\textit{SD1: Server goes down},\ \mathsf{rl}(\bot),\ \{ig_3, og_6\},\ \emptyset)$

$SD2 = \text{rr}(\textit{SD2: Server goes down},\ \mathsf{rl}(\bot),\ \{ig_3, og_5\},\ \emptyset)$

$\quad is = \text{da}(\textit{Integrity of server},\ \text{av}(\bot))$

$\quad ci = \text{da}(\textit{Confidentiality of information},\ \text{av}(\bot))$

$\quad as = \text{da}(\textit{Availability of server},\ \text{av}(\bot))$

$\quad im = \text{tms}(\textit{Install monitoring software})$

$\quad ir = \text{tms}(\textit{Improve routines for patching SQL server})$

$\quad iv = \text{tms}(\textit{Improve input validation})$

$\quad ip = \text{rtms}(\textit{Improve virus protection},\ \{tog_3\})$

$\quad ic = \text{rtms}(\textit{SQL injection countermeasures},\ \{tog_4\})$

The syntax of the high-level diagram is:

$$(\{cv, vn, hs, si, HA, VB, SD1, SD2, is, ci, as, im, ip, ic\},$$
$$\{cv \xrightarrow{\{vn\}} si.ig_2,\ si.og_3 \to hs.ig_1,\ si.og_4 \to SD1.ig_3,$$
$$si.og_4 \to SD2.ig_3,\ hs.og_1 \to HA,\ hs.og_2 \to VB,\ VB \to HA,$$
$$HA \to ci,\ VB \to is,\ SD1.og_6 \to as,\ SD2.og_5 \to is,$$
$$im \to hs.tig_1,\ ip.tog_3 \to vn,\ ic.tog_4 \to si.tig_2\})$$

The syntax of the referenced treatment scenario *SQL injection countermeasures* is:

$$(SQL\ injection\ countermeasures, \{tog_4\}, \{ir, iv\},$$
$$\{ir \rightarrow tog_4, iv \rightarrow tog_4\})$$

The syntax of the referenced threat scenario *Server is infected by computer virus* is:

$$(Server\ is\ infected\ by\ computer\ virus, \{ig_2, og_3, og_4\}, \{tig_2\},$$
$$\{la, st, se, es, ss, so\}, \{ig_2 \rightarrow st, ig_2 \rightarrow se, ig_2 \rightarrow ss,$$
$$se \xrightarrow{\{la\}} es, ss \rightarrow so, es \rightarrow so, so \rightarrow og_4, st \rightarrow og_3, tig_2 \rightarrow ss\})$$

A.2.3.4 High-level Treatment Overview Diagram

In this section, we give the textual syntax of the high-level treatment overview diagram found in Fig. 14.13 on p. 254 with its referenced diagrams found in Fig. 14.11 on p. 253 and Fig. 14.14 on p. 254. We use the following shorthand notation:

$h = \mathsf{htd}(Hacker)$

$cv = \mathsf{nht}(Computer\ virus)$

$ig_3 = \mathsf{ig}(i_3, \mathsf{l}(\bot))$

$og_5 = \mathsf{og}(o_5, \mathsf{l}(\bot))$

$og_6 = \mathsf{og}(o_6, \mathsf{l}(\bot))$

$tog_3 = \mathsf{tog}(t_3)$

$tog_4 = \mathsf{tog}(t_4)$

$tig_5 = \mathsf{tig}(t_5)$

$tig_6 = \mathsf{tig}(t_6)$

$tig_7 = \mathsf{tig}(t_7)$

$tig_8 = \mathsf{tig}(t_8)$

$HA = \mathsf{r}(HA\text{: }Hacker\ gets\ access\ to\ server, \mathsf{rl}(\bot))$

$VB = \mathsf{r}(VB\text{: }Virus\ creates\ back\ door\ to\ server, \mathsf{rl}(\bot))$

$SD1 = \mathsf{rr}(SD1\text{: }Server\ goes\ down, \mathsf{rl}(\bot), \{ig_3, og_6\}, \{tig_5, tig_7\})$

$SD2 = \mathsf{rr}(SD2\text{: }Server\ goes\ down, \mathsf{rl}(\bot), \{ig_3, og_5\}, \{tig_6, tig_8\})$

$SC2 = \mathsf{r}(SC2\text{: }Server\ crashes, \mathsf{rl}(\bot))$

$is = \mathsf{da}(Integrity\ of\ server, \mathsf{av}(\bot))$

$ci = \mathsf{da}(Confidentiality\ of\ information, \mathsf{av}(\bot))$

$as = \mathsf{da}(Availability\ of\ server,\ \mathsf{av}(\bot))$

$im = \mathsf{tms}(Install\ monitoring\ software)$

$ir = \mathsf{tms}(Improve\ routines\ for\ patching\ SQL\ server)$

$iv = \mathsf{tms}(Improve\ input\ validation)$

$ip = \mathsf{rtms}(Improve\ virus\ protection,\ \{tog_3\})$

$ic = \mathsf{rtms}(SQL\ injection\ countermeasures,\ \{tog_4\})$

The syntax of the high-level diagram is:

$(\{h, cv, HA, VB, SD1, SD2, is, ci, as, im, ip, ic\},$
$\{h \rightarrow HA, cv \rightarrow VB, cv \rightarrow SD1.ig_3, cv \rightarrow SD2.i_3, VB \rightarrow HA,$
$HA \rightarrow ci, VB \rightarrow is, SD1.og_6 \rightarrow as, SD2.og_5 \rightarrow is, im \rightarrow HA,$
$ip.tog_3 \rightarrow HA, ip.tog_3 \rightarrow VB, ip.tog_3 \rightarrow SD1.tig_5,$
$ip.tog_3 \rightarrow SD2.tig_6, ic.tog_4 \rightarrow SD1.tig_7, ic.tog_4 \rightarrow SD2.tig_8\})$

The syntax of the referenced treatment scenario *SQL injection countermeasures* is:

$(SQL\ injection\ countermeasures,\ \{tog_4\},\ \{ir, iv\},$
$\{ir \rightarrow tog_4, iv \rightarrow tog_4\})$

The syntax of the referenced risk *SD2: Server goes down* is:

$(SD2:\ Server\ goes\ down,\ \{ig_3, og_5\},\ \{tig_6, tig_8\},\ \{SC2\},$
$\{ig_3 \rightarrow SC2, SC2 \rightarrow og_5, tig_6 \rightarrow SC2, tig_8 \rightarrow SC2\})$

A.3 Dependent CORAS

This section defines the syntax of dependent CORAS diagrams, first by means of an OMG style meta-model in Sect. A.3.1, and then by means of an EBNF grammar in Sect. A.3.2. In Sect. A.3.3, we give an example of a dependent diagram using the EBNF grammar.

A.3.1 Meta-model

The meta-model of dependent CORAS extends the meta-model of basic CORAS with the Border package as shown in Fig. A.8. The diagram depicts only the extension of the packages of basic CORAS as shown in Fig. A.1. The Border package

Fig. A.8 Packages of the CORAS meta-model

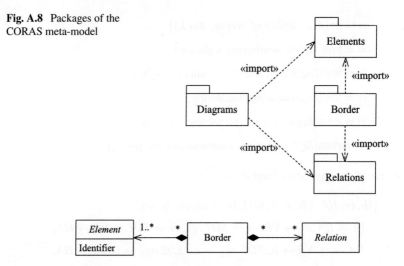

Fig. A.9 Meta-model for CORAS border

defines the border construct of the dependent CORAS diagram that comes in addition to the language elements and relations. Notice that dependent CORAS can also make use of referring elements, in which case the Elements package of Fig. A.8 is replaced by the Referring elements package of Fig. A.5.

A.3.1.1 Border

The class diagram of Fig. A.9 defines the meta-model for the border construct of the dependent CORAS language. The abstract class *Element* is defined in Fig. A.2 and Fig. A.6, whereas the abstract class *Relation* is defined in Fig. A.3.

Notice that Border is defined in the same way as the abstract class *Diagram* in Fig. A.4. The border can hence be understood as specifying a diagram, namely the diagram that is within the perimeter of the border.

There is one restriction that applies to the border:

- The diagram that is specified by a border must be a threat diagram or a treatment diagram.

A.3.1.2 Dependent Diagrams

A dependent diagram is simply a CORAS diagram that in addition to elements and relations consists of a border that defines which part of the diagram that makes the target of analysis. The class diagram of Fig. A.10 defines the meta-model for dependent CORAS diagrams by showing the extension of the class diagram of Fig. A.4. Notice that by the multiplicities in the meta-model, the border is optional, and there can be at most one border in a diagram.

A.3 Dependent CORAS

Fig. A.10 Meta-model for CORAS diagrams

The elements and relations of the border are subsets of the elements and relations, respectively, of the diagram. These subsets of the diagram make the target. The diagram elements and relations that remain when removing the border elements and relations make the assumptions.

In the graphical representation of dependent diagrams, elements can be placed on the border and relations can cross the border. In those cases, these elements and relations belong to the target, that is, they are within the border. While the graphical dependent diagrams explicitly show which elements are on the border and which relations cross the border, this is only implicit when elements and relations are classified as sets that belong either to the target or to the assumptions: If element e_1 belongs to the assumptions, element e_2 belongs to the target and relation $e_1 \to e_2$ belongs to the assumptions, then e_2 is placed on the border; if element e_1 belongs to the assumptions, element e_2 belongs to the target and relation $e_1 \to e_2$ belongs to the assumptions, then $e_1 \to e_2$ crosses the border.

The following restrictions apply to dependent CORAS diagrams:

- Dependent diagrams, that is, diagrams with a border, must be threat diagrams or treatment diagrams.
- All elements inside the border must be elements of the diagram containing the border.
- All relations inside the border must be relations of the diagram containing the border.
- All assets of the diagram must be inside the border.
- All impacts relation of the diagram must be inside the border.
- Threats cannot occur on the border.
- Relations cannot cross the border from the inside to the outside.
- Initiates relations cannot cross the border.
- If elements e_1 and e_2 belong to the target (i.e., are inside the border), then each relation $e_1 \to e_2$ belongs to the target.
- For each element e_2, if element e_1 and relation $e_1 \to e_2$ belong to the target, then e_2 belongs to the target.
- If elements e_1 and e_2 belong to the assumptions (i.e., are outside the border), then each relation $e_1 \to e_2$ belongs to the assumptions.
- For each element e_1, if element e_2 and relation $e_1 \to e_2$ belong to the assumptions, then e_1 belongs to the assumptions.

A.3.2 EBNF Grammar

In this section, we define a textual syntax for dependent CORAS in EBNF. The only new language construct is the border, the introduction of which requires a redefinition of the term *diagram*.

A.3.2.1 Border

The textual syntax of the border of a dependent CORAS diagram is defined as follows:

$$border = (\{element\}^-, \{relation\});$$

A.3.2.2 Dependent Diagrams

In the setting of basic CORAS, we define dependent diagrams by redefining the term *diagram* from Sect. A.1.2.3 to allow the specification of a border:

$$diagram = (\{element\}^-, \{relation\}, [border]);$$

The square brackets around *border* mean that the border is optional. If the border is omitted, the diagram is a set of elements and relations the syntax of which is as defined in Sect. A.1.2. If the border is included, the restrictions from Sect. A.3.1.2 on dependent diagrams apply. Notice that corresponding to the graphical specifications of dependent CORAS diagrams, the removal of the border yields a basic CORAS diagram.

In the setting of high-level CORAS, we correspondingly redefine the term *diagram* from Sect. A.2.2.3:

$$diagram = (\{element\}^-, \{relation\}, [border]) \mid referenced\ threat\ scenario \mid$$
$$referenced\ unwanted\ incident \mid referenced\ risk \mid$$
$$referenced\ treatment\ scenario;$$

A.3.3 Example

As an example, we give the textual syntax of the dependent CORAS diagram of Fig. 16.1 on p. 297. We use the following shorthand notation, where o is a name we have chosen for the output gate of the referring threat scenario:

$$lr = \mathsf{nht}(Lack\ of\ rain)$$
$$og = \mathsf{og}(o,\ \mathsf{I}(0.0))$$
$$fi = \mathsf{rts}(Fire,\ \{og\},\ \emptyset)$$
$$lh = \mathsf{ts}(Low\ hydro\ availability,\ \mathsf{I}(\bot))$$
$$hi = \mathsf{ts}(High\ import\ of\ power,\ \mathsf{I}(\bot))$$
$$hl = \mathsf{v}(High\ load\ on\ transmission\ corridor)$$
$$go = \mathsf{ui}(Grid\ overload,\ \mathsf{I}(1:5\ years))$$

$$pp = \mathsf{da}(\textit{Power production},\ \mathsf{av}(\bot))$$
$$cr = \mathsf{c}(\textit{critical})$$
$$ip = \mathsf{tms}(\textit{Increase power self-sufficiency rate})$$

There are two elements that are completely outside the border, namely *lr* and *fi*. The initiates relation from *lr* to *lh* is also outside the border. The textual syntax of the diagram is therefore represented as follows:

$$(\{lr, lh, hi, fi, hl, go, pp, ip\},$$
$$\{lr \xrightarrow{I(1.0)} lh, lh \to hi, hi \xrightarrow{\{hl\}} go, fi.og \xrightarrow{cl(1.0)} go, go \xrightarrow{cr} pp, ip \to hi\},$$
$$(\{lh, hi, hl, go, pp, ip\},$$
$$\{lh \to hi, hi \xrightarrow{\{hl\}} go, fi.og \xrightarrow{cl(1.0)} go, go \xrightarrow{cr} pp, ip \to hi\}))$$

Notice that by removing the border from the dependent diagram of Fig. 16.1, the result is a (high-level) treatment diagram. Correspondingly, by removing the third element of the textual representation, that is, removing the pair of elements and relations that represents the border, the result is the textual representation of the diagram of Fig. 16.1 without the border.

Since the border represents the target and the elements and relations that remain when removing the border from the diagram represents the assumptions, the following two pairs are the textual syntax of the target and the assumptions:

$$\textit{Target} = (\{lh, hi, hl, go, pp, ip\},$$
$$\{lh \to hi, hi \xrightarrow{\{hl\}} go, fi.og \xrightarrow{cl(1.0)} go, go \xrightarrow{cr} pp, ip \to hi\})$$
$$\textit{Assumptions} = (\{lr, fi\}, \{lr \xrightarrow{I(1.0)} lh\})$$

A.4 Legal CORAS

This section defines the syntax of legal CORAS, first by means of an OMG style meta-model in Sect. A.4.1, and then by means of an EBNF grammar in Sect. A.4.2. In Sect. A.4.3, we give an example of a legal CORAS diagram using the EBNF grammar.

A.4.1 Meta-model

Figure A.11 shows the extension of the meta-model in Fig. A.2 of basic CORAS with the elements that are introduced by legal CORAS.

Fig. A.11 Meta-model for CORAS elements

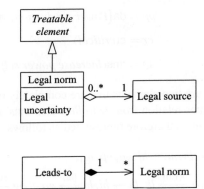

Fig. A.12 Meta-model for CORAS relations

Figure A.12 shows the meta-model for the relations in legal CORAS. The meta-model is an extension of the meta-model in Fig. A.3 of the relations in basic CORAS. This extension carries over to high-level CORAS and dependent CORAS, which means that legal norms can be specified also in high-level and dependent diagrams.

A.4.2 EBNF Grammar

We first extend the list of the types of values that are used in the definition of the grammar:

$$legal\ uncertainty = \mathsf{lu}(value);$$

Next, we introduce the concrete elements of legal CORAS that extend basic CORAS:

$$legal\ source = \mathsf{ls}(identifier);$$

$$legal\ norm = \mathsf{ln}(identifier, legal\ source, legal\ uncertainty);$$

Legal norms are classified as treatable elements, and we therefore redefine the definition from Sect. A.1.2.1 as follows:

$$treatable\ element = asset\ |\ threat\ |\ core\ element\ |\ legal\ norm;$$

The legal norm is the only new language element. A legal norm is specified as an annotation on a leads-to relation, so we need to redefine the leads-to relation of basic CORAS. However, before we redefine the leads-to relation we introduce a new term for the annotations, namely *legal norms*, that represents a set of legal norms.

$$legal\ norms = \{legal\ norm\};$$

A.4 Legal CORAS

The leads-to relation is then redefined to allow the annotation of one or more legal norms as follows:

$$leads\text{-}to = core\ element \xrightarrow{vulnerabilities, conditional\ likelihood, legal\ norms} core\ element;$$

This definition of the leads-to relation also applies when legal CORAS is combined with dependent CORAS. When using legal CORAS in the setting of high-level CORAS, we redefine the leads-to relation from Sect. A.2.2.2 as follows:

$$leads\text{-}to = core\ element\ |\ in\text{-}gate\ |\ referring\ core\ element.out\text{-}gate$$
$$\xrightarrow{vulnerabilities,\ conditional\ likelihood,\ legal\ norms} core\ element\ |\ out\text{-}gate\ |$$
$$referring\ core\ element.in\text{-}gate;$$

A.4.3 Example

As an example, we give the textual syntax of the legal CORAS diagram of Fig. 17.5 on p. 325. We use the following shorthand notation:

$e = \mathsf{hta}(Employee)$

$lr = \mathsf{v}(Lack\ of\ routines\ for\ updating\ personal\ data)$

$li = \mathsf{l}(likely)$

$pos = \mathsf{l}(possible)$

$lli = \mathsf{lu}(likely)$

$pd = \mathsf{ts}(Personal\ data\ is\ not\ held\ accurate\ and\ up\ to\ date,\ li)$

$dpa = \mathsf{ls}(Data\ Protection\ Act)$

$cli = \mathsf{ln}(Company\ may\ be\ held\ liable\ for\ irregular\ processing\ of\ personal\ data,$
$\qquad dpa,\ lli)$

$cp = \mathsf{ui}(Company\ is\ prosecuted\ for\ irregular\ processing\ of\ personal\ data,\ pos)$

$r = \mathsf{da}(Reputation,\ \mathsf{av}(\bot))$

In the textual syntax, the diagram is represented as follows:

$$\left(\{e, lr, pd, cli, cp, r\}, \{e \xrightarrow{\{lr\}} pd, pd \xrightarrow{cl(0.3),\ \{cli\}} cp, cp \to r\}\right)$$

Appendix B
The CORAS Language Semantics

In this appendix, we provide semantics for the CORAS language, adapted from [14]. The semantics is a systematic translation of CORAS diagrams into English. Each element or relation in a CORAS diagram is translated into a sentence in English, such that the translation of a diagram results in a textual paragraph.

The semantics is defined by a function $[\![\_]\!]$ that takes diagrams and fragments of diagrams, expressed in the textual syntax defined with EBNF in Appendix A, and returns strings of English text. We start by defining the semantics of basic CORAS diagrams in Sect. B.1. In Sect. B.2, we extend the semantics to high-level CORAS diagrams, in Sect. B.3 we extend the semantics to dependent CORAS diagrams, and in Sect. B.4 we extend the semantics to legal CORAS diagrams.

B.1 Basic CORAS

This section gives the semantics of basic CORAS diagrams. In the following, we give the semantics of elements in Sect. B.1.1, relations in Sect. B.1.2 and diagrams in Sect. B.1.3. In Sect. B.1.4, we provide examples of the application of the semantics.

We start, however, by defining the semantics of the different types of values that are used in the syntax. We let v range over *value*. The undefined value is represented by $\bot$ and ε represents the empty string. The semantics of the values are then:

$$[\![\mathsf{av}(\bot)]\!] = \text{value}$$

$$[\![\mathsf{av}(v)]\!] = \text{the value } v$$

$$[\![\mathsf{l}(\bot)]\!] = \text{with undefined likelihood}$$

$$[\![\mathsf{l}(v)]\!] = \text{with likelihood } v$$

$$[\![\mathsf{cl}(\bot)]\!] = \text{with undefined conditional likelihood}$$

$$[\![\mathsf{cl}(v)]\!] = \text{with conditional likelihood } v$$

$$\llbracket \mathsf{c}(\bot) \rrbracket = \varepsilon$$
$$\llbracket \mathsf{c}(v) \rrbracket = \text{with consequence } v$$
$$\llbracket \mathsf{rl}(\bot) \rrbracket = \text{with undefined risk level}$$
$$\llbracket \mathsf{rl}(v) \rrbracket = \text{with risk level } v$$

B.1.1 Elements

Below, we give the semantics of each of the elements of basic CORAS diagrams. In these definitions, *id* ranges over *identifier*, *av* over *asset value*, *l* over *likelihood*, and *rl* over *risk level*. As can be observed, each element is translated into an English sentence.

$$\llbracket \mathsf{p}(id) \rrbracket = id \text{ is a party.}$$
$$\llbracket \mathsf{da}(id, av) \rrbracket = id \text{ is a direct asset.}$$
$$\llbracket \mathsf{ia}(id, av) \rrbracket = id \text{ is an indirect asset.}$$
$$\llbracket \mathsf{htd}(id) \rrbracket = id \text{ is a deliberate human threat.}$$
$$\llbracket \mathsf{hta}(id) \rrbracket = id \text{ is an accidental human threat.}$$
$$\llbracket \mathsf{nht}(id) \rrbracket = id \text{ is a non-human threat.}$$
$$\llbracket \mathsf{v}(id) \rrbracket = id \text{ is a vulnerability.}$$
$$\llbracket \mathsf{ts}(id, l) \rrbracket = \text{Threat scenario } id \text{ occurs } \llbracket l \rrbracket.$$
$$\llbracket \mathsf{ui}(id, l) \rrbracket = \text{Unwanted incident } id \text{ occurs } \llbracket l \rrbracket.$$
$$\llbracket \mathsf{r}(id, rl) \rrbracket = \text{Risk } id \text{ occurs } \llbracket rl \rrbracket.$$
$$\llbracket \mathsf{tms}(id) \rrbracket = id \text{ is a treatment scenario.}$$

B.1.2 Relations

Relations can be decorated with consequences, likelihoods or conditional likelihoods, as well as sets of vulnerabilities and treatment categories. The semantics of consequences, likelihoods and conditional likelihoods are already defined above, but before we can define the semantics of relations we also need the semantics of vulnerability sets and treatment categories. They are defined as follows:

$$\llbracket \{\mathsf{v}(id)\} \rrbracket = \text{vulnerability } id$$
$$\llbracket \{\mathsf{v}(id_1), \mathsf{v}(id_2), \ldots, \mathsf{v}(id_n)\} \rrbracket = \text{vulnerabilities } id_1, id_2, \ldots, id_{n-1} \text{ and } id_n$$
$$\llbracket \mathsf{none} \rrbracket = \text{treats}$$

B.1 Basic CORAS

$$\llbracket \mathsf{avoid} \rrbracket = \text{avoids}$$
$$\llbracket \mathsf{reduce\ consequence} \rrbracket = \text{decreases the consequence of}$$
$$\llbracket \mathsf{reduce\ likelihood} \rrbracket = \text{decreases the likelihood of}$$
$$\llbracket \mathsf{transfer} \rrbracket = \text{transfers}$$
$$\llbracket \mathsf{retain} \rrbracket = \text{retains}$$

Below, we give the semantics of each of the relations. In these definitions, we make use of a function $id(\_)$ that returns the identity of an element such that $id(\mathsf{da}(id, \mathsf{av}(v))) = id$, and so forth. In the definitions, a ranges over *asset*, c over *consequence*, t over *threat*, r over *risk*, l over *likelihood*, e over *core element*, V over *vulnerabilities* (vulnerability sets), cl over *conditional likelihood*, tm over *treatment scenario*, tc over *treatment category*, and te over *treatable element*.

$$\llbracket a_1 \xrightarrow{c} a_2 \rrbracket = \text{Harm to } id(a_1) \text{ may result in harm to } id(a_2)\ \llbracket c \rrbracket.$$
$$\llbracket t \xrightarrow{\emptyset,\ \mathsf{l}(\bot)} r \rrbracket = id(t) \text{ initiates } id(r).$$
$$\llbracket t \xrightarrow{\emptyset,\ l} e \rrbracket = id(t) \text{ initiates } id(e)\ \llbracket l \rrbracket.$$
$$\llbracket t \xrightarrow{V,\ \mathsf{l}(\bot)} r \rrbracket = id(t) \text{ exploits } \llbracket V \rrbracket \text{ to initiate } id(r).$$
$$\llbracket t \xrightarrow{V,\ l} e \rrbracket = id(t) \text{ exploits } \llbracket V \rrbracket \text{ to initiate } id(e)\ \llbracket l \rrbracket.$$
$$\llbracket e \xrightarrow{\emptyset,\ \mathsf{cl}(\bot)} r \rrbracket = id(e) \text{ leads to } id(r).$$
$$\llbracket e_1 \xrightarrow{\emptyset,\ cl} e_2 \rrbracket = id(e_1) \text{ leads to } id(e_2)\ \llbracket cl \rrbracket.$$
$$\llbracket e \xrightarrow{V,\ \mathsf{cl}(\bot)} r \rrbracket = id(e) \text{ leads to } id(r), \text{ due to } \llbracket V \rrbracket.$$
$$\llbracket e_1 \xrightarrow{V,\ cl} e_2 \rrbracket = id(e_1) \text{ leads to } id(e_2)\ \llbracket cl \rrbracket, \text{ due to } \llbracket V \rrbracket.$$
$$\llbracket e \xrightarrow{c} a \rrbracket = id(e) \text{ impacts } id(a)\ \llbracket c \rrbracket.$$
$$\llbracket tm \xrightarrow{tc} te \rrbracket = id(tm)\ \llbracket tc \rrbracket\ id(te).$$

As with the elements, the semantics of a relation is given as a sentence in English. Note the special cases with initiates and leads-to relations going to risks. These relations are not allowed to have (conditional) likelihood values, so it does not make any sense to give them a translation with sentences containing "undefined (conditional) likelihood".

B.1.3 Diagrams

The semantics of a threat, risk, treatment or treatment overview diagram is simply the semantics of all of its elements and relations. Let *el* range over *element* and *rel* over *relation*. The semantics of a diagram is then

$$[\![(\{el_1, el_2, \ldots, el_n\}, \{rel_1, rel_2, \ldots, rel_m\})]\!]$$
$$= [\![el_1]\!][\![el_2]\!] \cdots [\![el_n]\!][\![rel_1]\!][\![rel_2]\!] \cdots [\![rel_m]\!]$$

Since each element and relation results in a sentence, the semantics of a diagram constitutes a paragraph in English.

In asset diagrams, there exists an implicit relation between the party of the diagram and the assets. In the semantics, we want to make that relation explicit, and asset diagrams therefore require special treatment when the semantics is defined. Let p range over *party*, a over *asset* and h over *harm*. Further, let $val(e)$ be a function returning the value of an element such that, for example, $val(\mathsf{da}(id, \mathsf{av}(v))) = \mathsf{av}(v)$. The semantics of an asset diagram is then defined as follows.

$$[\![(\{p, a_1, a_2, \ldots, a_n\}, \{h_1, h_2, \ldots, h_m\})]\!] = [\![p]\!][\![a_1]\!][\![a_2]\!] \cdots [\![a_n]\!]$$
$$id(p) \text{ assigns } [\![val(a_1)]\!] \text{ to } id(a_1).$$
$$id(p) \text{ assigns } [\![val(a_2)]\!] \text{ to } id(a_2).$$
$$\vdots$$
$$id(p) \text{ assigns } [\![val(a_n)]\!] \text{ to } id(a_n).$$
$$[\![h_1]\!][\![h_2]\!] \cdots [\![h_m]\!]$$

B.1.4 Examples

In this section, we provide examples of translations of each kind of CORAS diagram using the semantics defined above.

B.1.4.1 Asset Diagram

In this section, we give the semantics of the asset diagram found in Fig. 4.8 on p. 55. The textual syntax of the diagram is found in Sect. A.1.3.1. The semantics of the diagram is:

- *Company* is a party.
- *Integrity of server* is a direct asset.
- *Confidentiality of information* is a direct asset.
- *Availability of server* is a direct asset.
- *Company's reputation* is an indirect asset.
- *Company* assigns the value *high* to *Integrity of server*.
- *Company* assigns the value *critical* to *Confidentiality of information*.
- *Company* assigns the value *critical* to *Availability of server*.

B.1 Basic CORAS

- *Company* assigns value to *Company's reputation*.
- Harm to *Integrity of server* may result in harm to *Company's reputation*.
- Harm to *Integrity of server* may result in harm to *Confidentiality of information*.
- Harm to *Confidentiality of information* may result in harm to *Company's reputation*.
- Harm to *Availability of server* may result in harm to *Company's reputation*.

B.1.4.2 Threat Diagram

In this section, we give the semantics of the threat diagram found in Fig. 4.9 on p. 57. The textual syntax of the diagram is found in Sect. A.1.3.2. The semantics of the diagram is:

- *Hacker* is a deliberate human threat.
- *Computer virus* is a non-human threat.
- *Virus protection not up to date* is a vulnerability.
- Threat scenario *Server is infected by computer virus* occurs with likelihood *possible*.
- Unwanted incident *Hacker gets access to server* occurs with likelihood *unlikely*.
- Unwanted incident *Virus creates back door to server* occurs with likelihood *possible*.
- Unwanted incident *Server goes down* occurs with likelihood *unlikely*.
- *Integrity of server* is a direct asset.
- *Confidentiality of information* is a direct asset.
- *Availability of server* is a direct asset.
- *Hacker* initiates *Hacker gets access to server* with likelihood 0.1.
- *Computer virus* exploits vulnerability *Virus protection not up to date* to initiate *Server is infected by computer virus* with undefined likelihood.
- *Server is infected by computer virus* leads to *Virus creates back door to server* with undefined conditional likelihood.
- *Server is infected by computer virus* leads to *Server goes down* with conditional likelihood 0.2.
- *Virus creates back door to server* leads to *Hacker gets access to server* with undefined conditional likelihood.
- *Hacker gets access to server* impacts *Confidentiality of information* with consequence *high*.
- *Virus creates back door to server* impacts *Integrity of server* with consequence *high*.
- *Server goes down* impacts *Integrity of server* with consequence *low*.
- *Server goes down* impacts *Availability of server* with consequence *high*.

B.1.4.3 Risk Diagram

In this section, we give the semantics of the risk diagram found in Fig. 4.10 on p. 58. The textual syntax of the diagram is found in Sect. A.1.3.3. The semantics of the diagram is:

- *Hacker* is a deliberate human threat.
- *Computer virus* is a non-human threat.
- Risk *HA: Hacker gets access to server* occurs with risk level *unacceptable*.
- Risk *VB: Virus creates back door to server* occurs with risk level *unacceptable*.
- Risk *SD1: Server goes down* occurs with risk level *unacceptable*.
- Risk *SD2: Server goes down* occurs with risk level *acceptable*.
- *Integrity of server* is a direct asset.
- *Confidentiality of information* is a direct asset.
- *Availability of server* is a direct asset.
- *Hacker* initiates *HA: Hacker gets access to server*.
- *Computer virus* initiates *VB: Virus creates back door to server*.
- *Computer virus* initiates *SD1: Server goes down*.
- *Computer virus* initiates *SD2: Server goes down*.
- *VB: Virus creates back door to server* leads to *HA: Hacker gets access to server*.
- *HA: Hacker gets access to server* impacts *Confidentiality of information*.
- *VB: Virus creates back door to server* impacts *Integrity of server*.
- *SD1: Server goes down* impacts *Availability of server*.
- *SD2: Server goes down* impacts *Integrity of server*.

B.1.4.4 Treatment Diagram

In this section, we give the semantics of the treatment diagram found in Fig. 4.11 on p. 60. The textual syntax of the diagram is found in Sect. A.1.3.4. The semantics of the diagram is:

- *Hacker* is a deliberate human threat.
- *Computer virus* is a non-human threat.
- *Virus protection not up to date* is a vulnerability.
- Threat scenario *Server is infected by computer virus* occurs with likelihood *possible*.
- Risk *HA: Hacker gets access to server* occurs with risk level *unacceptable*.
- Risk *VB: Virus creates back door to server* occurs with risk level *unacceptable*.
- Risk *SD1: Server goes down* occurs with risk level *unacceptable*.
- Risk *SD2: Server goes down* occurs with risk level *acceptable*.
- *Integrity of server* is a direct asset.
- *Confidentiality of information* is a direct asset.
- *Availability of server* is a direct asset.
- *Install monitoring software* is a treatment scenario.
- *Implement new routines for updating virus protection* is a treatment scenario.

B.1 Basic CORAS

- *Hacker* initiates *HA: Hacker gets access to server*.
- *Computer virus* exploits vulnerability *Virus protection not up to date* to initiate *Server is infected by computer virus* with undefined likelihood.
- *Server is infected by computer virus* leads to *VB: Virus creates back door to server*.
- *Server is infected by computer virus* leads to *SD1: Server goes down*.
- *Server is infected by computer virus* leads to *SD2: Server goes down*.
- *VB: Virus creates back door to server* leads to *HA: Hacker gets access to server*.
- *HA: Hacker gets access to server* impacts *Confidentiality of information*.
- *VB: Virus creates back door to server* impacts *Integrity of server*.
- *SD1: Server goes down* impacts *Availability of server*.
- *SD2: Server goes down* impacts *Integrity of server*.
- *Install monitoring software* decreases the consequence of *HA: Hacker gets access to server*.
- *Implement new routines for updating virus protection* decreases the likelihood of *Virus protection not up to date*.

B.1.4.5 Treatment Overview Diagram

In this section, we give the semantics of the treatment overview diagram found in Fig. 4.12 on p. 62. The textual syntax of the diagram is found in Sect. A.1.3.5. The semantics of the diagram is:

- *Hacker* is a deliberate human threat.
- *Computer virus* is a non-human threat.
- Risk *HA: Hacker gets access to server* occurs with risk level *unacceptable*.
- Risk *VB: Virus creates back door to server* occurs with risk level *unacceptable*.
- Risk *SD1: Server goes down* occurs with risk level *unacceptable*.
- Risk *SD2: Server goes down* occurs with risk level *acceptable*.
- *Integrity of server* is a direct asset.
- *Confidentiality of information* is a direct asset.
- *Availability of server* is a direct asset.
- *Install monitoring software* is a treatment scenario.
- *Implement new routines for updating virus protection* is a treatment scenario.
- *Hacker* initiates *HA: Hacker gets access to server*.
- *Computer virus* initiates *VB: Virus creates back door to server*.
- *Computer virus* initiates *SD1: Server goes down*.
- *Computer virus* initiates *SD2: Server goes down*.
- *VB: Virus creates back door to server* leads to *HA: Hacker gets access to server*.
- *HA: Hacker gets access to server* impacts *Confidentiality of information*.
- *VB: Virus creates back door to server* impacts *Integrity of server*.
- *SD1: Server goes down* impacts *Availability of server*.
- *SD2: Server goes down* impacts *Integrity of server*.
- *Install monitoring software* decreases the consequence of *HA: Hacker gets access to server*.

- *Implement new routines for updating virus protection* decreases the likelihood of *VB: Virus creates back door to server.*
- *Implement new routines for updating virus protection* decreases the likelihood of *SD1: Server goes down.*
- *Implement new routines for updating virus protection* decreases the likelihood of *SD2: Server goes down.*

B.2 High-level CORAS

This section gives the semantics of high-level CORAS diagrams as an extension to the semantics of basic CORAS diagrams given in Sect. B.1. In Sect. B.2.1, we give the semantics of referring elements, in Sect. B.2.2 we extend the semantics of the relations and in Sect. B.2.3 we give the semantics of referenced diagrams. In Sect. B.2.4, we provide examples of the application of the semantics on high-level CORAS diagrams and the referenced diagrams to which they refer.

B.2.1 Referring Elements

When we define the semantics of the referring elements, we start by giving the translations of sets of gates. These translations are used to list the gates that belong to referring elements, and do therefore not show the likelihoods of the gates. Let g range over *gate* and *treatment gate*. The semantics for sets of gates are then:

$$[\![\{g\}]\!] = \text{gate } id(g)$$
$$[\![\{g_1, g_2, \ldots, g_n\}]\!] = \text{gates } id(g_1), id(g_2), \ldots, id(g_{n-1}) \text{ and } id(g_n)$$

The translation of the individual gates include their kind, and also takes into account that in-gates and out-gates may be assigned likelihoods. Let id range over *identifier* and let v range over *value*. The semantics for gates are then:

$$[\![\text{ig}(id, \text{l}(\bot))]\!] = id \text{ is an in-gate.}$$
$$[\![\text{ig}(id, \text{l}(v))]\!] = id \text{ is an in-gate with likelihood } v.$$
$$[\![\text{og}(id, \text{l}(\bot))]\!] = id \text{ is an out-gate.}$$
$$[\![\text{og}(id, \text{l}(v))]\!] = id \text{ is an out-gate with likelihood } v.$$
$$[\![\text{tig}(id)]\!] = id \text{ is a treatment gate.}$$
$$[\![\text{tog}(id)]\!] = id \text{ is a treatment gate.}$$

The semantics for the referring elements are similar to the semantics of basic elements, but must also take the gates of the referring elements into consideration. We do, however, not make any difference between gates and treatment gates in the

B.2 High-level CORAS

definitions. Let *id* range over *identifier*, *rl* over *risk level*, *g* over *gate* and *tg* over *treatment gate*. The translation scheme takes into account that referring core elements may or may not have gates. The semantics of the referring elements are given by the following.

$\llbracket \text{rts}(id, \emptyset, \emptyset) \rrbracket = id$ is a referring threat scenario with no gates.

$\llbracket \text{rts}(id, \{g_1, g_2, \ldots, g_n\}, \{tg_1, tg_2, \ldots, tg_m\}) \rrbracket$
 $= id$ is a referring threat scenario with $\llbracket \{g_1, g_2, \ldots, g_n\} \cup \{tg_1, tg_2, \ldots, tg_m\} \rrbracket$.
 $\llbracket g_1 \rrbracket \llbracket g_2 \rrbracket \cdots \llbracket g_n \rrbracket \llbracket tg_1 \rrbracket \llbracket tg_2 \rrbracket \cdots \llbracket tg_m \rrbracket$

$\llbracket \text{rui}(id, \emptyset, \emptyset) \rrbracket = id$ is a referring unwanted incident with no gates.

$\llbracket \text{rui}(id, \{g_1, g_2, \ldots, g_n\}, \{tg_1, tg_2, \ldots, tg_m\}) \rrbracket$
 $= id$ is a referring unwanted incident with
 $\llbracket \{g_1, g_2, \ldots, g_n\} \cup \{tg_1, tg_2, \ldots, tg_m\} \rrbracket$.
 $\llbracket g_1 \rrbracket \llbracket g_2 \rrbracket \cdots \llbracket g_n \rrbracket \llbracket tg_1 \rrbracket \llbracket tg_2 \rrbracket \cdots \llbracket tg_m \rrbracket$

$\llbracket \text{rr}(id, rl, \emptyset, \emptyset) \rrbracket = $ Referring risk *id*, with no gates, occurs $\llbracket rl \rrbracket$.

$\llbracket \text{rr}(id, rl, \{g_1, g_2, \ldots, g_n\}, \{tg_1, tg_2, \ldots, tg_m\}) \rrbracket$
 $= $ Referring risk *id*, with $\llbracket \{g_1, g_2, \ldots, g_n\} \cup \{tg_1, tg_2, \ldots, tg_m\} \rrbracket$, occurs $\llbracket rl \rrbracket$.
 $\llbracket g_1 \rrbracket \llbracket g_2 \rrbracket \cdots \llbracket g_n \rrbracket \llbracket tg_1 \rrbracket \llbracket tg_2 \rrbracket \cdots \llbracket tg_m \rrbracket$

$\llbracket \text{rtms}(id, \{tg_1, tg_2, \ldots, tg_m\}) \rrbracket$
 $= id$ is a referring treatment scenario with $\llbracket \{tg_1, tg_2, \ldots, tg_m\} \rrbracket$.
 $\llbracket tg_1 \rrbracket \llbracket tg_2 \rrbracket \cdots \llbracket tg_m \rrbracket$

The semantics of likelihoods and risk levels are the same as in basic CORAS.

B.2.2 Relations

In basic CORAS, the relations are between basic elements. These definitions carry over to high-level CORAS, but we also get a number of new cases because of the introduction of gates. In high-level CORAS, relations can have gates (contained in referring elements or at the border of referenced diagrams) as source and/or target, and this needs to be reflected in the semantics.

Let *t* range over *threat*, *rr* over *referring risk*, *re* over *referring core element*, *ig* over *in-gate*, *og* over *out-gate*, *l* over *likelihood* and *V* over *vulnerabilities* (vulnerability sets). The semantics of the new cases of the initiates relation are then:

$\llbracket t \xrightarrow{\emptyset, l(\bot)} rr.ig \rrbracket = id(t)$ initiates $id(rr)$ via gate $id(ig)$.

$\llbracket t \xrightarrow{\emptyset, l} re.ig \rrbracket = id(t)$ initiates $id(re)$ via gate $id(ig)$ $\llbracket l \rrbracket$.

$[\![t \xrightarrow{V,\, \mathsf{l}(\bot)} rr.ig]\!] = id(t)$ exploits $[\![V]\!]$ to initiate $id(rr)$ via gate $id(ig)$.

$[\![t \xrightarrow{V,\, l} re.ig]\!] = id(t)$ exploits $[\![V]\!]$ to initiate $id(re)$ via gate $id(ig)$ $[\![l]\!]$.

$[\![t \xrightarrow{\emptyset,\, \mathsf{l}(\bot)} og]\!] = id(t)$ has an effect via gate $id(og)$.

$[\![t \xrightarrow{V,\, \mathsf{l}(\bot)} og]\!] = id(t)$ has an effect via gate $id(og)$, due to $[\![V]\!]$.

The semantics of (conditional) likelihoods and vulnerability sets are the same as in basic CORAS. Note that we in these new definitions have special rules for relations with gates contained in referring risks as target, similar to the special rules for risks in basic CORAS. The same is the case for the new cases of the leads-to relation.

Let e range over *core element*, r over *risk* and cl over *conditional likelihood*. We get the following semantics for the new cases of the leads-to relation.

$[\![e \xrightarrow{\emptyset,\, \mathsf{cl}(\bot)} rr.ig]\!] = id(e)$ leads to $id(rr)$ via gate $id(ig)$.

$[\![e \xrightarrow{\emptyset,\, cl} re.ig]\!] = id(e)$ leads to $id(re)$ via gate $id(ig)$ $[\![cl]\!]$.

$[\![e \xrightarrow{V,\, \mathsf{cl}(\bot)} rr.ig]\!] = id(e)$ leads to $id(rr)$ via gate $id(ig)$, due to $[\![V]\!]$.

$[\![e \xrightarrow{V,\, cl} re.ig]\!] = id(e)$ leads to $id(re)$ via gate $id(ig)$ $[\![cl]\!]$, due to $[\![V]\!]$.

$[\![re.og \xrightarrow{\emptyset,\, \mathsf{cl}(\bot)} r]\!] = id(re)$ leads to $id(r)$ via gate $id(og)$.

$[\![re.og \xrightarrow{\emptyset,\, cl} e]\!] = id(re)$ leads to $id(e)$ via gate $id(og)$ $[\![cl]\!]$.

$[\![re.og \xrightarrow{V,\, \mathsf{cl}(\bot)} r]\!] = id(re)$ leads to $id(r)$ via gate $id(og)$, due to $[\![V]\!]$.

$[\![re.og \xrightarrow{V,\, cl} e]\!] = id(re)$ leads to $id(e)$ via gate $id(og)$ $[\![cl]\!]$, due to $[\![V]\!]$.

$[\![re.og \xrightarrow{\emptyset,\, \mathsf{cl}(\bot)} rr.ig]\!] = id(re)$ leads to $id(rr)$ via gates $id(og)$ and $id(ig)$.

$[\![re_1.og \xrightarrow{\emptyset,\, cl} re_2.ig]\!] = id(re_1)$ leads to $id(re_2)$ via gates $id(og)$ and $id(ig)$ $[\![cl]\!]$.

$[\![re.og \xrightarrow{V,\, \mathsf{cl}(\bot)} rr.ig]\!] = id(re)$ leads to $id(rr)$ via gates $id(og)$ and $id(ig)$, due to $[\![V]\!]$.

$[\![re_1.og \xrightarrow{V,\, cl} re_2.ig]\!] = id(re_1)$ leads to $id(re_2)$ via gates $id(og)$ and $id(ig)$ $[\![cl]\!]$, due to $[\![V]\!]$.

$[\![ig \xrightarrow{\emptyset,\, \mathsf{cl}(\bot)} e]\!] = id(e)$ is initiated via gate $id(ig)$.

$[\![ig \xrightarrow{\emptyset,\, cl} e]\!] = id(e)$ is initiated via gate $id(ig)$ $[\![cl]\!]$.

B.2 High-level CORAS

$[\![ig \xrightarrow{V,\ cl(\bot)} e]\!] = id(e)$ is initiated via gate $id(ig)$, due to $[\![V]\!]$.

$[\![ig \xrightarrow{V,\ cl} e]\!] = id(e)$ is initiated via gate $id(ig)$ $[\![cl]\!]$, due to $[\![V]\!]$.

$[\![ig_1 \xrightarrow{\emptyset,\ cl(\bot)} re.ig_2]\!] = id(re)$ is initiated via gates $id(ig_1)$ and $id(ig_2)$.

$[\![ig_1 \xrightarrow{\emptyset,\ cl} re.ig_2]\!] = id(re)$ is initiated via gates $id(ig_1)$ and $id(ig_2)$ $[\![cl]\!]$.

$[\![ig_1 \xrightarrow{V,\ cl(\bot)} re.ig_2]\!] = id(re)$ is initiated via gates $id(ig_1)$ and $id(g_2)$,
due to $[\![V]\!]$.

$[\![ig_1 \xrightarrow{V,\ cl} re.ig_2]\!] = id(re)$ is initiated via gates $id(ig_1)$ and $id(g_2)$ $[\![cl]\!]$,
due to $[\![V]\!]$.

$[\![e \xrightarrow{\emptyset,\ cl(\bot)} og]\!] = id(e)$ has an effect via gate $id(og)$.

$[\![e \xrightarrow{\emptyset,\ cl} og]\!] = id(e)$ has an effect via gate $id(og)$ $[\![cl]\!]$.

$[\![e \xrightarrow{V,\ cl(\bot)} og]\!] = id(e)$ has an effect via gate $id(og)$, due to $[\![V]\!]$.

$[\![e \xrightarrow{V,\ cl} og]\!] = id(e)$ has an effect via gate $id(og)$ $[\![cl]\!]$, due to $[\![V]\!]$.

$[\![re.og_1 \xrightarrow{\emptyset,\ cl(\bot)} og_2]\!] = id(re)$ has an effect via gates $id(og_1)$ and $id(og_2)$.

$[\![re.og_1 \xrightarrow{\emptyset,\ cl} og_2]\!] = id(re)$ has an effect via gates $id(og_1)$ and $id(og_2)$ $[\![cl]\!]$.

$[\![re.og_1 \xrightarrow{V,\ cl(\bot)} og_2]\!] = id(re)$ has an effect via gates $id(og_1)$ and $id(og_2)$,
due to $[\![V]\!]$.

$[\![re.og_1 \xrightarrow{V,\ cl} og_2]\!] = id(re)$ has an effect via gates $id(og_1)$ and $id(og_2)$ $[\![cl]\!]$,
due to $[\![V]\!]$.

Notice that with respect to initiates and leads-to relations with gates at the border of a referenced diagram as source or target we get a different semantics ("is initiated", "has an effect") than in the other cases. The reason that these cases have a different (somewhat weaker) semantics than usual is that we do not have the information on what is connected to the gates at the outside of the referenced diagram.

The impacts relation has only one new case. Let c range over *consequence* and a over *asset*. The new case of the impacts relation has the following semantics.

$[\![re.og \xrightarrow{c} a]\!] = id(re)$ impacts $id(a)$ via gate $id(og)$ $[\![c]\!]$.

The treats relation also has a number of new cases due to the introduction of treatment gates. Let tm range over *treatment scenario*, tc over *treatment category*, *tg* over *treatment gate* and *rtm* over *referring treatment scenario*. The semantics of the new cases are:

$$[\![tm \xrightarrow{tc} re.tg]\!] = id(tm) \; [\![tc]\!] \; id(re) \text{ via gate } id(tg).$$

$$[\![rtm.tg \xrightarrow{tc} e]\!] = id(rtm) \; [\![tc]\!] \; id(e) \text{ via gate } id(tg).$$

$$[\![rtm.tg_1 \xrightarrow{tc} re.tg_2]\!] = id(rtm) \; [\![tc]\!] \; id(re) \text{ via gates } id(tg_1) \text{ and } id(tg_2).$$

$$[\![tm \xrightarrow{tc} tg]\!] = id(tm) \; \text{ provides treatment via gate } id(tg).$$

$$[\![rtm.tg_1 \xrightarrow{tc} tg_2]\!] = id(rtm) \; \text{ provides treatment via gates } id(tg_1) \text{ and } id(tg_2).$$

$$[\![tg \xrightarrow{tc} e]\!] = id(e) \text{ is treated via gate } id(tg).$$

$$[\![tg_1 \xrightarrow{tc} re.tg_2]\!] = id(re) \text{ is treated via gates } id(tg_1) \text{ and } id(tg_2).$$

The semantics of treatment categories are the same as for basic CORAS.

B.2.3 Referenced Diagrams

Finally, we give the semantics for referenced diagrams. The semantics of a referenced diagram is similar to the semantics of a regular diagram in that it consists of the semantics of all its elements and relations. Additionally, it consists of the semantics of all the gates. The semantics of a referenced diagram moreover contains a kind of heading that gives the identifier of the diagram and lists the names of its gates.

Let *id* range over *identifier*, *g* over *gate*, *tg* over *treatment gate*, *el* over *element* (including referring elements) and *rel* over *relation*. The following semantics of referenced threat scenarios, referenced unwanted incidents and referenced risks (referenced core diagrams) takes into account that these diagrams may or may not have gates.

$$[\![(id, \emptyset, \emptyset, \{el_1, el_2, \ldots, el_n\}, \{rel_1, rel_2, \ldots, rel_m\})]\!]$$
$$= id \text{ is a referenced diagram with no gates}:$$
$$[\![el_1]\!][\![el_2]\!] \cdots [\![el_n]\!][\![rel_1]\!][\![rel_2]\!] \cdots [\![rel_m]\!]$$

$$[\![(id, \{g_1, g_2, \ldots, g_i\}, \{tg_1, tg_2, \ldots, tg_j\}, \{el_1, el_2, \ldots, el_n\}, \{rel_1, rel_2, \ldots, rel_m\})]\!]$$
$$= id \text{ is a referenced diagram with } [\![\{g_1, g_2, \ldots, g_i\} \cup \{tg_1, tg_2, \ldots, tg_j\}]\!]:$$
$$[\![g_1]\!][\![g_2]\!] \cdots [\![g_i]\!][\![tg_1]\!][\![tg_2]\!] \cdots [\![tg_j]\!]$$
$$[\![el_1]\!][\![el_2]\!] \cdots [\![el_n]\!][\![rel_1]\!][\![rel_2]\!] \cdots [\![rel_m]\!]$$

The semantics of gates and gate sets are the same as given for referring elements in Sect. B.2.1.

B.2 High-level CORAS

Referenced treatment scenarios are somewhat separate from the other referenced diagrams, and has therefore has a separate, but similar, semantic definition. Let tg range over *treatment gate*, tms range over *treatment scenario* and *referring treatment scenario*, and tr over *treats*. The semantics of referenced treatment scenarios is then:

$$[\![(id, \{tg_1, tg_2, \ldots, tg_i\}, \{tms_1, tms_2, \ldots, tms_n\}, \{tr_1, tr_2, \ldots, tr_m\})]\!]$$
$$= id \text{ is a referenced treatment scenario with } [\![\{tg_1, tg_2, \ldots, tg_i\}]\!] :$$
$$[\![tg_1]\!][\![tg_2]\!] \cdots [\![tg_i]\!][\![tms_1]\!][\![tms_2]\!] \cdots [\![tms_n]\!][\![tr_1]\!][\![tr_2]\!] \cdots [\![tr_m]\!]$$

The semantics of basic CORAS diagrams carry over to high-level CORAS diagrams because of the redefinition of *element*. Hence, there is no need to provide semantics specifically to high-level diagrams.

B.2.4 Examples

In this section, we provide examples of translations of high-level diagram and the referenced diagrams they refer to using the semantics defined above.

B.2.4.1 High-level Threat Diagram

In this section, we give the semantics of the high-level threat diagram found in Fig. 14.18 on p. 258 with its referenced diagrams found in Figs. 14.16, 14.19 and 14.22 on pp. 256, 260 and 263. The textual syntax of the diagrams is found in Sect. A.2.3.1. The semantics of the high-level diagram is:

- *Computer virus* is a non-human threat.
- *Virus protection not up to date* is a vulnerability.
- *Server is infected by computer virus* is a referring threat scenario with gates i_2, o_3 and o_4.
  - i_2 is an in-gate with likelihood 0.1.
  - o_3 is an out-gate with likelihood 0.05.
  - o_4 is an out-gate with likelihood 0.1.
- *Hacker attack on server* is a referring threat scenario with gates i_1, o_1 and o_2.
  - i_1 is an in-gate with likelihood 0.01.
  - o_1 is an out-gate with likelihood 0.05.
  - o_2 is an out-gate with likelihood 0.1.
- *Server goes down* is referring unwanted incident with gates i_3, o_5 and o_6.
  - i_3 is an in-gate with likelihood 0.05.
  - o_5 is an out-gate with likelihood 0.01.
  - o_6 is an out-gate with likelihood 0.05.
- *Integrity of server* is a direct asset.
- *Confidentiality of information* is a direct asset.

- *Availability of server* is a direct asset.
- *Computer virus* exploits vulnerability *Virus protection not up to date* to initiate *Server is infected by computer virus* via gate i_2 with likelihood 0.1.
- *Server is infected by computer virus* leads to *Hacker attack on server* via gates o_3 and i_1 with conditional likelihood 0.2.
- *Server is infected by computer virus* leads to *Server goes down* via gates o_4 and i_3 with conditional likelihood 0.2.
- *Hacker attack on server* impacts *Confidentiality of information* via gate o_1 with consequence *high*.
- *Hacker attack on server* impacts *Integrity of server* via gate o_2 with consequence *high*.
- *Server goes down* impacts *Integrity of server* via gate o_5 with consequence *low*.
- *Server goes down* impacts *Availability of server* via gate o_6 with consequence *medium*.

The semantics of the referenced threat scenario *Hacker attack on server* is:

- *Hacker attack on server* is a referenced diagram with gates i_1, o_1 and o_2:
  - i_1 is an in-gate with likelihood 0.02.
  - o_1 is an out-gate with likelihood $[0.02, 0.06]$.
  - o_2 is an out-gate with likelihood 0.02.
  - *Hacker* is a deliberate human threat.
  - Unwanted incident *Hacker gets access to server* occurs with likelihood $[0.02, 0.06]$.
  - Unwanted incident *Virus creates back door to server* occurs with likelihood 0.02.
  - *Hacker* initiates *Hacker gets access to server* with likelihood *unlikely*.
  - *Virus creates back door to server* leads to *Hacker gets access to server* with conditional likelihood 0.5.
  - *Virus creates back door to server* is initiated via gate i_1 with conditional likelihood 1.0.
  - *Hacker gets access to server* has an effect via gate o_1 with conditional likelihood 1.0.
  - *Virus creates back door to server* has an effect via gate o_2 with conditional likelihood 1.0.

The semantics of the referenced threat scenario *Server is infected by computer virus* is:

- *Server is infected by computer virus* is a referenced diagram with gates i_2, o_3 and o_4:
  - i_2 is an in-gate.
  - o_3 is an out-gate with likelihood 0.05.
  - o_4 is an out-gate.
  - *Low awareness of email worms* is a vulnerability.
  - Threat scenario *Server is infected by Trojan horse* occurs with undefined likelihood.

B.2 High-level CORAS

- Threat scenario *Server is infected by email worm* occurs with likelihood 0.1.
- Threat scenario *Email worm is spreading* occurs with likelihood 0.08.
- Threat scenario *Server is infected by SQL worm* occurs with likelihood 0.1.
- Threat scenario *Server is overloaded* occurs with undefined likelihood.
- *Server is infected by Trojan horse* is initiated via gate i_2 with conditional likelihood 1.0.
- *Server is infected by email worm* is initiated via gate i_2 with conditional likelihood 1.0.
- *Server is infected by SQL worm* is initiated via gate i_2 with conditional likelihood 1.0.
- *Server is infected by email worm* leads to *Email worm is spreading* with conditional likelihood 0.8, due to vulnerability *Low awareness of email worms*.
- *Server is infected by SQL worm* leads to *Server is overloaded* with conditional likelihood 0.8.
- *Email worm is spreading* leads to *Server is overloaded* with conditional likelihood 1.0.
- *Server is overloaded* has an effect via gate o_4 with conditional likelihood 1.0.
- *Server is infected by Trojan horse* has an effect via gate o_3 with conditional likelihood 1.0.

The semantics of the referenced unwanted incident *Server goes down* is:

- *Server goes down* is a referenced diagram with gates i_3, o_5 and o_6:
  - i_3 is an in-gate with likelihood 0.05.
  - o_5 is an out-gate with likelihood 0.01.
  - o_6 is an out-gate with likelihood 0.05.
  - Unwanted incident *Server crashes* occurs with likelihood 0.01.
  - Unwanted incident *Server does not respond* occurs with likelihood 0.05.
  - *Server crashes* is initiated via gate i_3 with conditional likelihood 0.2.
  - *Server does not respond* is initiated via gate i_3 with conditional likelihood 1.0.
  - *Server crashes* has an effect via gate o_5 with conditional likelihood 1.0.
  - *Server crashes* has an effect via gate o_6 with conditional likelihood 1.0.
  - *Server does not respond* has an effect via gate o_6 with conditional likelihood 1.0.

B.2.4.2 High-level Risk Diagram

In this section we give the semantics of the high-level risk diagram found in Fig. 14.23 on p. 265 with its referenced diagrams found in Figs. 14.8 and 14.9 on pp. 250–251. The textual syntax of the diagrams is found in Sect. A.2.3.2. The semantics of the high-level diagram is:

- *Hacker* is a deliberate human threat.
- *Computer virus* is a non-human threat.
- Risk *HA: Hacker gets access to server* occurs with risk level *unacceptable*.
- Risk *VB: Virus creates back door to server* occurs with risk level *unacceptable*.

- Referring risk *SD1: Server goes down*, with gates i_3 and o_6, occurs with risk level *unacceptable*.
  - i_3 is an in-gate.
  - o_6 is an out-gate.
- Referring risk *SD2: Server goes down*, with gates i_3 and o_5, occurs with risk level *acceptable*.
  - i_3 is an in-gate.
  - o_5 is an out-gate.
- *Integrity of server* is a direct asset.
- *Confidentiality of information* is a direct asset.
- *Availability of server* is a direct asset.
- *Hacker* initiates *HA: Hacker gets access to server*.
- *Computer virus* initiates *VB: Virus creates back door to server*.
- *Computer virus* initiates *SD1: Server goes down* via gate i_3.
- *Computer virus* initiates *SD2: Server goes down* via gate i_3.
- *VB: Virus creates back door to server* leads to *HA: Hacker gets access to server*.
- *HA: Hacker gets access to server* impacts *Confidentiality of information*.
- *VB: Virus creates back door to server* impacts *Integrity of server*.
- *SD1: Server goes down* impacts *Availability of server* via gate o_6.
- *SD2: Server goes down* impacts *Integrity of server* via gate o_5.

The semantics of the referenced risk *SD1: Server goes down* is:

- *SD1: Server goes down* is a referenced diagram with gates i_3 and o_6:
  - i_3 is an in-gate.
  - o_6 is an out-gate.
  - Risk *SC1: Server crashes* occurs with undefined risk level.
  - Risk *SN1: Server does not respond* occurs with undefined risk level.
  - *SC1: Server crashes* is initiated via gate i_3.
  - *SN1: Server does not respond* is initiated via gate i_3.
  - *SC1: Server crashes* has an effect via gate o_6.
  - *SN1: Server does not respond* has an effect via gate o_6.

The semantics of the referenced risk *SD2: Server goes down* is:

- *SD2: Server goes down* is a referenced diagram with gates i_3 and o_5:
  - i_3 is an in-gate.
  - o_5 is an out-gate.
  - Risk *SC2: Server crashes* occurs with undefined risk level.
  - *SD2: Server goes down* is initiated via gate i_3.
  - *SD2: Server goes down* has an effect via gate o_5.

B.2.4.3 High-level Treatment Diagram

In this section, we give the semantics of the high-level treatment diagram found in Fig. 14.10 on p. 252 with its referenced diagrams found in Figs. 14.11 and 14.12

B.2 High-level CORAS

on p. 253. The textual syntax of the diagrams is found in Sect. A.2.3.3. The semantics of the high-level diagram is:

- *Computer virus* is a non-human threat.
- *Virus protection not up to date* is a vulnerability.
- *Hacker attack on server* is a referring threat scenario with gates i_1, o_1, o_2 and t_1.
  - i_1 is an in-gate.
  - o_1 is an out-gate.
  - o_2 is an out-gate.
  - t_1 is a treatment gate.
- *Server is infected by computer virus* is a referring threat scenario with gates i_2, o_3, o_4 and t_2.
  - i_2 is an in-gate.
  - o_3 is an out-gate.
  - o_4 is an out-gate.
  - t_2 is a treatment gate.
- Risk *HA: Hacker gets access to server* occurs with undefined risk level.
- Risk *VB: Virus creates back door to server* occurs with undefined risk level.
- Referring risk *SD1: Server goes down*, with gates i_3 and o_6, occurs with undefined risk level.
  - i_3 is an in-gate.
  - o_6 is an out-gate.
- Referring risk *SD2: Server goes down*, with gates i_3 and o_5, occurs with undefined risk level.
  - i_3 is an in-gate.
  - o_5 is an out-gate.
- *Integrity of server* is a direct asset.
- *Confidentiality of information* is a direct asset.
- *Availability of server* is a direct asset.
- *Install monitoring software* is a treatment scenario.
- *Improve virus protection* is a referring treatment scenario with gate t_3.
  - t_3 is a treatment gate.
- *SQL injection countermeasures* is a referring treatment scenario with gate t_4.
  - t_4 is a treatment gate.
- *Computer virus* exploits vulnerability *Virus protection not up to date* to initiate *Server is infected by computer virus* via gate i_2 with undefined likelihood.
- *Server is infected by computer virus* leads to *Hacker attack on server* via gates o_3 and i_1 with undefined conditional likelihood.
- *Server is infected by computer virus* leads to *SD1: Server goes down* via gates o_4 and i_3.
- *Server is infected by computer virus* leads to *SD2: Server goes down* via gates o_4 and i_3.
- *Hacker attack on server* leads to *HA: Hacker gets access to server* via gate o_1.
- *Hacker attack on server* leads to *VB: Virus creates back door to server* via gate o_2.
- *VB: Virus creates back door to server* leads to *HA: Hacker gets access to server*.

- *HA: Hacker gets access to server* impacts *Confidentiality of information.*
- *VB: Virus creates back door to server* impacts *Integrity of server.*
- *SD1: Server goes down* impacts *Availability of server* via gate o_6.
- *SD2: Server goes down* impacts *Integrity of server* via gate o_5.
- *Install monitoring software* treats *Hacker gets access to server* via gate t_1.
- *Improve virus protection* treats *Virus protection not up to date* via gate t_3.
- *SQL injection countermeasures* treats *Server is infected by computer virus* via gates t_4 and t_2.

The semantics of the referenced treatment scenario *SQL injection countermeasures* is:

- *SQL injection countermeasures* is a referenced treatment scenario with gate t_4:
  - t_4 is a treatment gate.
  - *Improve routines for patching SQL server* is a treatment scenario.
  - *Improve input validation* is a treatment scenario.
  - *Improve routines for patching SQL server* provides treatment via gate t_4.
  - *Improve input validation* provides treatment via gate t_4.

The semantics of the referenced threat scenario *Server is infected by computer virus* is:

- *Server is infected by computer virus* is a referenced diagram with gates i_2, o_3, o_4 and t_2:
  - i_2 is an in-gate.
  - o_3 is an out-gate.
  - o_4 is an out-gate.
  - t_4 is a treatment gate.
  - *Low awareness of email worms* is a vulnerability.
  - Threat scenario *Server is infected by Trojan horse* occurs with undefined likelihood.
  - Threat scenario *Server is infected by email worm* occurs with undefined likelihood.
  - Threat scenario *Email worm is spreading* occurs with undefined likelihood.
  - Threat scenario *Server is infected by SQL worm* occurs with undefined likelihood.
  - Threat scenario *Server is overloaded* occurs with undefined likelihood.
  - *Server is infected by Trojan horse* is initiated via gate i_2.
  - *Server is infected by email worm* is initiated via gate i_2.
  - *Server is infected by SQL worm* is initiated via gate i_2.
  - *Server is infected by email worm* leads to *Email worm is spreading* with undefined conditional likelihood, due to vulnerability *Low awareness of email worms.*
  - *Server is infected by SQL worm* leads to *Server is overloaded* with undefined conditional likelihood.
  - *Email worm is spreading* leads to *Server is overloaded* with undefined conditional likelihood.

B.2 High-level CORAS

- *Server is overloaded* has an effect via gate o_4.
- *Server is infected by Trojan horse* has an effect via gate o_3.
- *Server is infected by SQL worm* is treated via gate t_2.

B.2.4.4 High-level Treatment Overview Diagram

In this section, we give the semantics of the high-level treatment overview diagram found in Fig. 14.13 on p. 254 with its referenced diagrams found in Figs. 14.11 and 14.14 on pp. 253 and 254. The textual syntax of the diagrams is found in Sect. A.2.3.4. The semantics of the high-level diagram is:

- *Hacker* is a deliberate human threat.
- *Computer virus* is a non-human threat.
- Risk *HA: Hacker gets access to server* occurs with undefined risk level.
- Risk *VB: Virus creates back door to server* occurs with undefined risk level.
- Referring risk *SD1: Server goes down*, with gates i_3, o_6, t_5 and t_7, occurs with undefined risk level.
  - i_3 is an in-gate.
  - o_6 is an out-gate.
  - t_5 is a treatment gate.
  - t_7 is a treatment gate.
- Referring risk *SD2: Server goes down*, with gates i_3, o_5, t_6 and t_8, occurs with undefined risk level.
  - i_3 is an in-gate.
  - o_5 is an out-gate.
  - t_6 is a treatment gate.
  - t_8 is a treatment gate.
- *Integrity of server* is a direct asset.
- *Confidentiality of information* is a direct asset.
- *Availability of server* is a direct asset.
- *Install monitoring software* is a treatment scenario.
- *Improve virus protection* is a referring treatment scenario with gate t_3.
  - t_3 is a treatment gate.
- *SQL injection countermeasures* is a referring treatment scenario with gate t_4.
  - t_4 is a treatment gate.
- *Hacker* initiates *HA: Hacker gets access to server*.
- *Computer virus* initiates *VB: Virus creates back door to server*.
- *Computer virus* initiates *SD1: Server goes down* via gate i_3.
- *Computer virus* initiates *SD2: Server goes down* via gate i_3.
- *VB: Virus creates back door to server* leads to *HA: Hacker gets access to server*.
- *HA: Hacker gets access to server* impacts *Confidentiality of information*.
- *VB: Virus creates back door to server* impacts *Integrity of server*.
- *SD1: Server goes down* impacts *Availability of server* via gate o_6.
- *SD2: Server goes down* impacts *Integrity of server* via gate o_5.
- *Install monitoring software* treats *HA: Hacker gets access to server*.

- *Improve virus protection* treats *HA: Hacker gets access to server* via gate t_3.
- *Improve virus protection* treats *VB: Virus creates back door to server* via gate t_3.
- *Improve virus protection* treats *SD1: Server goes down* via gates t_3 and t_5.
- *Improve virus protection* treats *SD2: Server goes down* via gates t_3 and t_6.
- *SQL injection countermeasures* treats *SD1: Server goes down* via gates t_4 and t_7.
- *SQL injection countermeasures* treats *SD2: Server goes down* via gates t_4 and t_8.

The semantics of the referenced treatment scenario *SQL injection countermeasures* is:

- *SQL injection countermeasures* is a referenced treatment scenario with gate t_4:
  - t_4 is a treatment gate.
  - *Improve routines for patching SQL server* is a treatment scenario.
  - *Improve input validation* is a treatment scenario.
  - *Improve routines for patching SQL server* provides treatment via gate t_4.
  - *Improve input validation* provides treatment via gate t_4.

The semantics of the referenced risk *SD2: Server goes down* is:

- *SD2: Server goes down* is a referenced diagram with gates i_3, o_5, t_6 and t_8:
  - i_3 is an in-gate.
  - o_5 is an out-gate.
  - t_6 is a treatment gate.
  - t_8 is a treatment gate.
  - Risk *SC2: Server crashes* occurs with undefined risk level.
  - *SC2: Server crashes* is initiated via gate i_3.
  - *SC2: Server crashes* has an effect via gate o_5.
  - *SC2: Server crashes* is treated via gate t_6.
  - *SC2: Server crashes* is treated via gate t_8.

B.3 Dependent CORAS

A dependent CORAS diagram consists of elements and relations. These can be both basic elements and relations, and referring elements and relations between referring elements. The semantics of these language construct is as defined in Sects. B.1 and B.2. What remains is therefore to define the semantics of the border construct of dependent diagrams, as well as the semantics of dependent diagrams.

B.3.1 Border

The border is a part of a diagram that in itself is a syntactically correct threat or treatment digram, and that consists of a set of elements and relations. The definition of the semantics is therefore precisely the same as the definition of the semantics of threat diagrams and treatment diagrams in Sect. B.1.3. This means that the semantics of a border is the semantics of all its elements and relations. The semantics of elements and relations are given in Sect. B.1 for basic CORAS and in Sect. B.2 for high-level CORAS.

B.3.2 Dependent Diagrams

The semantics of dependent diagrams captures the dependency of the target on the assumptions. The target is the fragment of the dependent diagram that is specified by the border, whereas the assumptions are what remains when removing the target from the diagram.

More formally, a dependent diagram is a triple $(E, R, (E_b, R_b))$ such that E are all the elements, R are all the relations, E_b are the elements of the border, and R_b are the relations of the border. By definition of dependent CORAS, $E_b \subseteq E$ and $R_b \subseteq R$. The pair (E_b, R_b) represents the target, whereas the pair $(E \setminus E_b, R \setminus R_b)$ of the remaining elements and relations represents the assumptions. The semantics of a dependent diagram $(E, R, (E_b, R_b))$ is then defined as follows:

$$[\![(E, R, (E_b, R_b))]\!] = [\![(E_b, R_b)]\!]$$

Assuming:

$$[\![(E \setminus E_b, R \setminus R_b)]\!]$$

To the extent there are explicit dependencies.

B.3.3 Example

In this section, we use the above defined semantics for dependent CORAS diagrams to exemplify the translation of the diagram of Fig. 16.1 on p. 297. The textual syntax of the diagram is found in Sect. A.3.3. The semantics of the diagram is:

- Threat scenario *Low hydro availability* occurs with undefined likelihood.
- Threat scenario *High import of power* occurs with undefined likelihood.
- *High load on transmission corridor* is a vulnerability.
- Unwanted incident *Grid overload* occurs with likelihood 1 : 5 *years*.
- *Power production* is a direct asset.
- *Increase power self-sufficiency rate* is a treatment scenario.
- *Low hydro availability* initiates *High import of power*.
- *High import of power* leads to *Grid overload* with undefined conditional likelihood, due to vulnerability *High load on transmission corridor*.
- *Fire* leads to *Grid overload* via gate *o* with conditional likelihood 1.0.
- *Grid overload* impacts *Power production* with consequence *critical*.
- *Increase power self-sufficiency rate* treats *High import of power*.

Assuming:

- *Lack of rain* is a non-human threat.
- *Fire* is a referring threat scenario with gate *o*.
  – *o* is an out-gate with likelihood 0.0.
- *Lack of rain* initiates *Low hydro availability* with likelihood 1.0.

To the extent there are explicit dependencies.

B.4 Legal CORAS

We start by defining the semantics of the legal uncertainty value. We let v range over *value* as before.

$$[\![\mathsf{lu}(\bot)]\!] = \text{with undefined legal uncertainty}$$

$$[\![\mathsf{lu}(v)]\!] = \text{with legal uncertainty } v$$

B.4.1 Elements

Next, we give the semantics of the legal norm element that is specific for legal CORAS. In this definition, *id* ranges over *identifier*, *ls* over *legal source* and *lu* over *legal uncertainty*. We furthermore make use of the function *id*(_) that returns the identity of an element.

$$[\![\mathsf{ln}(id, ls, lu)]\!] = \text{Legal norm } id \text{ is stated in the legal source } id(ls).$$

B.4.2 Relations

Before we define the semantics of the leads-to relation annotated with one or more legal norms, we define the semantics of sets of legal norms.

$$[\![\{\mathsf{ln}(id, ls, lu)\}]\!] = id \text{ applies } [\![lu]\!]$$

$$[\![\{\mathsf{ln}(id_1, ls_1, lu_1), \mathsf{ln}(id_2, ls_2, lu_2), \ldots, \mathsf{ln}(id_n, ls_n, lu_n)\}]\!] = id_1 \text{ applies } [\![lu_1]\!],$$
$$id_2 \text{ applies } [\![lu_2]\!], \ldots,$$
$$id_{n-1} \text{ applies } [\![lu_{n-1}]\!]$$
$$\text{and } id_n \text{ applies } [\![lu_n]\!]$$

Below, we give the semantics of the leads-to relation possibly annotated with a set of legal norms. This is a redefinition of the semantics of the leads-to relation as defined in Sect. B.1.2 and gives the semantics of legal CORAS in the setting of basic

B.4 Legal CORAS

CORAS or dependent CORAS. Notice that the semantics is as for basic CORAS in case the set of legal norms is empty. In the definitions, e ranges over *core element*, r over *risk*, cl over *conditional likelihood*, V over *vulnerabilities*, and LN over *legal norms*.

$\llbracket e \xrightarrow{\emptyset,\ \mathsf{cl}(\bot),\ \emptyset} r \rrbracket = id(e)$ leads to $id(r)$.

$\llbracket e \xrightarrow{\emptyset,\ \mathsf{cl}(\bot),\ LN} r \rrbracket = id(e)$ leads to $id(r)$ because $\llbracket LN \rrbracket$.

$\llbracket e_1 \xrightarrow{\emptyset,\ cl,\ \emptyset} e_2 \rrbracket = id(e_1)$ leads to $id(e_2)$ $\llbracket cl \rrbracket$.

$\llbracket e_1 \xrightarrow{\emptyset,\ cl,\ LN} e_2 \rrbracket = id(e_1)$ leads to $id(e_2)$ $\llbracket cl \rrbracket$ because $\llbracket LN \rrbracket$.

$\llbracket e \xrightarrow{V,\ \mathsf{cl}(\bot),\ \emptyset} r \rrbracket = id(e)$ leads to $id(r)$, due to $\llbracket V \rrbracket$.

$\llbracket e \xrightarrow{V,\ \mathsf{cl}(\bot),\ LN} r \rrbracket = id(e)$ leads to $id(r)$, due to $\llbracket V \rrbracket$ and because $\llbracket LN \rrbracket$.

$\llbracket e_1 \xrightarrow{V,\ cl,\ \emptyset} e_2 \rrbracket = id(e_1)$ leads to $id(e_2)$ $\llbracket cl \rrbracket$, due to $\llbracket V \rrbracket$.

$\llbracket e_1 \xrightarrow{V,\ cl,\ LN} e_2 \rrbracket = id(e_1)$ leads to $id(e_2)$ $\llbracket cl \rrbracket$, due to $\llbracket V \rrbracket$ and because $\llbracket LN \rrbracket$.

In the setting of high-level CORAS, we redefine the semantics of the leads-to relations as defined in Sect. B.2.2 as follows:

$\llbracket e \xrightarrow{\emptyset,\ \mathsf{cl}(\bot),\ \emptyset} rr.ig \rrbracket = id(e)$ leads to $id(rr)$ via gate $id(ig)$.

$\llbracket e \xrightarrow{\emptyset,\ \mathsf{cl}(\bot),\ LN} rr.ig \rrbracket = id(e)$ leads to $id(rr)$ via gate $id(ig)$ because $\llbracket LN \rrbracket$.

$\llbracket e \xrightarrow{\emptyset,\ cl,\ \emptyset} re.ig \rrbracket = id(e)$ leads to $id(re)$ via gate $id(ig)$ $\llbracket cl \rrbracket$.

$\llbracket e \xrightarrow{\emptyset,\ cl,\ LN} re.ig \rrbracket = id(e)$ leads to $id(re)$ via gate $id(ig)$ $\llbracket cl \rrbracket$ because $\llbracket LN \rrbracket$.

$\llbracket e \xrightarrow{V,\ \mathsf{cl}(\bot),\ \emptyset} rr.ig \rrbracket = id(e)$ leads to $id(rr)$ via gate $id(ig)$, due to $\llbracket V \rrbracket$.

$\llbracket e \xrightarrow{V,\ \mathsf{cl}(\bot),\ LN} rr.ig \rrbracket = id(e)$ leads to $id(rr)$ via gate $id(ig)$, due to $\llbracket V \rrbracket$ and because $\llbracket LN \rrbracket$.

$\llbracket e \xrightarrow{V,\ cl,\ \emptyset} re.ig \rrbracket = id(e)$ leads to $id(re)$ via gate $id(ig)$ $\llbracket cl \rrbracket$, due to $\llbracket V \rrbracket$.

$\llbracket e \xrightarrow{V,\ cl,\ LN} re.ig \rrbracket = id(e)$ leads to $id(re)$ via gate $id(ig)$ $\llbracket cl \rrbracket$, due to $\llbracket V \rrbracket$ and because $\llbracket LN \rrbracket$.

$\llbracket re.og \xrightarrow{\emptyset,\ \mathsf{cl}(\bot),\ \emptyset} r \rrbracket = id(re)$ leads to $id(r)$ via gate $id(og)$.

$\llbracket re.og \xrightarrow{\emptyset,\ \mathsf{cl}(\bot),\ LN} r \rrbracket = id(re)$ leads to $id(r)$ via gate $id(og)$ because $\llbracket LN \rrbracket$.

$[\![re.og \xrightarrow{\emptyset,\ cl,\ \emptyset} e]\!] = id(re)$ leads to $id(e)$ via gate $id(og)$ $[\![cl]\!]$.

$[\![re.og \xrightarrow{\emptyset,\ cl,\ LN} e]\!] = id(re)$ leads to $id(e)$ via gate $id(og)$ $[\![cl]\!]$ because $[\![LN]\!]$.

$[\![re.og \xrightarrow{V,\ cl(\bot),\ \emptyset} r]\!] = id(re)$ leads to $id(r)$ via gate $id(og)$, due to $[\![V]\!]$.

$[\![re.og \xrightarrow{V,\ cl(\bot),\ LN} r]\!] = id(re)$ leads to $id(r)$ via gate $id(og)$, due to $[\![V]\!]$ and because $[\![LN]\!]$.

$[\![re.og \xrightarrow{V,\ cl,\ \emptyset} e]\!] = id(re)$ leads to $id(e)$ via gate $id(og)$ $[\![cl]\!]$, due to $[\![V]\!]$.

$[\![re.og \xrightarrow{V,\ cl,\ LN} e]\!] = id(re)$ leads to $id(e)$ via gate $id(og)$ $[\![cl]\!]$, due to $[\![V]\!]$ and because $[\![LN]\!]$.

$[\![re.og \xrightarrow{\emptyset,\ cl(\bot),\ \emptyset} rr.ig]\!] = id(re)$ leads to $id(rr)$ via gates $id(og)$ and $id(ig)$.

$[\![re.og \xrightarrow{\emptyset,\ cl(\bot),\ LN} rr.ig]\!] = id(re)$ leads to $id(rr)$ via gates $id(og)$ and $id(ig)$ because $[\![LN]\!]$.

$[\![re_1.og \xrightarrow{\emptyset,\ cl,\ \emptyset} re_2.ig]\!] = id(re_1)$ leads to $id(re_2)$ via gates $id(og)$ and $id(ig)$ $[\![cl]\!]$.

$[\![re_1.og \xrightarrow{\emptyset,\ cl,\ LN} re_2.ig]\!] = id(re_1)$ leads to $id(re_2)$ via gates $id(og)$ and $id(ig)$ $[\![cl]\!]$ because $[\![LN]\!]$.

$[\![re.og \xrightarrow{V,\ cl(\bot),\ \emptyset} rr.ig]\!] = id(re)$ leads to $id(rr)$ via gates $id(og)$ and $id(ig)$, due to $[\![V]\!]$.

$[\![re.og \xrightarrow{V,\ cl(\bot),\ LN} rr.ig]\!] = id(re)$ leads to $id(rr)$ via gates $id(og)$ and $id(ig)$, due to $[\![V]\!]$ and because $[\![LN]\!]$.

$[\![re_1.og \xrightarrow{V,\ cl,\ \emptyset} re_2.ig]\!] = id(re_1)$ leads to $id(re_2)$ via gates $id(og)$ and $id(ig)$ $[\![cl]\!]$, due to $[\![V]\!]$.

$[\![re_1.og \xrightarrow{V,\ cl,\ LN} re_2.ig]\!] = id(re_1)$ leads to $id(re_2)$ via gates $id(og)$ and $id(ig)$ $[\![cl]\!]$, due to $[\![V]\!]$ and because $[\![LN]\!]$.

$[\![ig \xrightarrow{\emptyset,\ cl(\bot),\ \emptyset} e]\!] = id(e)$ is initiated via gate $id(ig)$.

$[\![ig \xrightarrow{\emptyset,\ cl(\bot),\ LN} e]\!] = id(e)$ is initiated via gate $id(ig)$ because $[\![LN]\!]$.

$[\![ig \xrightarrow{V,\ cl(\bot),\ \emptyset} e]\!] = id(e)$ is initiated via gate $id(ig)$, due to $[\![V]\!]$.

B.4 Legal CORAS

$[\![ig \xrightarrow{V,\ \text{cl}(\bot),\ LN} e]\!] = id(e)$ is initiated via gate $id(ig)$, due to $[\![V]\!]$ and because $[\![LN]\!]$.

$[\![ig_1 \xrightarrow{\emptyset,\ \text{cl}(\bot),\ \emptyset} re.ig_2]\!] = id(re)$ is initiated via gates $id(ig_1)$ and $id(ig_2)$.

$[\![ig_1 \xrightarrow{\emptyset,\ \text{cl}(\bot),\ LN} re.ig_2]\!] = id(re)$ is initiated via gates $id(ig_1)$ and $id(ig_2)$ because $[\![LN]\!]$.

$[\![ig_1 \xrightarrow{V,\ \text{cl}(\bot),\ \emptyset} re.ig_2]\!] = id(re)$ is initiated via gates $id(ig_1)$ and $id(g_2)$, due to $[\![V]\!]$.

$[\![ig_1 \xrightarrow{V,\ \text{cl}(\bot),\ LN} re.ig_2]\!] = id(re)$ is initiated via gates $id(ig_1)$ and $id(g_2)$, due to $[\![V]\!]$ and because $[\![LN]\!]$.

$[\![e \xrightarrow{\emptyset,\ \text{cl}(\bot),\ \emptyset} og]\!] = id(e)$ has an effect via gate $id(og)$.

$[\![e \xrightarrow{\emptyset,\ \text{cl}(\bot),\ LN} og]\!] = id(e)$ has an effect via gate $id(og)$ because $[\![LN]\!]$.

$[\![e \xrightarrow{V,\ \text{cl}(\bot),\ \emptyset} og]\!] = id(e)$ has an effect via gate $id(og)$, due to $[\![V]\!]$.

$[\![e \xrightarrow{V,\ \text{cl}(\bot),\ LN} og]\!] = id(e)$ has an effect via gate $id(og)$, due to $[\![V]\!]$ and because $[\![LN]\!]$.

$[\![re.og_1 \xrightarrow{\emptyset,\ \text{cl}(\bot),\ \emptyset} og_2]\!] = id(re)$ has an effect via gates $id(og_1)$ and $id(og_2)$.

$[\![re.og_1 \xrightarrow{\emptyset,\ \text{cl}(\bot),\ LN} og_2]\!] = id(re)$ has an effect via gates $id(og_1)$ and $id(og_2)$ because $[\![LN]\!]$.

$[\![re.og_1 \xrightarrow{V,\ \text{cl}(\bot),\ \emptyset} og_2]\!] = id(re)$ has an effect via gates $id(og_1)$ and $id(og_2)$, due to $[\![V]\!]$.

$[\![re.og_1 \xrightarrow{V,\ \text{cl}(\bot),\ LN} og_2]\!] = id(re)$ has an effect via gates $id(og_1)$ and $id(og_2)$, due to $[\![V]\!]$ and because $[\![LN]\!]$.

B.4.3 Example

In this section, we use the above defined semantics for legal CORAS diagrams to exemplify the translation of the diagram in Fig. 17.5 on p. 325. The textual syntax of the diagram is found in Sect. A.4.3. The semantics of the diagram is:

- *Employee* is an accidental human threat.

- *Lack of routines for updating personal data* is a vulnerability.
- Threat scenario *Personal data is not held accurate and up to date* occurs with likelihood *likely*.
- Legal norm *Company may be held liable for irregular processing of personal data* is stated in the legal source *Data Protection Act*.
- Unwanted incident *Company is prosecuted for irregular processing of personal data* occurs with likelihood *possible*.
- *Reputation* is a direct asset.
- *Employee* exploits vulnerability *Lack of routines for updating personal data* to initiate *Personal data is not held accurate and up to date*.
- *Personal data is not held accurate and up to date* leads to *Company is prosecuted for irregular processing of personal data* with conditional likelihood 0.3 because *Company may be held liable for irregular processing of personal data* applies with legal uncertainty *likely*.
- *Company is prosecuted for irregular processing of personal data* impacts *Reputation*.

Appendix C
The CORAS Guidelines

In this appendix, we provide the guidelines for the CORAS method. The detailed guidelines explain how to fulfil the goals of the eight steps and their tasks in general, and how to fulfil these goals by making efficient use of the CORAS language, in particular.

In the following sections, we provide the guidelines for each of the eight steps of the CORAS method.

The eight steps of the CORAS method

---

1. **Preparations for the analysis**

    The customer briefly informs the analysis team about the target it wishes to have analysed, and the analysis team prepares for the analysis

2. **Customer presentation of the target**

    The customer presents the system or organisation it wishes to have analysed; the focus and scope of the analysis is identified and an analysis plan is set up

3. **Refining the target description using asset diagrams**

    The analysis team presents its understanding of the target of analysis; the assets are identified, as well as the most important related threats and vulnerabilities

4. **Approval of the target description**

    The analysis team presents the documentation of the target of analysis for finalisation and approval by the customer; values are assigned to the identified assets, and the risk evaluation criteria are established

5. **Risk identification using threat diagrams**

    Risks are identified through a structured brainstorming

6. **Risk estimation using threat diagrams**

    The likelihoods and consequences for the identified risks are estimated

7. **Risk evaluation using risk diagrams**

    The risks are evaluated against the risk evaluation criteria

8. **Risk treatment using treatment diagrams**

    Treatments for the mitigation of unacceptable risks are identified and evaluated

---

C.1 Step 1: Preparations for the Analysis

Preparation for the analysis

- **Objective:** Gather basic information about the customer, the purpose and domain of the analysis, and the size of the analysis; ensure that the customer and other involved parties are prepared for their roles and responsibilities; appoint a contact person from the customer; agree on a tentative time schedule
- **How conducted:** Held as an interaction between the analysis team and customer representatives; the interaction is preferably held as a face-to-face meeting, but may be conducted by, for example, phone or other forms of interaction; the analysis team may subsequently make preparations by gathering data about the target of analysis, such as information and statistics about typical risks
- **Input documentation:** None
- **Output documentation:** Any relevant information about the target provided by the customer; information that is gathered by the analysis team about the domain and the target that is addressed; a tentative time schedule

Subtasks:	People that should participate:
• No subtasks in this step	• Analysis leader (required)
	• Analysis secretary (optional)
	• Representatives of the customer

Modelling guideline:
- Modelling is not part of this step

C.2 Step 2: Customer Presentation of the Target

Customer presentation of the target

- **Objective:** Achieve an initial understanding of what the parties wish to have analysed and what they are most concerned about; decide on the focus, scope and assumptions of the analysis; establish a detailed plan for the analysis
- **How conducted:** The analysis step is conducted as an interaction between the analysis team and representatives of the parties, preferably as a face-to-face meeting; the analysis team introduces the CORAS method before representatives of the parties present the goal and target of the analysis; through plenary discussions, the focus, scope and assumptions are determined, and the plan for the analysis is decided
- **Input documentation:** Any relevant information about the target provided by the customer or the parties in advance; information that is gathered by the analysis team about the type of system that is addressed
- **Output documentation:** Informal drawings and sketches describing the target, the focus and the scope of the analysis; a plan for the analysis with meeting dates, list of participants and dates for delivery of documentation and analysis report

C.2 Step 2: Customer Presentation of the Target

Customer presentation of the target

Subtasks:	**People that should participate:**
a. Presentation of the CORAS terminology and method	• Analysis leader (required)
b. Presentation of the goals and target of the analysis	• Analysis secretary (required)
c. Setting the focus and scope of the analysis	• Representatives of the customer:
d. Determining the meeting plan	– Decision makers (required)
	– Technical expertise (optional)
	– Users (optional)

Modelling guideline:
1. At this early stage of the analysis, it can be useful to describe the target with informal drawings, pictures or sketches on a blackboard
2. The presentation can later be supplemented with more formal modelling techniques such as the UML or data flow diagrams

C.2.1 Step 2a: Presentation of the CORAS Terminology and Method

Presentation of the CORAS terminology and method

- **Objective:** The target team understands the CORAS terms and their relationships, and is familiarised with the steps of the CORAS method
- **How conducted:** Presentation given by the analysis team; the precise definition of each of the CORAS terms is presented and the relationships between the terms are explained; the CORAS symbols illustrating the terms are displayed and a basic CORAS diagram is presented and explained; an overview of the CORAS method is displayed and each step of the method is briefly presented; the analysis team explains to the target team what can be expected from the analysis, and emphasises the responsibilities of the target team with respect to providing the necessary information and documentation, as well as gathering personnel with suitable background to participate in the various steps of the analysis
- **Input documentation:** None
- **Output documentation:** None

C.2.2 Step 2b: Presentation of the Goals and Target of the Analysis

Presentation of the goals and target of the analysis

- **Objective:** The analysis team understands what the parties wish to have analysed and what the goals of the analysis are
- **How conducted:** Presentation given by representative(s) of the target team explaining what the parties wishes to have analysed and what kind of incidents they are most worried about occurring; the target presentation typically includes business goals, work processes, users and roles, contracts and policies, hardware and software specifications and network layout

Presentation of the goals and target of the analysis

- **Input documentation:** Information about the target that may have been provided by the parties in advance, as well as information that is gathered by the analysis team about the kind of target that is addressed
- **Output documentation:** Informal drawings and sketches describing the target, as well as other relevant target documentation provided to the analysis team

C.2.3 Step 2c: Setting the Focus and Scope of the Analysis

Setting the focus and scope of the analysis

- **Objective:** The customer and the analysis team reach an agreement and common understanding of what is to be the focus, scope and assumptions of the analysis; a clear characterisation of the focus and scope facilitates the optimisation of the use of the available time and resources for the analysis
- **How conducted:** Plenary discussion guided by the analysis leader with the aim of determining which aspects or parts of the target that is of chief concern; the discussion must also determine what should be held outside of the analysis, including parts of the system or organisation under consideration, aspects of security such as integrity or confidentiality of information, or particular threats and incidents; the analysis team should be careful to clarify uncertainties or open questions in order to avoid misunderstandings later on
- **Input documentation:** The drawings and sketches, as well as other target information, gathered during the customer presentation of the target
- **Output documentation:** Notes and informal sketches produced on-the-fly during the discussion

C.2.4 Step 2d: Determining the Meeting Plan

Determining the meeting plan

- **Objective:** The target team and the analysis team agree on meeting dates, meeting participants and date of delivery of the analysis report
- **How conducted:** Plenary discussion lead by the analysis team; the plan must schedule sufficient time between the analysis steps for the analysis team to process and analyse the result from the previous step and to prepare for the next step; the analysis team must also ensure that personnel with adequate background are present at the various meetings and workshops
- **Input documentation:** The tentative time schedule that were put up during Step 1
- **Output documentation:** List of analysis participants and their roles; meeting plan with dates, tasks and participants; list of deadlines for delivery of documentation, including the final analysis report

C.3 Step 3: Refining the Target Description Using Asset Diagrams

Refining the target description using asset diagrams

- **Objective:** Ensure a common and more precise understanding of the target of analysis, including its scope, focus and main assets
- **How conducted:** The analysts present their understanding of the target of analysis; asset identification and high-level risk analysis are conducted as interactions between analysts and representatives from the customer, preferable in a face-to-face meeting
- **Input documentation:** Target models in a suitable formal or semi-formal language prepared by the analysts based on information gathered during the previous step and other target information provided to the analysts
- **Output documentation:** Updated and corrected models of the target of analysis, including CORAS asset diagrams; high-level risk analysis table documenting the results of the high-level analysis

Subtasks:	People that should participate:
a. Presentation of the target by the analysis team	• Analysis leader (required)
b. Asset identification	• Analysis secretary (required)
c. High-level analysis	• Representatives of the customer:
	– Decision makers (required)
	– Technical expertise (required)
	– Users (optional)

Modelling guideline:
- Asset diagrams:
  1. Draw the diagram frame and place the name of the party in the compartment in the upper left corner
  2. Place the assets within the frame
  3. Indicate with arrows which assets may be harmed via other assets
  4. Assets that are found to be indirect are modelled using the white indirect asset symbol; indirect assets are with respect to the target of analysis harmed only as a consequence of another asset being harmed first
  5. Repeat the above steps for each party of the analysis
- Target descriptions:
  1. Use a formal or semi-formal notation such as the UML, but ensure that the notation is explained thoroughly so that the target team understands it; if the customer has specific preferences or an in-house notation, such notations should then be used
  2. Create models of both the static and the dynamic features of the target; static features may be hardware configurations, network design, and so forth, while dynamic features may be work processes, information flow, and so forth
  3. For the static part of the description, UML class diagrams and UML collaboration diagrams (or similar notations) are suitable
  4. For the dynamic parts of the description UML activity diagrams and UML sequence diagrams (or similar notations) are suitable

C.3.1 Step 3a: Presentation of the Target by the Analysis Team

Presentation of the target by the analysis team

- **Objective:** Ensure that the analysis team has correctly understood the target and the objectives of the analysis; establish a common understanding of the target among the analysis team and the target team; refine the target description and bring it closer to finalisation
- **How conducted:** Presentation given by the analysts of both static and dynamic features of the target; static features may be hardware configurations, network design and organisational structure, whereas dynamic features may be work processes and information flow; corrections and comments from the target team derived through a plenary discussion are implemented
- **Input documentation:** Models in a suitable formal or semi-formal language such as the UML prepared by the analysts; the models are based on the information gathered during the previous steps and on target documentation provided to the analysts; UML class diagrams and UML collaboration diagrams, for example, are suited to model static system features, whereas UML activity diagrams and UML sequence diagrams, for example, are suited to capture dynamic features
- **Output documentation:** Updated and corrected models describing the target of analysis

C.3.2 Step 3b: Asset Identification

Asset identification

- **Objective:** Identify precisely with respect to which assets the analysis is to be conducted; increase the understanding of the appropriate scope and focus of the analysis
- **How conducted:** Plenary discussion based on suggestions prepared by the analysts
- **Input documentation:** CORAS asset diagram prepared by the analysts based on information gathered during the previous step
- **Output documentation:** Updated and corrected CORAS asset diagram

C.3.3 Step 3c: High-level Analysis

High-level analysis

- **Objective:** Identify what the parties are most worried about happening, thus increasing the understanding of the correct and appropriate focus and scope for the analysis; establish an initial risk picture at enterprise level and from a decision maker perspective
- **How conducted:** Threats, vulnerabilities, threat scenarios and unwanted incidents are identified through plenary discussion or brainstorming lead by the analysis team
- **Input documentation:** Asset diagram and target description
- **Output documentation:** High-level risk analysis table

C.4 Step 4: Approval of the Target Description

Approval of the target description

- **Objective:** Ensure that the background documentation for the rest of the analysis, including the target, focus and scope, is correct and complete as seen by the customer; decide a ranking of the assets according to importance in order to prioritise the time and resources during the analysis; establish scales for estimating risk and criteria for evaluating risks
- **How conducted:** The step is conducted as a structured walk-through of the target description, preferably at a face-to-face meeting; any errors or omissions in the documentation are identified and corrected; based on plenary discussions and guidance of the analysts, the target team decides on a ranking of the assets, a risk scale and the risk evaluation criteria
- **Input documentation:** Full target description prepared by the analysts based on information and models derived through the previous steps
- **Output documentation:** Correct and complete target description, as well as risk evaluation criteria with definitions of all scales; all the documentation must be approved by the parties before the next step of the analysis

Subtasks:	People that should participate:
a. Approval of the target description	• Analysis leader (required)
b. Ranking of assets	• Analysis secretary (required)
c. Setting the consequence scales	• Representatives of the customer:
d. Setting the likelihood scale	– Decision makers (required)
e. Defining the risk function	– Technical expertise (required)
f. Deciding the risk evaluation criteria	– Users (optional)

Modelling guideline:

- Modelling is not part of this step

C.4.1 Step 4a: Approval of the Target Description

Approval of the target description

- **Objective:** Finalise the documentation and characterisation of the target of analysis, including the assets, the focus and scope of the analysis as well as the assumptions made for the analysis
- **How conducted:** The analysis team gives a plenary walk-though of the documentation; errors or omissions that are identified are recorded for corrections; the task is terminated once the full documentation has been approved by the customer
- **Input documentation:** The target description consisting of models and other documentation of the target of analysis and the assets prepared by the analysis team on the basis of the previous steps
- **Output documentation:** Correct and complete target description; the models and other documentation are cleaned up and properly prepared by the analysis team and serve as the finalised target description to be used during the remaining analysis

C.4.2 Step 4b: Ranking of Assets

Ranking of assets

- **Objective:** Facilitate identification of the most important assets to support prioritising of risks and thereby also prioritising the time and resources for risk analysis and treatment identification
- **How conducted:** Plenary discussion lead by the analysis team
- **Input documentation:** CORAS asset diagrams from previous step
- **Output documentation:** A list of assets ordered by importance; each asset can be given a value and/or a number from a scale defined for ranking assets

C.4.3 Step 4c: Setting the Consequence Scales

Setting the consequence scales

- **Objective:** Establish consequence scales for describing the harm of identified unwanted incidents on direct assets; to be used when estimating risk values
- **How conducted:** Determined by the target team through plenary discussion possibly based on suggestions provided by the analysis team
- **Input documentation:** Consequence scales suggested by the analysts
- **Output documentation:** A consequence scale defined for each direct asset

C.4.4 Step 4d: Setting the Likelihood Scale

Setting the likelihood scale

- **Objective:** Establish a scale for describing the likelihood of unwanted incidents to occur; the scale is defined in terms of frequencies or probabilities; to be used when estimating risk values
- **How conducted:** Determined by the target team through plenary discussion possibly based on suggestions provided by the analysis team
- **Input documentation:** A likelihood scale suggested by the analysis team
- **Output documentation:** A likelihood scale

C.4.5 Step 4e: Defining the Risk Function

Defining the risk function

- **Objective:** Determine the level of risk an unwanted incident represents as a function of the likelihood and consequence of its occurrence

Defining the risk function

- **How conducted:** The analysis team will typically draw a matrix for each of the consequence scales with the likelihood scale horizontally and the consequence scale vertically; based on suggestions provided by the analysis team and on plenary discussions, the target team decides the risk value for each entry in the matrix; a risk value scale is set with an appropriate granularity for the analysis; if the risk matrix representation is not appropriate, an alternative risk function must be defined
- **Input documentation:** The consequence scales and the likelihood scale set during the previous tasks
- **Output documentation:** The decided risk functions

C.4.6 Step 4f: Deciding the Risk Evaluation Criteria

Deciding the risk evaluation criteria

- **Objective:** Determine what level of risk the customer is willing to accept; will later be used to determine which risks that must be evaluated for possible treatment
- **How conducted:** The customer determines for which combinations of consequence and likelihood the resulting risk is acceptable and for which combinations the risk must be evaluated for possible treatment
- **Input documentation:** The risk function defined in the previous task
- **Output documentation:** Risk evaluation criteria for each asset

C.5 Step 5: Risk Identification Using Threat Diagrams

Risk identification using threat diagrams

- **Objective:** Identify the risks that must be managed; determine where, when, why and how they may occur
- **How conducted:** Conducted as a brainstorming session involving a target team consisting of people of different backgrounds with different insight into the problem at hand; using the assets and the high-level analysis as starting point the risks are gradually identified by identifying unwanted incidents, threats, threat scenarios and vulnerabilities; the risk identification is conducted with respect to the target description, and the results are documented on-the-fly by drawing CORAS threat diagrams as the information is gathered
- **Input documentation:** Target description including CORAS asset diagrams; high-level risk analysis table
- **Output documentation:** CORAS threat diagrams and notes made by analysis secretary during the risk identification; the diagrams are cleaned up and completed by the analysis team offline before the next step of the analysis

Risk identification using threat diagrams

Subtasks:	People that should participate:
a. Categorising threat diagrams	• Analysis leader (required)
b. Identification of threats and unwanted incidents	• Analysis secretary (required)
c. Identification of threat scenarios	• Representatives of the customer:
d. Identification of vulnerabilities	– Decision makers (optional[*])
	– Technical expertise (required)
	– Users (required)

Modelling guideline:
- Threat diagrams:
  1. Decide how to structure the threat diagrams; a diagram may either focus on one asset at the time, a particular aspect of the target, or a specific kind of threat; for instance, deliberate sabotage in one diagram, mistakes in another, environmental in a third, and so forth; this makes it easier to generalise over the risks, for example "these risks all harm asset X", "these risks are caused by human errors", or "these risks are related to the network"
  2. Assets are listed to the very right
  3. Relevant threats are placed to the very left
  4. Unwanted incidents are placed in between with relations from the threats that may cause them, and relations to the assets they impact
  5. Assets that are not harmed by any incidents can be removed from the diagram
  6. Add threat scenarios between the threats and the unwanted incidents in the same order as they occur in real time (in other words in a logical consequence)
  7. Insert the vulnerabilities before the threat scenario or unwanted incident to which they lead; for example, a vulnerability called "poor backup solution" is typically placed before the threat scenario "the backup solution fails to run the application database correctly"

[*]This workshop usually has a technical focus; the competence of decision makers is more relevant in the next step

C.5.1 Step 5a: Categorising Threat Diagrams

Categorising threat diagrams

- **Objective:** Facilitate structuring of the risk identification brainstorming and a categorisation of the identified risks
- **How conducted:** Decision of analysis team based on the description of the target of analysis and the high-level analysis; threat diagrams may be categorised according to for example assets, threats, work processes, organisational domains, and so forth
- **Input documentation:** Target description and high-level risk analysis table
- **Output documentation:** A set of categories used by the analysis team to guide and structure the risk identification brainstorming session; for each category, the analysis team selects relevant target models such as UML class, sequence and activity diagrams describing features and aspects of the target of analysis at an appropriate level of detail for the analysis

C.5.2 Step 5b: *Identification of Threats and Unwanted Incidents*

Identification of threats and unwanted incidents

- **Objective:** Identify how assets may be harmed and the threats that cause the harm; facilitate the subsequent task of identifying threat scenarios
- **How conducted:** The analysis secretary presents on a whiteboard or some visual display a set of assets adjusted to the very right; the analysis leader has the target team to discuss and suggest how these may be harmed and which threats causes the harm by referring to the models of the target of analysis; the suggestions are added on-the-fly by the analysis secretary as unwanted incidents and threats, respectively; the threats are adjusted to the very left and the unwanted incidents are placed next to the assets; relations are added between threats and unwanted incidents and between unwanted incidents and assets; the analysis leader constantly probes the target team and, if necessary, stimulates the discussion by giving suggestions and by leading the attention to relevant parts of the target models; the activity is repeated for each category of threat diagram; assets for which unwanted incidents are not identified are removed from the diagrams
- **Input documentation:** Target description, high-level risk analysis table and sets of initial, preliminary threat diagram fragments structured according to the decided categorisation
- **Output documentation:** CORAS threat diagrams depicting which threats cause which unwanted incidents, and which unwanted incidents harm which assets

C.5.3 Step 5c: *Identification of Threat Scenarios*

Identification of threat scenarios

- **Objective:** Explain how threats may cause unwanted incidents by identifying the threat scenarios that are initiated by the threats and that may lead to unwanted incidents
- **How conducted:** The analysis leader has the target team to discuss and explain how the identified threats can initiate scenarios that lead to the identified unwanted incidents; the analysis secretary documents the suggestions on-the-fly by adding threat scenarios ordered by time from left to right; directed relations are added from threats to threat scenarios, from threat scenarios to threat scenarios and from threat scenarios to unwanted incidents; relations may also be added from unwanted incidents to unwanted incidents and to threat scenarios if relevant; the analysis leader constantly probes the target team and stimulates the discussion by referring to relevant parts of the target models and, if necessary, giving suggestions
- **Input documentation:** Target description, high-level risk analysis table and the CORAS threat diagrams from Step 5b showing threats, unwanted incidents and assets
- **Output documentation:** The CORAS threat diagrams from Step 5b extended with the relevant threat scenarios and relations

C.5.4 Step 5d: Identification of Vulnerabilities

Identification of vulnerabilities

- **Objective:** Explain and identify the vulnerabilities, which are the flaws or weaknesses of the target of analysis that opens for unwanted incidents to occur; this completes the risk identification step
- **How conducted:** The analysis leader has the target team to discuss and explain why threats can initiate threat scenarios, why one threat scenario can lead to another and why threat scenarios can lead to an unwanted incident by identifying the relevant system vulnerabilities; the analysis secretary documents the suggestions on-the-fly by inserting vulnerabilities on the relations leading to the unwanted incidents or to the threat scenarios; if necessary, the analysis leader again probes the target team and stimulates the discussion by giving suggestions
- **Input documentation:** Target description, high-level risk analysis table and the CORAS threat diagram from Step 5c showing threats, threat scenarios, unwanted incidents and assets
- **Output documentation:** The CORAS threat diagrams documenting the results of the risk identification brainstorming session; the diagrams are cleaned up by the analysis team offline before the next step of the analysis

C.6 Step 6: Risk Estimation Using Threat Diagrams

Risk estimation using threat diagrams

- **Objective:** Determine the risk level of the identified risks
- **How conducted:** Conducted as a workshop involving a target team representing various backgrounds, including technical expertise and decision making; the threat scenarios and the unwanted incidents are annotated with likelihoods based on input from the target team; each relation between an unwanted incident and an asset is annotated with the consequence describing the impact of the incident on the asset; risk levels are documented using CORAS risk diagrams modelling each of the identified risks and their risk values as calculated from the estimated likelihoods and consequences
- **Input documentation:** CORAS threat diagrams from the risk identification step; the likelihood scale, the consequence scales and the risk functions defined during Step 4
- **Output documentation:** CORAS threat diagrams completed with a likelihood assigned to each unwanted incident and a consequence assigned to each relation between an unwanted incident and an asset; CORAS risk diagrams modelling the risks and their estimated risk levels

Subtasks:	People that should participate:
a. Likelihood estimation	• Analysis leader (required)
b. Consequence estimation	• Analysis secretary (required)
c. Risk estimation	• Representatives of the customer:
	– Decision makers (required)
	– Technical expertise regarding the target (required)
	– Users (required)

C.6 Step 6: Risk Estimation Using Threat Diagrams

Risk estimation using threat diagrams

Modelling guideline:
- Risk estimation on threat diagrams:
    1. Add likelihood estimates to the threat scenarios if this kind of information is available
    2. Add likelihood estimates to each unwanted incident, either directly or based on the threat scenarios or unwanted incidents that lead up to it
    3. Annotate each impacts relation from an unwanted incident to an asset with a consequence from the respective asset's consequence scale
- Risk diagrams:
    1. Use the threat diagrams and replace each unwanted incident with one risk for each impacts relation, where each risk shows the risk description, the risk id, and the risk value
    2. Remove threat scenarios and vulnerabilities, but keep the relations between the threats and the risks

C.6.1 Step 6a: Likelihood Estimation

Likelihood estimation

- **Objective:** Estimate the likelihood of the unwanted incidents documented under the risk identification step to occur; to be used when estimating risk levels
- **How conducted:** The CORAS threat diagrams resulting from the risk identification step are taken as starting point; based on discussions between the target team members guided by the analysis leader, each of the identified unwanted incidents is assigned a likelihood estimate; the likelihood scale used is the one that were set during Step 4 of the analysis; if likelihoods cannot be assigned directly to the unwanted incidents, these are estimated for the threat scenarios leading up to them, providing data for calculating the resulting likelihood for the unwanted incident; likelihoods can also be estimated for the initiates relations and conditional likelihoods for the leads-to relations for further input data; support for the estimation should be gathered by the analysis team in advance, for example by consulting domain experts, historical data and statistical data; when a likelihood estimate is arrived at for a threat scenario, an unwanted incident or a relation, the analysis secretary annotates the threat diagram accordingly
- **Input documentation:** The CORAS threat diagrams resulting from the risk identification step; the likelihood scale defined during Step 4
- **Output documentation:** CORAS threat diagrams annotated with a likelihood estimate for each of the unwanted incidents and any number of the identified threat scenarios, initiates relations and leads-to relations

C.6.2 Step 6b: Consequence Estimation

Consequence estimation

- **Objective:** Estimate the consequence of the unwanted incidents on each of the assets they affect; to be used when estimating risk levels

Consequence estimation

- **How conducted:** The CORAS threat diagrams annotated with the likelihood estimates resulting from the previous subtask are taken as starting point; based on discussions between the target team members guided by the analysis leader, each relation between an unwanted incident and an asset is assigned a consequence estimate; the consequence scale used is the one that were set for the given asset during Step 4 of the analysis; when a consequence has been decided upon, the analysis secretary annotates the threat diagram accordingly
- **Input documentation:** The CORAS threat diagrams annotated with likelihood estimates resulting from the previous subtask; the consequence scales defined during Step 4
- **Output documentation:** CORAS threat diagrams annotated with a likelihood estimate for each unwanted incident and a consequence estimate for each relation between an unwanted incident and an asset

C.6.3 Step 6c: Risk Estimation

Risk estimation

- **Objective:** Calculate the risk level for each of the identified risks
- **How conducted:** For each relation between an unwanted incident and an asset, the risk level is calculated based on the estimated likelihood and consequence; the risks and their risk levels are documented using CORAS risk diagrams
- **Input documentation:** The CORAS threat diagrams with likelihood and consequence estimates resulting from the previous subtask; the risk functions defined during Step 4
- **Output documentation:** CORAS risk diagrams annotated with the estimated risk level of each of the identified risks

C.7 Step 7: Risk Evaluation Using Risk Diagrams

Risk evaluation using risk diagrams

- **Objective:** Decide which of the identified risks are acceptable and which of the risks that must be further evaluated for possible treatment
- **How conducted:** Likelihood and consequence estimates are confirmed or adjusted; adjustments of the risk evaluation criteria are made if needed; risks that each contributes to the same overall risk are grouped and their accumulated risk value calculated; using the threat diagrams with respect to the direct assets combined with the asset diagrams, the risks with respect to the indirect assets are identified and estimated; each of the identified risks is evaluated by comparing against the risk evaluation criteria
- **Input documentation:** CORAS threat diagrams with estimated likelihoods and consequences; CORAS risk diagrams; CORAS asset diagrams; the risk evaluation criteria
- **Output documentation:** CORAS threat diagrams for both direct and indirect assets; CORAS risk diagrams with evaluation results

C.7 Step 7: Risk Evaluation Using Risk Diagrams

Risk evaluation using risk diagrams

Subtasks:	People that should participate:
a. Confirming the risk estimates	• Analysis leader (required)
b. Confirming the risk evaluation criteria	• Analysis secretary (required)
c. Providing a risk overview	• Representatives of the customer:
d. Accumulating risks	– Decision makers (required)
e. Estimating risks with respect to indirect assets	– Technical expertise regarding the target (optional)
f. Evaluating the risks	– Users (optional)

Modelling guideline:
- Risk diagrams with evaluation results:
  1. For the direct assets, use the risk diagrams from Step 6 and replace the risk value with *acceptable* or *unacceptable*, depending on the result of the risk evaluation
  2. For the indirect assets, extend the risk diagrams for the direct assets by adding the implied risks for the indirect assets
  3. Give each indirect risk a unique name or id but keep the risk description; annotate the risks with *acceptable or unacceptable*, depending on the result of the risk evaluation

C.7.1 Step 7a: Confirming the Risk Estimates

Confirming the risk estimates

- **Objective:** Ensure that the estimates conducted during Step 6 are appropriate
- **How conducted:** The analysis team cleans up and quality checks the documentation resulting from the risk estimation step; before the risk evaluation step is conducted, the analysis team sends the documentation to the customer for internal review and quality check; any adjustments are implemented by correcting the threat diagrams and risk diagrams from the previous analysis step
- **Input documentation:** The CORAS threat diagrams with likelihood and consequence estimates from Step 6
- **Output documentation:** Finalised and confirmed CORAS threat diagrams and risk diagrams

C.7.2 Step 7b: Confirming the Risk Evaluation Criteria

Confirming the risk evaluation criteria

- **Objective:** Ensure that the risk evaluation criteria defined during Step 4 are appropriate
- **How conducted:** The target team, and especially the decision makers, decides whether the risk evaluation criteria are appropriate or whether they should be adjusted; new knowledge and insight that are gathered during the analysis process may result in adjustments

Confirming the risk evaluation criteria

- **Input documentation:** Risk evaluation criteria from Step 4 of the analysis
- **Output documentation:** Finalised and confirmed risk evaluation criteria

C.7.3 Step 7c: Providing a Risk Overview

Providing a risk overview

- **Objective:** Establish an overview of the general risk picture
- **How conducted:** The risk overview is presented as a walk-through of the CORAS risk diagrams
- **Input documentation:** The finalised CORAS threat diagrams with confirmed likelihoods and consequences; the finalised and confirmed CORAS risk diagrams
- **Output documentation:** None

C.7.4 Step 7d: Accumulating Risks

Accumulating risks

- **Objective:** Calculate and document the accumulated value of several risks that contribute to the same overall risk
- **How conducted:** The individual risks that contribute to the same general risk are modelled in a combined CORAS risk diagram; the accumulated likelihoods and consequences are calculated on the basis of the respective CORAS threat diagrams
- **Input documentation:** The finalised CORAS threat diagrams with confirmed likelihoods and consequences; the finalised CORAS risk diagrams
- **Output documentation:** CORAS risk diagrams documenting the accumulated risks

C.7.5 Step 7e: Estimating Risks with Respect to Indirect Assets

Estimating risks with respect to indirect assets

- **Objective:** Document and estimate risks with respect to the indirect assets
- **How conducted:** Risks towards the indirect assets are identified by considering the identified risks towards the direct assets; using the asset diagrams, determine the extent to which the indirect assets are harmed via harm to the direct assets and estimate the resulting risk levels; conducting the task requires establishing consequence scales, risk functions and risk evaluation criteria for each of the indirect assets

Estimating risks with respect to indirect assets

- **Input documentation:** The finalised CORAS threat diagrams for the direct assets; the CORAS asset diagrams
- **Output documentation:** Consequence scales, risk functions and risk evaluation criteria for indirect assets; CORAS threat diagrams and CORAS risk diagrams documenting risks towards indirect assets

C.7.6 Step 7f: Evaluating the Risks

Evaluating the risks

- **Objective:** Decide which of the identified risks are acceptable and which must be evaluated for possible treatment
- **How conducted:** Each of the identified risks with estimated risk value is compared against the risk evaluation criteria; risks that are not acceptable are considered for treatment in the next step of the analysis
- **Input documentation:** CORAS risk diagrams; risk evaluation criteria
- **Output documentation:** CORAS risk diagrams annotated with risk evaluation result

C.8 Step 8: Risk Treatment Using Treatment Diagrams

Risk treatment using treatment diagrams

- **Objective:** Identify cost effective treatments for the unacceptable risks
- **How conducted:** Conducted as a brainstorming session involving a target team consisting of personnel of various backgrounds; treatments are identified by a walk-through of the threat diagrams that document the unacceptable risks and their causes
- **Input documentation:** CORAS risk diagrams and CORAS threat diagrams documenting the unacceptable risks; preliminary treatment diagrams prepared by the analysis team
- **Output documentation:** CORAS treatment diagrams documenting the identified treatments for the risks with respect to direct and indirect assets

Subtasks:	People that should participate:
a. Grouping of risks	• Analysis leader (required)
b. Treatment identification	• Analysis secretary (required)
c. Treatment evaluation	• Representatives of the customer:
	– Decision makers (required)
	– Technical expertise (required)
	– Users (required)

Risk treatment using treatment diagrams

Modelling guideline:
- Treatment diagrams:
  1. Use the threat diagrams as a basis and replace each unwanted incident with the risks it represents; this yields one risk for each relation between an unwanted incident and an asset; acceptable risks may be removed along with the unwanted incidents, vulnerabilities, assets, threats and threat scenarios that do not constitute unacceptable risks
  2. Annotate the diagram with treatments, pointing to where they will be applied
  3. If several treatments point towards the same risks (a many-to-many relation) we recommend using "junction points" to avoid multiple, crossing lines
- Treatment overview diagrams:
  1. Use the risk diagrams as a basis; remove the acceptable risks (including threats and assets that are not associated with an unacceptable risk)
  2. Add treatments as specified in the treatment diagrams; treatments directed towards vulnerabilities or threat scenarios should point towards the risks they indirectly treat
  3. If several treatments point towards the same risks (a many-to-many relation), we recommend using "junction points" to avoid multiple, crossing lines

C.8.1 Step 8a: Grouping of Risks

Grouping of risks

- **Objective:** Identify several risks that may be treated by the same measures in order to facilitate identification of the most efficient treatments
- **How conducted:** Conducted by the analysis team as part of the preparations for Step 8; by going through the threat diagrams showing unacceptable risks, common elements of the various diagrams such as threats, vulnerabilities and threat scenarios are identified; such common elements may indicate that risks also may have common treatments; at the same time the analysis team prepares preliminary treatment diagrams documenting all the unacceptable risks and ready to be filled in with treatment scenarios
- **Input documentation:** CORAS threat diagrams showing only unacceptable risks and their causes
- **Output documentation:** An overview of common elements with respect to which risk may be grouped; preliminary treatment diagrams ready to be filled in with treatment scenarios

C.8.2 Step 8b: Treatment Identification

Treatment identification

- **Objective:** Identification of possible treatments for the unacceptable risks

C.8 Step 8: Risk Treatment Using Treatment Diagrams

Treatment identification

- **How conducted:** Conducted as a brainstorming involving a target team consisting of personnel of various backgrounds; preliminary treatment diagrams showing only the unacceptable risks are displayed, and possible treatments are identified through brainstorming; risks may be reduced or removed by reducing the likelihood and/or consequence of unwanted incidents, by transferring the risk to another party (for example, through insurance or outsourcing), or by avoiding the activity that leads to the risk
- **Input documentation:** Preliminary CORAS treatment diagrams showing only unacceptable risks and their causes ready to be filled in with treatment scenarios
- **Output documentation:** CORAS treatment diagrams showing the identified treatments for the unacceptable risks

C.8.3 Step 8c: Treatment Evaluation

Treatment evaluation

- **Objective:** Identify the treatments the implementation costs of which are lower than their benefit with respect to reducing the risk level
- **How conducted:** Cost-benefit analysis to determine the usefulness of the identified treatments
- **Input documentation:** CORAS treatment diagrams showing the identified treatments for the unacceptable risks
- **Output documentation:** Overview of all treatments with their cost and benefit estimates; CORAS treatment overview diagrams

Appendix D
The CORAS Terminology

Analysis leader – The leader of a *risk analysis*. The person responsible for leading the tasks of the risk analysis and guiding the *analysis participants*.

Analysis member – A member of an *analysis team* that assists the *analysis leader* and the *analysis secretary*.

Analysis participant – Participant in a *risk analysis*. Member of either the *analysis team* or the *target team*.

Analysis role – The role of an *analysis participant* in the *analysis team* or the *target team*.

Analysis secretary – The secretary of a *risk analysis*. The person responsible for documenting the results of the risk analysis and assisting the *analysis leader* when necessary.

Analysis team – A team of people responsible for conducting a *risk analysis*. An analysis team consists of an *analysis leader*, an *analysis secretary* and zero or more *analysis members*.

Antecedent of legal norm – The circumstances that triggers a *legal norm*.

Asset – Something to which a *party* assigns value and hence for which the party requires protection.

Asset diagram – *CORAS diagram* documenting the *parties* and *assets* in a *risk analysis*.

Asset value – The value, priority or criticality assigned to an *asset* by a *party*.

Asset-driven risk analysis – Approach to *risk analysis* based on identifying the *assets* to be protected early in the analysis as part of the characterisation of the *target of analysis*, and where the rest of the analysis is driven by these assets.

Assumption – Something we take as granted or accept as true (although it may not be so).

Avoid – Avoid *risk* by not continuing the activity that gives rise to it.

Basic CORAS – See *basic CORAS language*.

Basic CORAS language – The *CORAS language* excluding *high-level CORAS language*, *dependent CORAS language* and *legal CORAS language*.

Basic diagram – Diagram in the *basic CORAS language*.

Border – Container enclosing the *target* in a *dependent diagram*, separating the target from the *assumptions*.

Communicate and consult – The subprocess of *risk management* of interacting with internal and external *stakeholders*.

Consequence – The impact of an *unwanted incident* on an *asset* in terms of harm or reduced *asset value*.

Consequent of legal norm – Requirements imposed by a *legal norm*.

Context – See *context of analysis*.

Context of analysis – The premises for and the background of a *risk analysis*. This includes the purposes of the analysis and to whom the analysis is addressed.

CORAS – An approach to *risk analysis* consisting of the *CORAS method*, the *CORAS risk modelling language* and the *CORAS tool*.

CORAS analysis – *Risk analysis* with *CORAS*.

CORAS diagram – Diagram in the *CORAS risk modelling language*.

CORAS element – See *element*.

CORAS language – See *CORAS risk modelling language*.

CORAS method – A method for *asset-driven defensive risk analysis* formalised in eight steps. A part of *CORAS*.

CORAS process – The eight steps of the *CORAS method*.

CORAS risk modelling language – Language for modelling *threats*, *vulnerabilities*, *threat scenarios*, *unwanted incidents*, *risks*, *treatment scenarios*, *parties*, *assets*, *legal norms*, the *relations* between these, and the *border* between the *assumptions* and the *target*. A part of *CORAS*.

CORAS tool – Modelling tool for the *CORAS risk modelling language*. A part of *CORAS*.

Defensive risk analysis – *Risk analysis* concerned with protecting what is already there.

Dependent CORAS – See *dependent CORAS language*.

Dependent CORAS language – The *CORAS language* including the *border* element.

Dependent diagram – Diagram in the *dependent CORAS language*. A dependent diagram is a *threat diagram* or a *treatment diagram* with a *border* separating the *assumptions* from the *target*.

Diagram construct – *Element* or *relation* of a *CORAS diagram*.

Direct asset – An *asset* that is not *indirect*.

Element – Construct in the *CORAS language* representing a *threat*, a *vulnerability*, a *threat scenario*, an *unwanted incident*, a *risk*, a *treatment scenario*, a *party*, an *asset*, a *legal norm*, or the *border* between the *assumptions* and the *target*.

Environment – See *environment of target*.

Environment of target – The surrounding things of relevance that may affect or interact with the *target*; in the most general case, the rest of the world.

Evaluate risks – The subprocess of *risk management* of prioritising *risks* according to their severity or *risk level* in order to identify the ones that will be subject to *treatment*.

Establish the context – The subprocess of *risk management* of identifying *stakeholders* and *vulnerabilities*, and to decide what parts of the system, process or organisation will receive attention.

Estimate risks – The subprocess of *risk management* of estimating *likelihoods* and *consequences* for the *risks* that have been identified.

Factual uncertainty – The uncertainty of whether the circumstances of the *antecedent of a legal norm* will occur and thereby trigger the *legal norm*.

Factual risk – See *risk*.

Focus – See *focus of analysis*.

Focus of analysis – The main issue or central area of attention in a *risk analysis*. The focus is within the *scope of the analysis*.

Harm See *harm relation*.

Harm relation – e_1 *harm* e_2: harm to e_1 may result in harm to e_2.

High-level CORAS – See *high-level CORAS language*.

High-level CORAS language – The *CORAS language* including *referring elements* and *referenced diagrams*.

High-level diagram – Diagram in the *high-level CORAS language*.

Identify risks – The subprocess of *risk management* of identifying potential *threats*, *threat scenarios*, and *unwanted incidents* that may constitute *risks*.

Impacts – See *impacts relation*.

Impacts relation – e_1 *impacts* e_2: e_1 impacts e_2 with some *consequence*.

Indirect asset – An *asset* that, with respect to the *target* and *scope of the analysis*, is harmed only via harm to other *assets*.

Initiates – See *initiates relation*.

Initiates relation – e_1 *initiates* e_2: e_1 exploits some set of *vulnerabilities* to initiate e_2 with some *likelihood*.

Leads-to – See *leads-to relation*.

Leads-to relation – e_1 *leads-to* e_2: e_1 leads to e_2 with some *likelihood*, due to some set of *vulnerabilities*.

Legal CORAS – See *legal CORAS language*.

Legal CORAS language – The *CORAS language* including *legal norms*.

Legal diagram – Diagram in the *legal CORAS language*.

Legal norm – A norm that stems from a legal source, such as domestic or international law, contracts or legally binding agreements.

Legal risk – A *risk* that may be caused by one or more *legal norms*. The impact or significance of a legal norm is determined by estimating both the *factual uncertainty* of the circumstances that serve as the *antecedent of the legal norm* and the *legal uncertainty* of whether the norm applies to the circumstances.

Legal uncertainty – The uncertainty of whether a *legal norm* applies to given circumstances.

Likelihood – The frequency or probability of something to occur.

Monitor and review – The subprocess of *risk management* addressing the continuous reviewing and monitoring with respect to all aspects of *risks*.

Offensive risk analysis – *Risk analysis* concerned with balancing potential gain against risk of investment loss.

Party – An organisation, company, person, group or other body on whose behalf a *risk analysis* is conducted.

Path – A set of connected, consecutive *relations* through a *CORAS diagram*.

Reduce consequence – Reduce *risk level* by reducing the harm of *unwanted incidents* to *assets*.

Reduce likelihood – Reduce *risk level* by reducing the *likelihood* of *unwanted incidents* to occur.

Referenced diagram – A *CORAS diagram* that is contained inside a CORAS language *element* and is referred to by a *referring element*.

Referenced risk – See *risk* and *referenced diagram*.

Referenced threat scenario – See *threat scenario* and *referenced diagram*.

Referenced treatment scenario – See *treatment scenario* and *referenced diagram*.

Referenced unwanted incident – See *unwanted incident* and *referenced diagram*.

Referring element – An *element* that is a reference to a *referenced diagram*.

Referring risk – See *risk* and *referring element*.

Referring threat scenario – See *threat scenario* and *referring element*.

Referring treatment scenario – See *treatment scenario* and *referring element*.

Referring unwanted incident – See *unwanted incident* and *referring element*.

Relation – Specification of how *elements* are related to each other, shown as arrows in *CORAS diagrams*.

Retain – Keep *risk* at current level by informed decision.

Risk – The *likelihood* of an *unwanted incident* and its *consequence* for a specific *asset*.

Risk analysis – The process of *establishing the context*, *identifying risks*, *estimating risks*, *evaluating risks*, and *treating risks*.

Risk assessment – The process *identifying risks*, *estimating risks*, and *evaluating risks*.

Risk diagram – *CORAS diagram* documenting *risks*, the *threats* initiating them and the *assets* to which they are related.

Risk evaluation – The task of comparing the estimated *risk levels* against *risk evaluation criteria*; Step 7 of the *CORAS process*. See also *evaluate risks*.

Risk evaluation criterion – Specification of the *risk levels* the *parties* of a *risk analysis* are willing to accept.

Risk evaluation matrix – Table defining *risk evaluation criteria* by means of *likelihood* and *consequence*.

Risk estimation – The task of estimating the *risk levels* of the identified *risks* by means of estimating the *likelihoods* and *consequences* of *unwanted incidents*; Step 6 of the *CORAS process*. See also *estimate risks*.

Risk function – Function defining *risk level* by means of *likelihood* and *consequence*.

Risk identification – The task of identifying *threats*, *vulnerabilities*, *threat scenarios*, and *unwanted incidents* constituting *risks* towards *assets*; Step 5 of the *CORAS process*. See also *identify risks*.

Risk level – The level or value of a *risk* as derived from its *likelihood* and *consequence*.

Risk management – Coordinated activities to direct and control an organisation with regard to *risks*.

Risk reduction – Reduction in the *risk level* of a *risk*.

Risk treatment – The task of identifying *treatments* of identified *risks*; Step 8 of the *CORAS process*. See also *treat risks*.

Scope – See *scope of analysis*.

Scope of analysis – The extent or range of a *risk analysis*. The scope defines the border of the analysis, in other words what is held inside of and what is held outside of the analysis.

Stakeholder – See *party*.

Target – See *target of analysis*.

Target description – The documentation of all the information that serves as the input to and the basis for a *risk analysis*. This includes the documentation of the *target of analysis*, the *focus* and *scope of the analysis*, the *environment of the target*, the *assumptions* of the analysis, the *parties* and *assets* of the analysis, and the *context of the analysis*.

Target of analysis – The system, organisation, enterprise, or the like that is the subject of a *risk analysis*.

Target team – The *analysis participants* representing the customer or other *parties* in a *risk analysis*.

Threat – A potential cause of an *unwanted incident*.

Threat diagram – *CORAS diagram* documenting *threats*, *vulnerabilities*, *threat scenarios*, *unwanted incidents*, the *relations* between these, and their effect on *assets*.

Threat scenario – A chain or series of events that is initiated by a *threat* and that may lead to an *unwanted incident*.

Transfer – Share the *risk* with other another *party* or parties.

Treatment – An appropriate measure to reduce *risk level*.

Treatment category – A general approach to treating *risks*. The categories are *avoid*, *reduce consequence*, *reduce likelihood*, *transfer*, and *retain*.

Treatment diagram – *CORAS diagram* documenting *threats*, *vulnerabilities*, *threat scenarios*, *risks*, and *treatment scenarios* providing *treatment* to the former.

Treatment overview diagram – *CORAS diagram* documenting *risks*, the *threats* initiating them, the *assets* to which they are related, and *treatment scenarios* providing *treatment* to the *risks*.

Treatment scenario – The implementation, operationalisation or execution of appropriate measures to reduce *risk level*.

Treat risks – The subprocess of *risk management* of finding proper *treatments*.

Treats – See *treats relation*.

Treats relation – e_1 *treats* e_2: e_1 reduces *risk* by providing *treatment* to e_2.

Unwanted incident – An event that harms or reduces the value of an *asset*.

Vulnerability – A weakness, flaw or deficiency that opens for, or may be exploited by, a *threat* to cause harm to or reduce the value of an *asset*.

Appendix E
Glossary of Terms

E.1 Logic

$\neg A$	Negation
	Not A
$A \wedge B$	Logical conjunction
	A and B
$A \vee B$	Logical disjunction
	A or B
$A \Rightarrow B$	Logical implication
	A implies B
	If A then B

E.2 Sets

$\emptyset$	Empty set
$A \cup B$	Union of A and B
$A \subseteq B$	A subset of B
$A \setminus B$	Difference between A and B
	The elements in A not in B
$e \in A$	e element in A
$e \notin A$	e not element in A
$\{e \mid \varphi\}$	Set of e such that φ

E.3 Likelihoods

$l_1 + l_2$	Sum of l_1 and l_2
$l_1 - l_2$	Difference between l_1 and l_2

$l_1 \cdot l_2$	Product of l_1 and l_2
$l_1/l_2, \frac{l_1}{l_2}$	Quotient of l_1 and l_2
$l_1 = l_2$	l_1 equal to l_2
$l_1 \neq l_2$	l_1 different from l_2
$l_1 < l_2$	l_1 less than l_2
$l_1 \leq l_2$	l_1 less than or equal to l_2
$l_1 > l_2$	l_1 greater than l_2
$l_1 \geq l_2$	l_1 greater than or equal to l_2
$l\%$	l percent
$f : n\ y,\ f : n\ years$	f times per n years
$\max(l_1, l_2)$	Maximum of l_1 and l_2
$\min(l_1, l_2)$	Minimum of l_1 and l_2

E.4 Likelihood Intervals

$[l_1, l_2]$	Interval from l_1 to l_2 including endpoints $\{l \mid l_1 \leq l \leq l_2\}$
$\langle l_1, l_2 \rangle$	Interval from l_1 to l_2 excluding endpoints $\{l \mid l_1 < l < l_2\}$
$[l_1, l_2 \rangle$	Interval from l_1 to l_2 including l_1 but excluding l_2 $\{l \mid l_1 \leq l < l_2\}$
$\langle l_1, l_2]$	Interval from l_1 to l_2 including l_2 but excluding l_1 $\{l \mid l_1 < l \leq l_2\}$
$\langle l, \infty \rangle$	Open ended interval from but excluding l $\{l' \mid l < l'\}$
$[l, \infty \rangle$	Open ended interval from and including l $\{l' \mid l \leq l'\}$
$[f_1, f_2] : n\ y$	From f_1 to f_2 times per n years $[f_1 : n\ y, f_2 : n\ y]$
$[l_1, l_2] + [l'_1, l'_2]$	Sum of intervals $[l_1, l_2]$ and $[l'_1, l'_2]$ $[l_1 + l'_1, l_2 + l'_2]$
$[l_1, l_2] \cdot [l'_1, l'_2]$	Product of intervals $[l_1, l_2]$ and $[l'_1, l'_2]$ $[l_1 \cdot l'_1, l_2 \cdot l'_2]$
$[l_1, l_2] - [l'_1, l'_2]$	Difference between intervals $[l_1, l_2]$ and $[l'_1, l'_2]$ $[l_1 - l'_1, l_2 - l'_2]$
$[l_1, l_2] \cdot l$	Product of interval $[l_1, l_2]$ and l $[l_1 \cdot l, l_2 \cdot l]$
$l \in [l_1, l_2]$	l in interval $[l_1, l_2]$
$l \notin [l_1, l_2]$	l outside of interval $[l_1, l_2]$
$[l_1, l_2] \subseteq [l'_1, l'_2]$	Interval $[l_1, l_2]$ contained in interval $[l'_1, l'_2]$

E.5 Deductions

$\frac{P_1 \quad P_2 \cdots P_n}{C}$	Derivation from premises $P_1, P_2, \ldots, P_n$ to conclusion C
	C is valid follows from the validity of $P_1, P_2, \ldots,$ and P_n
$t \to e$	t initiates e
$t \xrightarrow{l} e$	t initiates e with likelihood l
$e_1 \to e_2$	e_1 leads-to e_2
$e_1 \xrightarrow{l} e_2$	e_1 leads-to e_2 with conditional likelihood l
$e \to a$	e impacts a
$e(l)$	e occurs with likelihood l
$e([l_1, l_2])$	e occurs with likelihood $[l_1, l_2]$
$t \sqcap e$	The subset of instances of e that are initiated by t
$e_1 \sqcap\!\mid e_2$	The subset of instances of e_2 that are preceded by e_1
$e_1 \sqcup e_2$	All instances of e_1 and e_2
$D = (E, R)$	Diagram consisting of elements E and relations R
$D \cup D'$	Conjoint diagram of D and D'
	$D = (E, R) \wedge D' = (E', R') \Rightarrow D \cup D' = (E \cup E', R \cup R')$
$e \in D$	e element in diagram D
	$D = (E, R) \wedge e \in E$
$e \notin D$	e not element in diagram D
	$D = (E, R) \Rightarrow e \notin E$
$P \subseteq D$	P path in D
	$P = \{e_1 \to e_2, e_2 \to e_3, \ldots, e_{n-1} \to e_n\} \wedge D = (E, R) \wedge$
	$e_1, e_2, \ldots, e_n \in E \wedge e_1 \to e_2, e_2 \to e_3, \ldots, e_{n-1} \to e_n \in R$
$e \to P$	P is commencing in e
	$P = \{e_1 \to e_2, e_2 \to e_3, \ldots, e_{n-1} \to e_n\} \wedge e = e_1$
$P \to e$	P is ending in e
	$P = \{e_1 \to e_2, e_2 \to e_3, \ldots, e_{n-1} \to e_n\} \wedge e = e_n$
$D \ddagger D'$	D' independent of D
	$e \in D' \wedge e' \in D \cup D' \wedge P \subseteq D \cup D' \wedge e' \to P \wedge P \to e \Rightarrow e' \notin D$
$A \triangleright T$	Dependent diagram of assumptions A and target T
	T assuming A to the extent there are explicit dependencies

E.6 Extended Backus-Naur Form

$T = E;$	Term T is defined by expression E
$\{T\}$	Set of instances of T
$\{T\}^-$	non-empty set of instances of T
$[T]$	Optional instance of T
$T_1 \mid T_2$	Instance of T_1 or instance of T_2
$(T_1, T_2, \ldots, T_n)$	Tuple of instance of T_1, instance of $T_2, \ldots,$ and instance of T_n
$\bot$	Undefined or unspecified value

E.7 Semantics

$[\![T]\!]$	Semantics of T
$id(e)$	Identity of element e
$val(e)$	Value of element e
ε	The empty string

E.8 Miscellaneous

$*$	Zero or more
$n..*$	n or more

Acronyms

AS/NZS	Standards Australia/Standards New Zealand
CCA	Cause-Consequence Analysis
CCTA	Central Computing and Telecommunications Agency
CRAMM	CCTA Risk analysis and Management Method
DFD	Data Flow Diagrams
EBNF	Extended Backus-Naur Form
EMF	Eclipse Modeling Framework
ETA	Event Tree Analysis
FMEA	Failure Mode Effect Analysis
FMECA	Failure Mode Effect and Criticality Analysis
FRAAP	Facilitated Risk Analysis and Assessment Process
FTA	Fault Tree Analysis
GP	General Practitioner
GUI	Graphical User Interface
HazOp	Hazard and Operability analysis
HMSC	High-level Message Sequence Chart
ICT	Information and Communication Technology
IEC	International Electrotechnical Commission
ISO	International Organization for Standardization
IT	Information Technology
ITU	International Telecommunication Union
MSC	Message Sequence Chart
NIST	U.S. National Institute of Standards and Technology
OCTAVE	Operationally Critical Threat, Asset, and Vulnerability Evaluation
OMG	Object Management Group
PHCC	Primary Health Care Centre
SDL	Specification and Description Language
STRIDE	Spoofing identity, Tampering with data, Repudiation, Information disclosure, Denial of service, Elevation of privilege

UML	Unified Modeling Language
UML QoS&FT	UML Profile for Modeling Quality of Service and Fault Tolerance Characteristics and Mechanisms
XML	eXtensible Markup Language

References

1. Alberts, C.J., Davey, J.: OCTAVE criteria version 2.0. Technical report CMU/SEI-2001-TR-016. Carnegie Mellon University (2004)
2. Aven, T.: Reliability and Risk Analysis. Springer, Berlin (1992)
3. Aven, T., Sklet, S., Vinnem, J.E.: Barrier and operational risk analysis of hydrocarbon releases (BORA-Release). Part I. Method description. J. Hazard. Mater. A **137**, 681–691 (2006)
4. Baker, W.H., Hylender, C.D., Valentine, J.A.: 2008 data breach investigations report. A study conducted by the Verizon business risk team, Verizon (2008)
5. Barber, B., Davey, J.: The use of the CCTA risk analysis and management methodology CRAMM in health information systems. In: 7th International Congress on Medical Informatics, MEDINFO'92, pp. 1589–1593. North-Holland, Amsterdam (1992)
6. Basel Committee on Banking Supervision, Bank for International Settlements: International Convergence of Capital Measurement and Capital Standards (2006)
7. Ben-Gal, I.: Bayesian networks. In: Ruggeri, F., Kenett, R.S., Faltin, F.W. (eds.) Encyclopedia of Statistics in Quality and Reliability. John Wiley & Sons, New York (2007)
8. Bing, J.: Trust and legal certainty in electronic commerce: An essay. In: Magnusson, C.S., Wahlgren, P. (eds.) Festskrift till Peter Seipel, pp. 27–49. Norstedts Juridik, Stockholm (2006)
9. Bouti, A., Kadi, A.D.: A state-of-the-art review for FMEA/FMECA. Int. J. Reliab. Qual. Saf. Eng. **1**, 515–543 (1994)
10. den Braber, F., Hogganvik, I., Lund, M.S., Stølen, K., Vraalsen, F.: Model-based security analysis in seven steps – a guided tour to the CORAS method. BT Technol. J. **25**(1), 101–117 (2007)
11. Computer Security Institute: 2008 CSI Computer Crime and Security Survey (2008)
12. Computer Security Institute: 2009 CSI Computer Crime and Security Survey (2009)
13. CRAMM – the total information security toolkit. http://www.cramm.com/. Accessed 25 March 2010
14. Dahl, H.E.I., Hogganvik, I., Stølen, K.: Structured semantics for the CORAS security risk modelling language. Technical report STF07 A970, SINTEF ICT (2007)
15. Department for Business, Enterprise & Regulatory Reform: 2008 Information Security Breaches Survey (2008)
16. EUROCONTROL: Methodology report for the 2005/2012 integrated risk picture for Air Traffic Management in Europa. EEC Technical/Scientific Report No. 2006-041 (2006)
17. Fenton, N., Neil, M.: Combining evidence in risk analysis using Bayesian networks. Agena White Paper W0704/01, Agena (2004)
18. Fenton, N.E., Krause, P., Neil, M.: Software measurement: Uncertainty and causal modeling. IEEE Softw. **19**(4), 116–122 (2002)
19. Giese, H., Tichy, M.: Component-based hazard analysis: Optimal designs, product lines, and online-reconfiguration. In: 25th International Conference on Computer Safety, Security and

Reliability (SAFECOMP'06). Lecture Notes in Computer Science, vol. 4166, pp. 156–169. Springer, Berlin (2006)
20. Giese, H., Tichy, M., Schilling, D.: Compositional hazard analysis of UML component and deployment models. In: 23rd International Conference on Computer Safety, Reliability and Security (SAFECOMP'04). Lecture Notes in Computer Science, vol. 3219, pp. 166–179. Springer, Berlin (2004)
21. Goel, S., Chen, V.: Can business process reengineering lead to security vulnerabilities: Analyzing the reengineered process. Int. J. Prod. Econ. **115**(1), 104–112 (2008)
22. Harel, D.: Statecharts: A visual formalism for complex systems. Sci. Comput. Program. **8**(3), 231–274 (1987)
23. Hogganvik, I.: A graphical approach to security risk analysis. Ph.D. thesis, University of Oslo (2007)
24. Hogganvik, I., Stølen, K.: Risk analysis terminology for IT-systems: Does it match intuition? In: 4th International Symposium on Empirical Software Engineering, ISESE'05, pp. 13–23. IEEE Computer Society, Los Alamitos (2005)
25. Hogganvik, I., Stølen, K.: A graphical approach to risk identification, motivated by empirical investigations. In: 9th International Conference on Model Driven Engineering Languages and Systems, MoDELS'06. Lecture Notes in Computer Science, vol. 4199, pp. 574–588. Springer, Berlin (2006)
26. Howard, M., LeBlanc, D.: Writing Secure Code, 2nd edn. Microsoft Press, Redmond (2003)
27. Howard, R.A.: Dynamic Probabilistic Systems, Volume I: Markov Models. John Wiley & Sons, New York (1971)
28. Howard, R.A., Matheson, J.E.: Influence diagrams. Decis. Anal. **2**(3), 127–143 (2005)
29. Innerhofer-Oberperfler, F., Breu, R.: Using an enterprise architecture for IT risk management. In: Information Security South Africa Conference (ISSA'06) (2006)
30. International Electrotechnical Commission: IEC 61025 Fault Tree Analysis (FTA) (1990)
31. International Electrotechnical Commission: IEC 60300-3-9 Dependability management – Part 3: Application guide – Section 9: Risk analysis of technological systems – Event Tree Analysis (ETA) (1995)
32. International Electrotechnical Commission: IEC 61165 Application of Markov Techniques (1995)
33. International Electrotechnical Commission: IEC 61882 Hazard and Operability studies (HAZOP studies) – Application guide (2001)
34. International Organization for Standardization: ISO 27001 Information technology – Security techniques – Information security management systems – Requirements (2005)
35. International Organization for Standardization: ISO/IEC 17799 Information technology – Security techniques – Code of practice for information security management (ISO27002) (2005)
36. International Organization for Standardization: ISO 31000 Risk management – Principles and guidelines (2009)
37. International Organization for Standardization: ISO Guide 73 Risk management – Vocabulary (2009)
38. International Telecommunication Union: Message Sequence Chart (MSC), ITU-T Recommendation Z.120 (1999)
39. International Telecommunication Union: Specification and description language (SDL), ITU-T Recommendation Z.100 (2000)
40. Jones, C.B.: Development methods for computer programmes including a notion of interference. Ph.D. thesis, Oxford University (1981)
41. Jürjens, J.: Secure Systems Development with UML. Springer, Berlin (2005)
42. Kaiser, B., Liggesmeyer, P., Mäckel, O.: A new component concept for fault trees. In: 8th Australian Workshop on Safety Critical Systems and Software (SCS'03), pp. 37–46. Australian Computer Society, Darlinghurst (2003)
43. Kemeny, J.G., Snell, J.L.: Finite Markov Chains. Springer, Berlin (1976)
44. Lee, E., Park, Y., Shin, J.G.: Large engineering project risk management using a Bayesian belief network. Expert Syst. Appl. **36**(3), 5880–5887 (2009)

45. Lodderstedt, T., Basin, D.A., Doser, J.: SecureUML: A UML-based modeling language for model-driven security. In: 5nd International Conference on the Unified Modeling Language, UML'02. Lecture Notes in Computer Science, vol. 2460, pp. 426–441. Springer, Berlin (2004)
46. Lund, M.S., den Braber, F., Stølen, K.: Maintaining results from security assessments. In: 7th European Conference on Software Maintenance and Reengineering (CSMR'03), pp. 341–350. IEEE Computer Society, Los Alamitos (2003)
47. Lund, M.S., den Braber, F., Stølen, K., Vraalsen, F.: A UML profile for the identification and analysis of security risks during structured brainstorming. Technical report STF40 A03067, SINTEF ICT (2004)
48. Mahler, T.: Defining legal risk. In: Commercial Contracting for Strategic Advantage – Potentials and Prospects, pp. 10–31. Turku University of Applied Sciences, Turku (2007)
49. Mahler, T.: Legal risk management. Developing and evaluating elements of a method for proactive legal analyses, with a particular focus on contracts. Ph.D. thesis, University of Oslo (2010)
50. Mannan, S. (ed.): Lees' Loss Prevention the Process Industries. Hazard Identification, Assessment and Control, vol. 1, 3rd edn. Elsevier/Butterworth/Heinemann, Amsterdam/Stoneham/London (2005)
51. McCormick, R.: Legal Risk in the Financial Markets. Oxford University Press, London (2006)
52. McDermott, J., Fox, C.: Using abuse case models for security requirements analysis. In: 15th Annual Computer Security Applications Conference (ACSAC'99), pp. 55–64. IEEE Computer Society, Los Alamitos (1999)
53. McGraw, G.: Software Security: Building Security in. Addison-Wesley, Reading (2006)
54. Microsoft Corporation: Security Development Lifecycle. Version 5.0 (2010)
55. Microsoft Solutions for Security and Compliance and Microsoft Security Center of Excellence: The Security Risk Management Guide (2006)
56. Misra, J., Chandy, K.M.: Proofs of networks of processes. IEEE Trans. Softw. Eng. **7**(4), 417–426 (1981)
57. Nielsen, D.S.: The cause/consequence diagram method as basis for quantitative accident analysis. Technical report RISO-M-1374, Danish Atomic Energy Commission (1971)
58. Object Management Group: UML Profile for Modeling Quality of Service and Fault Tolerance Characteristics and Mechanisms. Version 1.1 (2008). OMG Document: formal/2008-04-05
59. Object Management Group: OMG Unified Modeling Language (OMG UML), Superstructure. Version 2.2 (2009). OMG Document: formal/2009-02-02
60. Papadoupoulos, Y., McDermid, J., Sasse, R., Heiner, G.: Analysis and synthesis of the behaviour of complex programmable electronic systems in conditions of failure. Reliab. Eng. Syst. Saf. **71**(3), 229–247 (2001)
61. Peltier, T.R.: Information Security Risk Analysis, 2nd edn. Auerbach Publications, Boca Raton (2005)
62. Peltier, T.R.: How to Complete a Risk Assessment in 5 Days or Less. Auerback Publications, Boca Raton (2009)
63. Poulsen, K.: Slammer worm crashed Ohio nuke plant network. SecurityFocus (2003)
64. Poulsen, K.: Sluggish movement on power grid cyber security. SecurityFocus (2004)
65. PriceWaterhouseCoopers: The global state of information security 2010 (2009)
66. Redmill, F., Chudleigh, M., Catmur, J.: System Safety: HAZOP and Software HAZOP. Wiley, New York (1999)
67. Refsdal, A., Stølen, K.: Employing key indicators to provide a dynamic risk picture with a notion of confidence. In: Trust Management III. Third IFIP WG 11.11 International Conference (IFIPTM'09), pp. 215–233. Springer, Berlin (2009)
68. Reid, K.: Risk-e-business: A framework for legal risk management developed through an analysis of selected legal risk in Internet commerce. Ph.D. thesis, University of New South Wales (2000)
69. Schneier, B.: Attack trees: Modeling security threats. Dr. Dobb's J. **24**(12), 21–29 (1999)
70. Schneier, B.: Secrets & Lies: Digital Security in a Networked World. Wiley, New York (2000)
71. Sherer, S.A.: Using risk analysis to manage software maintenance. J. Softw. Maint.: Res. Pract. **9**(6), 345–364 (1997)

72. Sindre, G., Opdahl, A.L.: Eliciting security requirements by misuse cases. In: 37th International Conference on Technology of Object-Oriented Languages and Systems (TOOLS Pacific'00), pp. 120–131. IEEE Computer Society, Los Alamitos (2000)
73. SOPHOS: Security threat report: 2010 (2010)
74. Stålhane, T., Wedde, K.J.: Practical experience with the application of HazOp to a software intensive system. In: Project Control for 2000 and Beyond (ESCOM-ENCRESS'98), pp. 271–281. Shaker Pub (1998)
75. Standards Australia/Standards New Zealand: AS/NZS 4360 Risk Management (2004)
76. Standards Australia/Standards New Zealand: HB 296 Legal risk management (2007)
77. Stoneburner, G., Goguen, A., Feringa, A.: Risk Management Guide for Information Technology Systems. National Institute of Standards and Technology, Gaithersburg (2002). NIST Special Publication 800-30
78. Susskind, R.E.: The Future of Law. Oxford University Press, London (1996)
79. Swiderski, F., Snyder, W.: Threat Modeling. Microsoft Press, Redmond (2004)
80. Symantec: Symantec global internet security threat report – trends for 2008 (2009)
81. Vose, D.: Risk Analysis – A Quantitative Guide. Wiley, New York (2008)
82. Vraalsen, F., Mahler, T., Lund, M.S., Hogganvik, I., den Braber, F., Stølen, K.: Assessing enterprise risk level: The CORAS approach. In: Khadraoui, D., Herrmann, F. (eds.) Advances in Enterprise Information Technology Security, chap. XVIII, pp. 311–333. Information Science Reference, Hershey (2007)
83. Wahlgren, P.: Juridisk riskanalys: Mot en säkrare juridisk metod. Jure, Stockholm (2003) (in Swedish)

Index

A
Abuse-cases, 346
Analysis
  assumptions of, **88**
  context, 48, **75**, *436*
  focus of, 21, 71, 87, **88**, *437*
  high-level, 10, 25, 94, 104
  leader, 22, 72, 337, *435*
  member, 72, *435*
  of legal aspects, 328
  participant, 72, *435*
  role, 72, *435*
  scope of, 21, 47, 87, **88**, *440*
  secretary, 22, 72, *435*
  target, 5, 47, **84**, 205, *440*
  team, 7, 71, 311, 339, *435*
Antecedent of legal norm, 318, *435*
Approval of target description, 10, 29, 109
AS/NZS 4360, 8, 349
Asset, 5, 15, 45, **53**, 71, 101, 298, *435*
  change to, 284
  diagram, 24, 54, 94, *435*
  direct, 27, **56**, 102, 291, *437*
  identification, 4, 24, 99
  indirect, 27, 54, **56**, 102, 171, 291, *438*
  ranking of, 110, 113
  value, 53, 284, *435*
Asset-driven
  risk analysis, 5, 168, *435*
  risk identification, 123
Assumption, 6, 23, 80, *435*
  change to, 283
  consequence, **308**
  environment, 21, 295

  independence, **305**
  of analysis, **88**
  simplification, **306**
Avoid risk, **61**, 191, *436*

B
Basic CORAS, *436*
  diagram, 54, 82, 244, *436*
  element, 252
  language, 6, *436*
    semantics, *389*
    syntax, *357*
Bayesian
  network, 17, 347
  reasoning, 350
Because relation, 325
Border, 297, *436*

C
Cause-consequence
  analysis (CCA), 17
  diagram, 347
CCA, 17
CCTA, 16
CCTA Risk analysis and Management
    Method (CRAMM), 16, 348
Central Computing and
    Telecommunications Agency
    (CCTA), 16
Change, 281
  classification, 281
  perspective, 281, 285
    before-after, 288

Change *(cont.)*
  continuous evolution, 292
  maintenance, 286
 process, 286
 to knowledge, 285
Change management, 11, 281
Communicate and consult, 14, *436*
Composition
 leads-to relation, 213
 threat scenario, 212, 235
 unwanted incident, 212, 235
Conditional likelihood, 147, 207
Conjunction, 231
Consequence, 11, 30, 53, **57**, 152, *436*
 accumulate, 171
 aggregate, 263
 estimation, 35, 37, 114, 152
 in high-level CORAS, 262
 interval, 115
 reduce, **61**, 191, *438*
 scale, 29, 111, 114, 263
Consequent of legal norm, 318, *436*
Consistency checking, 11, 227
 in high-level CORAS, 257
Construct, 11, 54, 126, 207, *437*
 high-level, 247
Context of analysis, 48, **75**, *436*
 change to, 284
CORAS, 3, 5, *436*
 analysis, 80, *436*
 basic, *436*
 dependent, *436*
 diagram, 6, 45, 205, *436*
  asset, 24, 54, *435*
  basic, 54, 82, 244, *436*
  dependent, 296, *437*
  high-level, 244, *437*
  legal, 324, *438*
  referenced, 243, *438*
  risk, 37, 58, 163, *439*
  threat, 11, 31, 35, 56, 123, 145, 235, *440*
  treatment, 39, 60, 185, *440*
  treatment overview, 40, 62, 199, *440*
 element, 46, 54, *436*, *437*
 high-level, 243, *437*
 language, 6, 45, 125, *436*
  basic, 6, *436*
  construct, 11, 54, 126, 207, *437*

  dependent, 6, *436*
  high-level, 6, 243, *437*
  legal, 6, 317, 324, *438*
  likelihoods in, 206
  semantics, *389*
  syntax, *357*
 legal, 317, *438*
 method, 5, 6, 21, 45, 75, 80, 317, *436*
 process, 289, 354, *436*
 risk modelling language, 9, 25, 45, 324, *436*
 tool, 6, 128, 337, *436*
Cost-benefit, 11, 185, 194
CRAMM, 16, 348

D

Data flow diagram (DFD), 347
Decomposition
 threat scenario, 237, 247
 unwanted incident, 237
Defensive risk analysis, 3, 346, *436*
Dependent CORAS, *436*
 diagram, 296, *437*
  combine, 311
  threat, 309
  treatment, 297
 language, 6, *436*
  semantics, 301, *408*
  syntax, *381*
DFD, 347
Diagram
 asset, 24, 54, 94, *435*
 basic, 54, 82, 244, *436*
 dependent, 296, *437*
 element, 46, 54, *437*
 high-level, 244, *437*
 legal, 324, *438*
 path, 51, 136, 218, 304, *438*
 referenced, 243, *438*
 risk, 37, 58, 163, *439*
 threat, 11, 31, 35, 56, 123, 145, 235, *440*
 treatment, 39, 60, 185, *440*
 treatment overview, 40, 62, 199, *440*
Dimension
 factual, 317
 legal, 317
Direct asset, 27, **56**, 102, 291, *437*
Disjunction, 214, 231

E

EBNF, 12, *357*, *445*
Element, 46, 54, *437*
  basic, 252
  likelihood of, 206
  referring, 243, *439*
Environment, 75, *437*
  assumption, 21, 295
  change to, 283
  of target, 11, 21, **88**, 205, 295, *437*
Establish the context, 13, *437*
Estimate risks, 14, *437*
ETA, 17, 346
Evaluate risks, 14, *437*
Event tree, 17, 346
  analysis (ETA), 17, 346
Evolution, 281, 353
Extended Backus-Naur Form (EBNF), 12, *357*, *445*

F

Facilitated Risk Analysis and Assessment Process (FRAAP), 348
Factual
  dimension, 317
  risk, 12, *437*
  uncertainty, 319, *437*
Failure Mode Effect Analysis/Failure Mode Effect and Criticality Analysis (FMEA/FMECA), 16, 354
Fault tree, 16, 346
  analysis (FTA), 16, 346
FMEA/FMECA, 16, 354
Focus of analysis, 21, 71, 87, **88**, 296, *437*
  change to, 283
FRAAP, 348
Frequency, 30, 57, 117, 148, 206, 224
  calculating, 220
  rule for initiates relation, **220**
  rule for leads-to relation, **222**
  rule for scenarios and incidents, **222**

G

Gate, 246
  in-gate, 246
  out-gate, 246
  treatment, 252
Grammar
  basic CORAS language, 357

  dependent CORAS language, *381*
  high-level CORAS language, *368*
  legal CORAS language, *385*

H

Harm relation, **56**, 101, *437*
Hazard and Operability analysis (HazOp), 16
HazOp, 16
High-level analysis, 10, 25, 94, 104
  table, 32, 94
High-level CORAS, 243, *437*
  consequences in, 262
  consistency checking, 257
  construct, 247
  diagram, 244, *437*
    risk, 249
    threat, 249
    treatment, 251
    treatment overview, 253
  language, 6, 243, *437*
    semantics, 265, *396*
    syntax, *368*
  likelihoods in, 255
  risk levels in, 264
Human threat, 105
  accidental, 32, 47
  deliberate, 32, 47

I

Identify risks, 14, *437*
Impacts relation, **56**, 147, 262, 298, *437*
In-gate, 246
Indicator, 293
Indirect asset, 27, 54, **56**, 171, 291, *438*
Influence diagrams, 353
Initiates relation, **56**, 298, *438*
  likelihood of, 205
  rule for frequency, **220**
  rule for probability, **209**
International Organization for Standardization (ISO), 8, 349
Interval
  calculation, 226
  consequence, 115
  likelihood, 30, 117, 206, 225, 444
ISO 31000, 8, 349

L

Leads-to relation, **56**, 147, *438*
  composition, 213
  likelihood of, 205
  mutually exclusive, 213
  rule for frequency, **222**
  rule for probability, **209**, **213**, **216**
  statistically independent, 213
Legal
  aspect, 6, 317, 356
    analysis, 328
  dimension, 317
  judgement, 320
  norm, 7, 317, **318**, *438*
    antecedent of, 318, *435*
    consequent of, 318, *436*
  qualification, 318
  risk, 3, 317, **324**, 355, *438*
  risk management, 354
  source, 318
  uncertainty, 319, *438*
    scale, 330
Legal CORAS, 317, *438*
  diagram, 324, *438*
  language, 6, 317, 324, *438*
    semantics, 326, *410*
    syntax, *385*
Likelihood, 11, 30, **57**, 117, 148, *438*, 443
  accumulate, 170
  aggregate, 36, 239
  analysing, 205
  calculating, 205
  conditional, 147, 207
  consistency checking, 11, 227
  distribution, 206, 225
  estimation, 35, 148, 205
  in high-level CORAS, 255
  interval, 30, 117, 206, 225, 444
  reasoning, 205
  reduce, **61**, 191, *438*
  rule for initiates relation, **209**, **220**
  rule for leads-to relation, **209**, **213**, **216**, **222**
  rule for scenarios and incidents, **209**, **210**, **222**
  scale, 29, 111, 116
  specifying, 206

M

Markov
  analysis, 18, 347
  model, 18, 347
Message Sequence Charts (MSC), 351
Microsoft
  security development lifecycle, 349
  security risk management process, 348
  threat modeling, 347
Misuse-cases, 346
Monitor and review, 14, *438*
MSC, 351
Mutually exclusive
  leads-to relations, 213
  scenarios and incidents, 209

N

NIST, 349
Non-human threat, 33, 47, 105

O

Object Management Group (OMG), 12, 357
OCTAVE, 15, 347
Offensive risk analysis, 3, *438*
OMG, 12, 357
On-the-fly modelling, 6, 32, 128, 337
Operationally Critical Threat, Asset, and
    Vulnerability Evaluation
    (OCTAVE), 15, 347
Out-gate, 246

P

Party, 5, 15, 53, **54**, 72, 100, *438*
  change to, 284
Path, 51, 136, 218, 304, *438*
Probability, 30, 57, 117, 148, 206, 224
  calculating, 208
  rule for initiates relation, **209**
  rule for leads-to relation, **209**, **213**, **216**
  rule for scenarios and incidents, **209**, **210**
ProSecO, 352

R

Ranking of assets, 110, 113
Reduce consequence, **61**, 191, *438*
Reduce likelihood, **61**, 191, *438*
Referenced diagram, 243, *438*
  risk, 249, *438*
  threat scenario, 245, *438*
  treatment scenario, 252, *439*

unwanted incident, 248, *439*
Referring element, 243, *439*
    expand, 244
    risk, 249, *439*
    threat scenario, 244, *439*
    treatment scenario, 252, *439*
    unwanted incident, 244, *439*
Relation, 6, 54, 298, *439*
    because, 325
    harm, **56**, 101, *437*
    impacts, **56**, 147, 262, 298, *437*
    initiates, **56**, 298, *438*
    leads-to, **56**, 147, *438*
    treats, **61**, *441*
Retain risk, **61**, 191, *439*
Risk, 3, 15, 45, **58**, 158, 206, *439*
    accumulate, 168
    avoid, **61**, 191, *436*
    diagram, 37, 58, 163, *439*
    estimation, 11, 35, 54, 145, 155, 205, *439*
    evaluation, 11, 37, 54, 163, 335, *439*
        criteria, 29, 85, 120, 166, 291, *439*
        matrix, 31, *439*
    factual, 12, *437*
    function, 59, 118, 275, *439*
    grouping, 187
    identification, 10, 31, 54, 123, 289, *439*
        asset-driven, 123
    legal, 3, 317, **324**, 355, *438*
    level, 14, 21, 55, **59**, 118, 158, 205, *439*
        in high-level CORAS, 264
    matrix, 118
    modelling, 5, 45, 206, 345
        on-the-fly, 6, 32, 128, 337
    monitoring, 353
    overview, 167
    reduction, 196, *440*
    referenced, 249, *438*
    referring, 249, *439*
    retain, **61**, 191, *439*
    transfer, **61**, 191, *440*
    treatment, 11, 39, 111, 185, **189**, 335, *440*
Risk analysis, 3, 45, 243, 281, *439*
    asset-driven, 5, 168, *435*
    defensive, 3, 346, *436*
    high-level, 94
    method, 7, 14, 131, 345

    offensive, 3, *438*
    preparations, 71
    technique, 6, 14
Risk assessment, 4, 14, *439*
Risk management, 7, 13, 348, *439*
    legal, 354

S
Scale
    consequence, 29, 111, 114, 263
    legal uncertainty, 330
    likelihood, 29, 111, 116
Scope of analysis, 21, 47, 87, **88**, *440*
    change to, 283
SDL, 351
SecureUML, 346
Semantics
    basic CORAS language, *389*
    dependent CORAS language, 301, *408*
    high-level CORAS language, 265, *396*
    legal CORAS language, 326, *410*
Separate scenarios and incidents, 222
Specification and Description Language (SDL), 351
Stakeholder, 3, 53, 283, *440*
Standards Australia/Standards New Zealand (AS/NZS), 8, 349
Statechart, 351
Statistically independent
    leads-to relations, 213
    scenarios and incidents, 209
STRIDE, 347
Syntax
    basic CORAS language, *357*
    dependent CORAS language, *381*
    high-level CORAS language, *368*
    legal CORAS language, *385*

T
Target, 295, *440*
    change to, 282
    description, 10, 24, 93, **94**, 281, *440*
        approval of, 10, 29, 109
    environment, 205
    environment of, 11, 21, **88**, *437*
    of analysis, 5, 47, **84**, 205, *440*
    presentation, 10, 23, 79
    simplification, **307**
    team, 74, *440*

Threat, 11, 45, **46**, 129, 298, *440*
  diagram, 11, 31, 35, 56, 123, 145, 235, *440*
  human, 105
    accidental, 32, 47
    deliberate, 32, 47
    identification, 127
    non-human, 33, 47, 105
Threat scenario, 10, **48**, 131, *440*
  composition, 212, 235
  decomposition, 237, 247
  high-level, 243
  identification, 131
  likelihood of, 205
  mutually exclusive, 209
  referenced, 245, *438*
  referring, 244, *439*
  rule for frequency, **222**
  rule for probability, **209, 210, 216**
  rule for probability of relation, **216**
  separate, 222
  statistically independent, 209
Transfer risk, **61**, 191, *440*
Treat risks, 14, *440*
Treatment, 11, 45, **189**, *440*
  category, **61**, 191, *440*
    avoid, **61**, 191, *436*
    reduce consequence, **61**, 191, *438*
    reduce likelihood, **61**, 191, *438*
    retain, **61**, 191, *439*
    transfer, **61**, 191, *440*
  cost-benefit, 11, 185, 194
  diagram, 39, 60, 185, *440*
  evaluation, 194
  gate, 252
  identification, 39, 54, 189
  implementation, 185, 282
  option, 40, 192
  overview diagram, 40, 62, 199, *440*
  plan, 11, 187
Treatment scenario, **61**, *440*
  high-level, 251
  referenced, 252, *439*
  referring, 252, *439*
Treats relation, **61**, *441*

U

U.S. National Institute of Standards and Technology (NIST), 349
UML, 10, 94, 345
  activity diagram, 25, 98
  class diagram, 14, 72
  collaboration diagram, 25, 97
  profile, 346
  Profile for Modeling Quality of Service and Fault Tolerance Characteristics and Mechanisms (UML QoS&FT), 346
  QoS&FT, 346
  use-case, 346
UMLSec, 346
Uncertainty
  factual, 319, *437*
  legal, 319, *438*
  scale, 330
Unified Modeling Language (UML), 10, 94, 345
Unwanted incident, 11, 15, 46, **52**, 128, *441*
  composition, 212, 235
  decomposition, 237
  high-level, 243
  identification, 127
  likelihood of, 205
  mutually exclusive, 209
  referenced, 248, *439*
  referring, 244, *439*
  rule for frequency, **222**
  rule for probability, **209, 210, 216**
  separate, 222
  statistically independent, 209

V

Vulnerability, 10, 45, **50**, 135, *441*
  identification, 135